THE FRONTIERS COLLECTION

THE FRONTIERS COLLECTION

Series Editors:

A.C. Elitzur L. Mersini-Houghton M. Schlosshauer M.P. Silverman R. Vaas H.D. Zeh
J. Tuszynski

The books in this collection are devoted to challenging and open problems at the forefront of modern science, including related philosophical debates. In contrast to typical research monographs, however, they strive to present their topics in a manner accessible also to scientifically literate non-specialists wishing to gain insight into the deeper implications and fascinating questions involved. Taken as a whole, the series reflects the need for a fundamental and interdisciplinary approach to modern science. Furthermore, it is intended to encourage active scientists in all areas to ponder over important and perhaps controversial issues beyond their own speciality. Extending from quantum physics and relativity to entropy, consciousness and complex systems – the Frontiers Collection will inspire readers to push back the frontiers of their own knowledge.

For a full list of published titles, please see back of book or springer.com/series 5342

Series home page – springer.com

Bernd Hoefflinger

Editor

Chips 2020

A Guide to the Future of Nanoelectronics

 Springer

Editor
Bernd Hoefflinger
Leonberger Str. 5
71063 Sindelfingen
Germany

Series Editors:
Avshalom C. Elitzur
Bar-Ilan University, Unit of Interdisciplinary Studies, 52900 Ramat-Gan, Israel
email: avshalom.elitzur@weizmann.ac.il

Laura Mersini-Houghton
Dept. Physics, University of North Carolina, Chapel Hill, NC 27599-3255, USA
email: mersini@physics.unc.edu

Maximilian A. Schlosshauer
Institute for Quantum Optics and Quantum Information, Austrian Academy of Sciences,
Boltzmanngasse 3, A-1090 Vienna, Austria
email: schlosshauer@nbi.dk

Mark P. Silverman
Trinity College, Dept. Physics, Hartford CT 06106, USA
email: mark.silverman@trincoll.edu

Jack A. Tuszynski
University of Alberta, Dept. Physics, Edmonton AB T6G 1Z2, Canada
email: jtus@phys.ualberta.ca

Rüdiger Vaas
University of Giessen, Center for Philosophy and Foundations of Science, 35394 Giessen,
Germany
email: ruediger.vaas@t-online.de

H. Dieter Zeh
Gaiberger Straße 38, 69151 Waldhilsbach, Germany
email: zeh@uni-heidelberg.de

ISSN 1612-3018
ISBN 978-3-642-22399-0   e-ISBN 978-3-642-23096-7
DOI 10.1007/978-3-642-23096-7
Springer Heidelberg Dordrecht London New York

Library of Congress Control Number: 2011943001

Springer is part of Springer Science+Business Media (www.springer.com)

Preface

On July 30, 1959, Robert Noyce filed his Integrated Circuit patent. Its fiftieth anniversary in 2009 became the origin of this book and its central question: After the unparalleled progress of microchips over half a century, can this story continue, and what will it take to make this happen?

To greet the year 2000, I had written an invited paper "Chips 2020" for the German magazine *Elektronik* (with a strong vote for 3D chip integration), and it looked attractive to check the status at half-time, in 2010. However, the central issue of this book emerged owing to more and more signs of the imminent end of the nano(meter) *roadmap*: The law that halving the transistor size every 18 months would bring automatic performance and market growth is about to end in 2015. When the billions of 10 nm × 10 nm transistors packed side-by-side on a chip are hardly useful because of their fundamental statistical variance, we face the most important turning point in the history of microelectronics: Declining growth in markets and services and an energy crisis on top, because, with the chip technology of 2010 and with the present annual doubling of video and TV on the internet, this service alone would require the total worldwide electrical power in 2015.

Chips 2020 explains the background to the 20–10 nm transistor limits in different applications, and it focuses on the new strategy for the sustainable growth of a nanoelectronics ecosystem with a focus on ultra-low energy of all chip functionalities, femtojoule electronics, enabled by 3D integration on-chip and of chip-systems incorporating new architectures as well as new lithography and silicon technologies.

At the critical time of this new fundamental energy orientation, I have been very fortunate that several world leaders with their teams agreed to contribute to this book: Greg Delagi of Texas Instruments, on intelligent mobile companions, the new lead products; Georg Kimmich of ST-Ericsson, on 3D integration for wireless applications; Burn Lin of TSMC, on nanolithography; Jiri Marek and Udo Gomez of Robert Bosch GmbH, on MEMS (micro-electro-mechanical systems) for automotive and consumer applications; Barry Pangrle of Mentor Graphics, on power-efficient design; Peter Roth and colleagues at IBM, on superprocessors; Yiannos Manoli with his team at the University of Freiburg, together with Boris Murmann of Stanford University, on analog–digital interfaces and energy harvesting; Albrecht

Rothermel of the University of Ulm, on retinal implants for blind patients; and Ben Spaanenburg with his co-author at Lund University, on digital neural networks. This book would be unthinkable without their contributions, and, in this critical situation for nanoelectronics, I am even more obliged to them for their involvement in this project.

Despite our broad agenda, we could cover only selected topics, which, we hope, are exemplary for the potential and challenges for 2020 chips.

My sincere thanks go to Claus Ascheron at Springer in Heidelberg, who constantly pursued the idea for this book and who finally convinced me in 2009 to realize it. I thank the team at Springer Science+Business Media for their confidence in this publication and for its professional production. I thank Stefanie Krug for her perfect execution of many of the illustrations and Deborah Marik for her professional editing of the manuscript.

With experience as the first MOS product manager at Siemens and of co-founding a Technical University and establishing several microchip research facilities in the USA and Germany, I have included educational, research, and business aspects of the nanoelectronics ecosystem. I hope that, with this scope, this book will be helpful to all those who have to make decisions associated with future electronics, from students to graduates, educators, and researchers, as well as managers, investors, and policy makers.

Sindelfingen   Bernd Hoefflinger

Contents

Contributors

Greg Delagi Texas Instruments, 12,500 TI Boulevard, MS 8723, Dallas, TX, 75243, USA, t-wright@ti.com

Dr. Udo-Martin Gómez Robert Bosch GmbH, Automotive Electronics (AE/NE4), Postfach 1,342, 72703 Reutlingen, Germany, Udo-Martin.Gomez@de.bosch.com

Thorsten Hehn IMTEK Microelectronics, University of Freiburg, Georges-Koehler-Allee 102, 79106 Freiburg, Germany

Prof. Bernd Hoefflinger Leonberger Strasse 5, 71063 Sindelfingen, Germany, bhoefflinger@t-online.de

Dr. Daniel Hoffmann HSG-IMIT, Institute of Micromachining and Information Technology, Wilhelm-Schickard-Strasse 10, 78052 Villingen-Schwenningen, Germany

Dr. Christian Jacobi IBM Systems & Technology Group, Technology Development, Schoenaicher Strasse 220, 71032 Boeblingen, Germany, cjacobi@de.ibm.com

Dr. Matthias Keller IMTEK Microelectronics, University of Freiburg, Georges-Koehler-Allee 102, 79110 Freiburg, Germany, mkeller@imtek.de

Georg Kimmich ST-Ericsson 12, rue Jules Horovitz, B.P. 217, 38019 Grenoble Cedex, France, georg.kimmich@stericsson.com

Matthias Kuhl IMTEK Microelectronics, University of Freiburg, Georges-Koehler-Allee 102, 79110 Freiburg, Germany

Dr. Burn J. Lin TSMC, 153 Kuang Fu Road, Sec. 1, Lane 89, 1st Fl, Hsinchu, Taiwan, R.O.C. 300, burnlin@tsmc.com

Niklas Lotze HSG-IMIT, Institute of Micromachining and Information Technology, Wilhelm-Schickard-Strasse 10, 78052 Villingen-Schwenningen, Germany

Suleiman Malki Department of Electrical & Information Technology, Lund University, P.O. Box 118, Lund 2210, Germany

Prof. Yiannos Manoli Fritz Huettinger Chair of Microelectronics, University of Freiburg, IMTEK, Georges-Koehler-Allee 102, 79110 Freiburg, Germany, manoli@imtek.de

Dr. Jiri Marek Robert Bosch GmbH, Automotive Electronics, Postfach 1,342, 72703 Reutlingen, Germany, Jiri.Marek@de.bosch.com

Dominic Maurath IMTEK Microelectronics, University of Freiburg, Georges-Koehler-Allee 102, 79110 Freiburg, Germany, maurath@imtek.de

Christian Moranz IMTEK Microelectronics, University of Freiburg, Georges-Koehler-Allee 102, 79110 Freiburg, Germany

Prof. Boris Murmann Stanford University, 420 Via Palu Mall, Allen-208, Stanford, CA, 94305-40070, USA, murmann@stanford.edu

Dr. Barry Pangrle Mentor Graphics, 1,001 Ridder Park Drive, San Jose, CA, 95131-2314, USA, barry_pangrle@mentor.com

Daniel Rossbach HSG-IMIT, Institute of Micromachining and Information Technology, Wilhelm-Schickard-Strasse 10, 78052 Villingen-Schwenningen, Germany

Dr. Peter Hans Roth IBM Systems & Technology Group, Technology Development, Schoenaicher Strasse 220, 71032 Boeblingen, Germany, peharo@de.ibm.com

Prof. Albrecht Rothermel University of Ulm, Institute of Microelectronics, Albert-Einstein-Allee 43, 89081 Ulm, Germany, info@albrecht-rothermel.de

Prof. Lambert Spaanenburg Dept. of Electrical & Information Technology, Lund University, PO Box 11822100 Lund, Sweden, Lambert.Spaanenburg@eit.lth.se

Dirk Spreemann HSG-IMIT, Institute of Micromachining and Information Technology, Wilhelm-Schickard-Strasse 10, 78052 Villingen-Schwenningen, Germany

Kai Weber System z Core Verification Lead, IBM Systems & Technology Group, Technology Development, Schoenaicher Strasse 220, 71032 Boeblingen, Germany, Kai.Weber@de.ibm.com

Biographies of Authors

Greg Delagi has been senior vice president and general manager of TI's Wireless Terminals Business since 2007, where he leads TI's worldwide development of semiconductors for mobile phones. Greg Delagi started his career with Texas Instruments in 1984 with the Materials and Controls Business. He later joined TI's Semiconductor Group. Delagi was responsible for several of the company's business units such as DSL, Cable, VoIP and Wireless LAN, creating systems-on-chip solutions. As vice president and manager of TI's Digital Signal Processing (DSP) business, he guided the company to its world leadership with a 62% share of the world market in 2006. He is a graduate of Nichols College in Dudley, MA, where he earned a BSBA in 1984.

Udo Gómez has been Director of Engineering since 2006 for Advanced Sensor Concepts at Bosch, Automotive Electronics Division, in Reutlingen, Germany, responsible for MEMS sensor predevelopment activities. He also serves as chief expert for MEMS sensors. From 2003 until 2005, he was responsible for the development of a novel automotive inertial sensor cluster platform for active safety applications. Dr. Gómez studied physics at the University of Stuttgart, Germany. In 1997, he received his Ph.D. from the University of Stuttgart for his work on molecular electronics. In January 1998, he joined the California Institute of Technology as a post-doctoral fellow. In 1999, he started his work at Bosch Corporate Research.

Bernd Hoefflinger started his career as an assistant professor in the Department of Electrical Engineering at Cornell University, Ithaca, NY, USA. Returning to Germany, he served as the first MOS product manager at Siemens, Munich. With that background, he became a co-founder of the University of Dortmund, Germany, and later head of the Electrical Engineering Departments at the University of Minnesota and at Purdue University, IN, USA. In 1985, he became the director of the newly established Institute for Microelectronics Stuttgart (IMS CHIPS), a public contract research and manufacturing institute. In 1993, IMS CHIPS became the world's first ISO 9000 certified research institute. He launched rapid prototyping with electron-beam lithography in 1989. He established the institute as a leader in high-dynamic-range CMOS imagers and video cameras from 1993 onwards. Among the developments in CMOS photosensors was the chip design and manufacturing for the first sub-retinal implants in humans in Europe in 2005. He retired in 2006.

Christian Jacobi received both his M.S. and Ph.D. degrees in computer science from Saarland University, Germany, in 1999 and 2002, respectively. He joined IBM Research and Development in 2002, working on floating-point implementation for various IBM processors. For IBM's System z10 mainframe, he worked on the L1.5 cache unit. From 2007 to 2010, Dr. Jacobi was on international assignment to Poughkeepsie, NY, where he worked on the Load-Store-Unit for the IBM zEnterprise 196 mainframe. Dr. Jacobi is now responsible for the cache design of IBM's mainframe processors, and for future architecture extensions.

Matthias Keller was born in Saarlouis, Germany, in 1976. He received his Diploma degree in electrical engineering from the University of Saarland, Saarbrücken, Germany, in 2003 and a Dr.-Ing. degree (summa cum laude) from the University of Freiburg, Germany, in 2010. From 2003 to 2009, he was a research assistant at the Fritz Huettinger Chair of Microelectronics at the University of Freiburg, Germany, in the field of analog CMOS integrated circuit design with an emphasis on continuous-time delta-sigma A/D converters. In 2009, he was awarded a tenured position as a research associate ("Akademischer Rat"). His research interests are analog integrated circuits based on CMOS technology, in particular delta-sigma A/D converters and patch-clamp readout circuits.

Georg Kimmich is product manager at ST-Ericsson in Grenoble, France. He is responsible for the product definition of digital baseband system-on-chip products within the 3G Multimedia and Platform Division. His focus is on the technology roadmap definition for chip packaging. Georg Kimmich graduated in computer engineering at HFU, Furtwangen, Germany in 1990. He held several design and management positions in the design automation and ASIC design domain with Thomson Multimedia and STMicroelectronics in Germany, France and the US. In 2006, he joined the application processor division of STMicroelectronics, where he was responsible for the system-on-chip development team of the Nomadik application processor family.

Burn J. Lin became a vice president at TSMC, Ltd., in 2011. He joined TSMC in 2000 as a senior director. Prior to that, he founded Linnovation, Inc. in 1992. Earlier he held various technical and managerial positions at IBM after joining IBM in 1970. He has been extending the limit of optical lithography for close to four decades. He pioneered many lithography techniques, among them deep-UV lithography starting from 1975, multi-layer resist from 1979, Exposure-Defocus methodology from 1980, k_1 reduction from 1987, attenuated phase-shifting mask from 1991, 193 nm immersion lithography from 2002, and polarization-dependent stray light from 2004. He is currently working on cost-effective optical lithography and multiple-electron-beam mask-less lithography for the 20 nm node and beyond.

Dr. Lin is the editor-in-chief of the *Journal of Micro/nanolithography, MEMS, and MOEMS*, life member of the US National Academy of Engineering, IEEE Life Fellow, and SPIE Fellow. He has received numerous awards, among them the 1st Semi IC Outstanding Achievement Award in 2010, the 2009 IEEE Cledo Brunetti Award, the 2009 Benjamin G. Lammé Meritorious Achievement Medal, and in 2004 the 1st SPIE Frits Zernike award. Throughout his career, he has received numerous TSMC and IBM Awards. He has written one book and three book chapters, published 121 articles, mostly first-authored, and holds 66 US patents.

Suleyman Malki received, in October 2008, a Ph.D. degree from the Department of Electrical and Information Technology at Lund University for work on the topic "Reconfigurable and Parallel Computing Structures in Structured Silicon". Before that he received a master's degree in computer science and engineering from Lund University. His research has focused mainly on the implementation and verification of highly intelligent systems on reconfigurable chips (FPGA). Currently he is devoting most of his time to his recent biological and industrial offspring.

Yiannos Manoli was born in Famagusta, Cyprus. As a Fulbright scholar, he received a B.A. degree (summa cum laude) in physics and mathematics from Lawrence University in Appleton, WI, in 1978 and a M.S. degree in electrical engineering and computer science from the University of California, Berkeley, in 1980. He obtained a Dr.-Ing. degree in electrical engineering from the University of

Duisburg, Germany, in 1987. From 1980 to 1984, he was a Research Assistant at the University of Dortmund, Germany. In 1985, he joined the newly founded Fraunhofer Institute of Microelectronic Circuits and Systems, Duisburg, Germany. From 1996 to 2001, he held the Chair of Microelectronics with the Department of Electrical Engineering, University of Saarbrücken, Germany. In 2001, he joined the Department of Microsystems Engineering (IMTEK) of the University of Freiburg, Germany, where he established the Chair of Microelectronics. Since 2005, he has additionally served as one of the three directors at the "Institute of Micromachining and Information Technology" (HSG-IMIT) in Villingen-Schwenningen, Germany. He spent sabbaticals with Motorola (now Freescale) in Phoenix, AZ, and with Intel, Santa Clara, CA. Prof. Manoli has received numerous best paper and teaching awards, and he has served on the committees of ISSCC, ESSCIRC, IEDM and ICCD. He was Program Chair (2001) and General Chair (2002) of the IEEE International Conference on Computer Design (ICCD). He is on the Senior Editorial Board of the IEEE *Journal on Emerging and Selected Topics in Circuits and Systems* and on the Editorial Board of the *Journal of Low Power Electronics*. He served as guest editor of *Transactions on VLSI* in 2002 and *Journal of Solid-State Circuits* in 2011.

Jiri Marek has been Senior Vice President of Engineering Sensors at Bosch, Automotive Electronics Division, since 2003, responsible for the MEMS activities at Bosch. He started his work at Bosch in 1986. From 1990 until 1999, he was responsible for the Sensor Technology Center. In 1999, he became Vice President Engineering of Restraint Systems and Sensors. Dr. Marek studied Electrical Engineering at the University of Stuttgart, Germany, and Stanford University, USA. In 1983, he received his Ph.D. from the University of Stuttgart for his work at the Max Planck Institute Stuttgart on the analysis of grain boundaries in large-grain polycrystalline solar cells. After a post-doctoral fellowship with IBM Research, San José, CA, he was a development engineer with Hewlett-Packard, Optical Communication Division.

Boris Murmann is an Associate Professor in the Department of Electrical Engineering, Stanford University, CA, USA. He received his Dipl.-Ing. (FH) degree in communications engineering from Fachhochschule Dieburg, Germany, in 1994, a M.S. degree in electrical engineering from Santa Clara University, CA, in 1999, and his Ph.D. degree in electrical engineering from the University of California, Berkeley, in 2003. From 1994 to 1997, he was with Neutron Mikrolektronik GmbH, Hanau, Germany, where he developed low-power and smart-power ASICs in automotive CMOS technology. Dr. Murmann's research interests are in the area of mixed-signal integrated circuit design, with special emphasis on data converters and sensor interfaces. In 2008, he was a co-recipient of the Best Student Paper Award at the VLSI Circuit Symposium and the recipient of the Best Invited Paper Award at the Custom Integrated Circuits Conference (CICC). In 2009, he received the Agilent Early Career Professor Award. Dr. Murmann is a member of the International Solid-State-Circuits Conference (ISSCC) program committee, an associate editor of the *IEEE Journal of Solid-State Circuits*, and a Distinguished Lecturer of the IEEE Solid-State Circuits Society.

Barry Pangrle is a Solutions Architect for Low Power in the Engineered Solutions Group at Mentor Graphics Corporation. He has a B.S. in computer engineering and a Ph.D. in computer science, both from the University of Illinois at Urbana-Champaign. He has been a faculty member at University of California, Santa Barbara and Penn State University, where he taught courses in computer architecture and VLSI design while performing research in high-level design automation. Barry has previously worked at Synopsys, initially on high-level design tools and then later as an R&D director for power optimization and analysis tools. He was the Director of Design Methodology for a fabless start-up company and has also worked at a couple of privately held EDA companies, where he focused on design automation tools for low power/energy designs. He has published over 25 reviewed works in high level design automation and low power design and served as a Technical Program Co-chair for the 2008 ACM/IEEE International Symposium on Low Power Electronics Design (ISLPED). He was also actively involved with the technical program committees for ISLPED and DAC for 2009 and was one of the General Co-Chairs for ISLPED 2010.

Peter H. Roth received his Dipl.-Ing. degree in electrical engineering and his Dr.-Ing. degree from the University of Stuttgart, Germany, in 1979 and 1985, respectively. In 1985, he joined the IBM Germany Research and Development Lab in Boeblingen, starting in the department of VLSI logic-chip development. Since 1987 he has been leading the VLSI test and characterization team of the Boeblingen lab. Later, Dr. Roth led several development projects in the area of IBM's Mainframe and Power microprocessors. He also was heavily involved in the development of gaming products such as the Cell processor for the Sony PlayStation. Today, Dr. Roth is responsible for the hardware strategy for the IBM Germany Research and Development Laboratory.

Albrecht Rothermel received his Dipl.-Ing. degree in electrical engineering from the University of Dortmund, and his Dr.-Ing. (Ph.D.) degree from the University of Duisburg, both Germany, in 1984 and 1989, respectively. From 1985 to 1989, he was with the Fraunhofer Institute of Microelectronic Circuits and Systems, Duisburg, Germany. From 1990 to 1993, he was with Thomson Consumer Electronics (Thomson Multimedia), Corporate Research, Villingen-Schwenningen, Germany. As manager of the IC design laboratory, he was involved in analog and mixed circuit design for audio and video. Since 1994, he has been with the Institute of Microelectronics, University of Ulm, Germany, as a Professor of Electrical Engineering. Dr. Rothermel was a guest scientist at Thomson Multimedia in Indianapolis, USA (1997), at the Edith-Cowan University in Perth, Western Australia (2003), and at the Shandong University in Jinan, P.R. China (2006). He has published more than 100 papers, book chapters, and patents and received numerous best-paper awards. He was an Associate Editor of the *IEEE Journal of Solid-State Circuits*, and he is a member of the program committees of ISSCC, ESSCIRC, and ICCE. He is a senior member and distinguished lecturer of the IEEE.

Lambert Spaanenburg received his master's degree in electrical engineering from Delft University and his doctorate in technical sciences from Twente University, both in The Netherlands. He started his academic journey at Twente University in the field of VLSI design, eventually serving as CTO of ICN, the commercial spin-off of the ESPRIT Nelsis project. At IMS in Stuttgart, he co-created the neural control for the Daimler OSCAR 1992 prototype, currently an upcoming standard for car safety measures. He became a full professor at the University of Groningen, The Netherlands. Further research of his group in neural image processing led in 2002 to Dacolian, which held a 60% market share for license-plate recognition before it merged into Q-Free ASA. Overall, Dr. Spaanenburg has produced more than 200 conference papers, 20 reviewed journal articles and seven books or chapters. He has served annually on several program committees and journal boards. He has been involved in many technology transfer projects, including eight university spin-offs. Currently a guest at Lund University, he is exploiting further research on embedded sensory networks with cloud connectivity under the umbrella of RaviteQ AB in Sweden.

Kai Weber received his Dipl.-Ing.(BA) in information technology from the University of Collaborative Education, Stuttgart, and a bachelor degree in computer science from the Open University, UK, in 2003. He joined IBM Research and Development in 2003, working on formal verification of floating-point units and leading the floating-point unit verification teams for IBM's POWER6 and z10 processors. Mr. Weber was the processor verification team leader for the IBM zEnterprise 196 mainframe and continues in this role for future generations of IBM System z processors. In addition Mr. Weber is participating in two joint research projects with the University of Tuebingen and Saarland University evaluating hybrid systems and business analytic workloads.

Acronyms and Abbreviations

AAEE	American Association for Engineering Education
AC	Alternating current
ACM	Association for Computing Machinery
ADC	Analog–digital converter
AI	Artificial intelligence
ALE	Atomic-layer epitaxy
ALU	Arithmetic logic unit
AMD	Age-related macula degeneration
ANN	Artificial neural network
AP	Action potential
APS	Active-pixel sensor
APSM	Advanced porous silicon membrane
AQL	Acceptable quality level
ARPA	Advanced Research Projects Agency
ASIC	Application-specific integrated circuit
ASIP	Algorithm-specific integrated processor
ASP	Application-specific processor
ASS	Application-specific system
AVG	Available voltage gain
AVS	Adaptive voltage scaling
BAN	Body-area network
BAW	Bulk acoustic wave
BEOL	Back-end of line
BI	Backside illumination
BiCMOS	Bipolar CMOS
BIST	Built-in self-test
BMI	Brain–machine interface
BST	Boundary-scan test
C4	Controlled-collapse chip connect
CAD	Computer-aided design
CAM	Content-addressable memory
CARE	Concerted Action for Robotics in Europe

CAT	Computer-aided test
CCD	Charge-coupled device
CDMA	Code-division multiple access
CE	Continuing education
CFC	Chip-integrated fuel cell
CIFB	Cascade of integrators in feedback
CIFF	Cascade of integrators in feed-forward
CIS	Charge-integration sensor
CISC	Complex-instruction-set computer
CMOS	Complementary metal–oxide–semiconductor
CNN	Cellular neural network
CNT	Carbon nanotube
COO	Cost of ownership
CORDIC	Coordinate rotation digital computer
CPU	Central processing unit
CSF	Contrast-sensitivity function
CT-CNN	Continuous-time CNN
DA	Design automation
DAC	Design-Automation Conference
DAC	Digital–analog converter
DARPA	Defense Advanced Research Projects Agency
DBB	Digital baseband
dBm	Power on a log scale relative to 1 mW
DC	Direct current
DCF	Digital cancelation filter
DDR2	Dual data rate RAM
DEM	Dynamic element matching
DFT	Design for test
DIBL	Drain-induced barrier lowering
DIGILOG	Digital logarithmic
DLP	Digital light processing
DMA	Direct memory access
DN	Digital number
DNN	Digital neural network
DPG	Digital pattern generator
DPT	Double-patterning technology liquid-immersion lithography
DRAM	Dynamic random-access memory
DRIE	Deep reactive-ion etching
DSP	Digital signal processor
DT-CNN	Discrete-time CNN
DTL	Diode–transistor logic
D-TLB	Data-cache translation lookaside buffer
DVFS	Dynamic voltage and frequency scaling
EBL	Electron-beam lithography

ECC	Error-correcting code
ECG	Electrocardiogram
ECL	Emitter-coupled logic
EDA	Electronic design automation
EEG	Electroencephalography
EITO	European Information Technology Organization
ELO	Epitaxial lateral overgrowth
ELTRAN	Epitaxial-layer transfer
EMI	Electromagnetic interference
ENOB	Effective number of bits
EOT	Equivalent oxide thickness
ERC	Engineering Research Center
ERD	Emerging research devices
ERM	Emerging research materials
ESD	Electrostatic discharge
ESL	Electronic-system level
ESP	Electronic safety package
ESP	Electronic Stability Program
ESPRIT	European Strategic Programme for Research in Information Technology
EUV	Extreme ultraviolet
EWS	Electrical wafer sort
FACETS	Fast Analog Computing with Emergent Transient States
FC	Fuel cell
FD SOI	Fully depleted silicon-on-inulator
FED	Future Electron Devices
FEOL	Front-end of line
FeRAM	Ferroelectric random-access memory
FET	Field-effect transistor
F^2	Square of minimum feature size
FFT	Fast Fourier transform
FIFO	First-in-first-out
FIPOS	Full isolation by porous silicon
FIR	Finite impulse response
FIT	Failure in 10^7 h
FLOP	Floating-point operation
FMEA	Failure mode and effect analysis
FOM	Figure-of-merit
FPAA	Field-programmable analog array
FPGA	Field-programmable gate array
fps	frames per second
FR	Floating-point register
FSM	Finite-state machine
FSR	Full signal range

GAPU	Global analogic programming unit
GBP	Gain–bandwidth product
GFC	Glucose fuel cell
GIPS	Giga instructions per second
GOPS	Giga operations per second
GPS	Global Positioning System
GPU	Graphics processing unit
GR	General-purpose register
GSM	Global System for Mobile Communication
HAC	Hardware accelerator
HD	High definition *or* high density
HDL	Hardware description language
HDR	High dynamic range
HDRC	High-Dynamic-Range CMOS
HDTV	High-definition television
HIPERLOGIC	High-performance logic
HKMG	High-k metal gate
HLS	High-level synthesis
HVM	High-volume manufacturing
IBL	Ion-beam lithography
IC	Integrated circuit
ICT	Information and communication technology
IEDM	International Electron Devices Meeting
IEEE	Institute of Electrical and Electronics Engineers
IFFT	Inverse fast Fourier transform
I^2L	Integrated injection logic
IO	Input–output
IOMMU	I/O memory mapping unit
IP-core	Intellectual property core
IS	Instruction set
ISO	International Organization for Standardization
ISSCC	International Solid-State Circuits Conference
IT	Information technology
I-TLB	Instruction-cache translation lookaside buffer
ITRS	International Technology Roadmap for Semiconductors
ITS	Intelligent transportation systems
JEDEC	Joint Electron Device Engineering Council
JESSI	Joint European Submicron Silicon Initiative
JIT compiler	Just-in-time compiler
JND	Joust noticeable difference
JPEG	Joint Photography Expert Group
KGD	Known-good die
LCA	Life-cycle analysis
LDO	Low-voltage dropout regulator

LOG	Localized epitaxial overgrowth
LPC	Linear predictive coding
LPDDR	Low-power DDR
LSB	Least significant bit
LSI	Large-scale integration
LTE	Long-term evolution
LUT	Look-up table
LVDL	Low-voltage differential logic
MASH	Multi-stage noise shaping
MBE	Molecular-beam epitaxy
MCC	Manchester carry chain
MCM	Multi-chip module
MCU	Multi-core processing units
MEBL	Multiple-electron-beam lithography
MEMS	Micro electro mechanical system
MIPS	Mega instructions per second
MITI	Ministry of International Trade and Industry (Japan)
MLC	Multi-level per cell
MMI	Man–machine interface
MMI	Machine–machine interface
MOPS	Million operations per second
MOS	Metal–oxide–semiconductor
MOSIS	MOS IC implementation system
MPEG	Motion-Picture Expert Group
MPU	Microprocessor unit
MRAM	Magnetic random-access memory
NA	Numerical aperture
NGL	Next-generation lithography
NHTSA	National Highway Traffic Safety Administration
NIR	Near infrared
NM	Noise margin
NMOS	Negative channel-charge MOS
NoC	Network on chip
NTF	Noise transfer function
NV	Nonvolatile
OCT	Optical-coherence tomography
ODM	Original-device manufacturer
OECD	Organisation for Economic Co-operation and Development
OECF	Optoelectronic conversion function
OFDM	Orthogonal frequency-division multiplexing
OpenCL	Open Compute Language – Standard established by the Khronos Group for platform independent description of highly parallel computations
ORTC	Overall roadmap technology characteristics

OSAT	Out-sourced assembly and test
OSCI	Open SystemC Initiative
OSR	Oversampling ratio
PAE	Power-added efficiency
PC	Personal computer
PCB	Printed-circuit board
PCM	Phase-change memory
PCS	Personal Communications Service
PDN	Power distribution network
PD	SOI partially depleted silicon-on-inulator
PE	Processing element
PEM	Polymer electrolyte membrane
PLL	Phase-locked loop
PMOS	Positive channel-charge MOS
PMU	Power management unit
POP	Package-on-package
PPA	Power, performance, and area
ppm	Parts per million
PVDF	Polyvinylidenefluoride
PZT	Lead zirconate titanate
QCIF	Quarter common intermediate format
QMS	Quality-management system
QXGA	Quad extended graphics array
RAM	Random-access memory
R&D	Research and development
REBL	Reflective electron-beam lithography
RET	Resolution-enhancement technique
RF	Radio frequency
RFID	Radio frequency identification
RISC	Reduced-instruction-set computer
ROI	Region of interest
ROM	Read-only memory
ROW	Rest-of-world
RP	Retinitis pigmentosa
RRAM	Resistive random-access memory
RTL	Register-transfer level
Rx	Receive
SAR	Successive approximation register
SBT	Strontium bismuth tantalite
SC	Switch capacitor
SD	Standard definition
SDR	Single data rate
SECE	Synchronous electric charge extraction
SEG	Selective epitaxial growth

SET	Single-electron transistor
SIA	Semiconductor industry association
SIMOX	Silicon isolation by implanting oxygen
SiO_2	Silicon dioxide
SLD	System-level design
SMASH	Sturdy multi-stage noise shaping
SMP	Symmetric multi-processing
SMT	Simultaneous multithreading
SNDR	Signal-to-noise-and-distortion ratio
SNM	Static noise margin
SNR	Signal-to-noise ratio
SNT	Silicon nanotube
SO	Switch operational amplifier
SOC	System-on-chip
SOI	Silicon-on-insulator
SPAD	Single-photon avalanche diode
SPI	Serial peripheral interface
SQNR	Signal-to-quantization-noise ratio
SRAM	Static random-access memory
SRC	Semiconductor Research Corporation
STF	Signal transfer function
STL	Subthreshold leakage
STT	Spin–torque transfer
SyNAPSE	Systems of Neuromorphic Adaptive Plastic Scalable Electronics
TAM	Total available market
TFC	Thin-film-on-CMOS
TLM	Transfer-level modeling
TOPS	Tera operations per second
TQM	Total quality management
TSV	Through-silicon via
TTL	Transistor–transistor logic
Tx	Transmit
UGBW	Unity-gain bandwidth
UL	Underwriters Laboratories
ULP	Ultralow power
UM	Universal model (CNN)
UMTS	Universal Mobile Telecommunications System
VCO	Voltage-controlled oscillator
VDE	Verband Deutscher Elektrotechniker
VDI	Verband Deutscher Ingenieure
VDC	Vehicle dynamics control
VDS	Vehicle dynamics system
VGA	Video Graphics Array
VHDL	VLSI (originally VHSIC) hardware description language

VHSIC	Very-high-speed integrated circuit
VLSI	Very-large-scale integration
VRD	Virtual retinal display
VSoC	Vision system on chip
wph	Wafers per hour
WSI	Wafer-scale integration
WSTS	World Semiconductor Trade Statistics
XML	Extensible Markup Language

Chapter 1
Introduction: Towards Sustainable 2020 Nanoelectronics

Bernd Hoefflinger

Abstract Faced with the immanent end of the nanometer roadmap at 10 nm, and with an electronics energy crisis, we have to engineer the largest strategy change in the 50-years history of microelectronics, renamed to nanoelectronics in 2000 with the first chips containing 100-nm transistors.

Accepting the 10 nm-limit, the new strategy for the future growth of chip functionalities and markets has to deliver, within a decade, another 1,000× improvement in the energy per processing operation as well as in the energy per bit of memory and of communication. As a team from industry and from research, we present expectations, requirements and possible solutions for this challenging energy scenario of femto- and atto-Joule electronics.

The introduction outlines the book's structure, which aims to describe the innovation eco-system needed for optimum-energy, sustainable nanoelectronics. For the benefit of the reader, chapters are grouped together into interest areas like transistors and circuits, technology, products and markets, radical innovations, as well as business and policy issues.

1.1 From Nanoelectronics to Femtoelectronics

In the year 2000, the first microchips were produced with gate lengths <100 nm, and microelectronics received the new label *nanoelectronics*. The drive in the industry along the *nano-roadmap* towards shorter transistors continued in order to build faster processors and to pack more memory bits on each chip. At the same time, the research community had widespread programs running on

B. Hoefflinger (✉)
Leonberger Strasse 5, 71063 Sindelfingen, Germany
e-mail: bhoefflinger@t-online.de

B. Hoefflinger (ed.), *CHIPS 2020*, The Frontiers Collection,
DOI 10.1007/978-3-642-23096-7_1, © Springer-Verlag Berlin Heidelberg 2012

quantum–nanometer-scale phenomena with considerable optimism for near-term practical success. By the year 2010, the planned 32 nm milestone (node) had been reached, including most of the expected chip specifications, but with a twist since 2005: the 3D integration of many thin chips on top of each other, connected through TSVs (through-silicon vias), to enhance the progress on the roadmap. But the year 2010 was also marked by the rapidly growing consensus that the *end of the roadmap is near* at 15 nm (2016) or 10 nm (2018?) at best, and that *none of the new quantum-nano devices will have any economic impact before 2025–2030*. This poses the serious question: Will the progression of chips come to a standstill, and with it the world's driving technology (information and communication)? The answer is: Not necessarily, if we accept the nanometer-limit and, at the same time, exchange the nanometer priority for a femtojoule priority: *Energy per function*, often called the *power efficiency*, *is the new yardstick*.

Between 2010 and 2020, the energy per chip function, such as processing, memory, or communication, has to be reduced by a factor of 1,000 if the nanoelectronics market is going to have enough to offer to six billion potential global customers. Remarkably, many of these chip functions run at picojoule (pJ) levels in 2010. The challenge is now to achieve femtojoule (fJ) levels. Therefore we set our focus on moving from *nano*(meter) *to femto*(joule) *electronics*.

The task is challenging indeed, because the nanometer roadmap only offers a final contribution of threefold at best, which may enter as $3 = 9$ in an energy figure-of-merit (FOM), leaving us with another factor of >100 in needed improvements within a decade. One approach to identifying a future strategy is to consider how the remarkable advances in the 100 years of electronics and, particularly, in the past 50 years of integrated circuits were achieved and which repertory of R&D results of the past 30 years could be put to the test in product developments of the present decade. This is the theme of Chap. 2, "From Micro-electronics to Nanoelectronics".

Our path to the 2020 goals will point out critical electronic functions, which are most challenging and intriguing, since we reach and go below the energy per operation in the synapses of natural brains: The typical energy per operation of a neuron's synapse is 10 fJ.

We pose five fundamental questions in this book, and we give answers, all of which point out tough energy requirements on future nanochips so that we conclude: *Sustainable nanoelectronics has to be femto(joule) electronics*.

Here are the five questions:

– Why do we hit the end of the nano-roadmap?
– Which femtoelectronic solutions can we find for critical functions?
– What are the requirements for new chip products and how can we meet them?
– Which radical femtoelectronic solutions should we seek for intelligent computing?
– What are the challenges for the femtoelectronics ecosystem of education, research, and business?
– How does the 2020 world benefit from femtoelectronic chips?

1.2 Why Do We Hit the End of the Nano-roadmap?

Because microelectronics has advanced so successfully over 50 years with a linear strategy of scaling down transistor sizes, it comes as a surprise that this mode of progression will hit its limits by 2015. We deal with these problems in Chaps. 3 (10 nm transistors), 7 (ITRS, the nano-roadmap), and 8 (nanolithography).

In Chap. 3, we evaluate the future of eight chip technologies, revealing that their development towards a fruitful future took between 25 and more than 35 years, an example being the present emphasis on the 3D integration of chips, which was taken on as a major research topic 30 years ago, only to be put aside in the late 1980s because the 2D scaling strategy did the job well and less *disruptively*. Transistors with gate lengths of <10 nm on the scaling roadmap were built in the lab and published before 2000. Why are their fundamental problems being considered only now? In Sect. 3.2, we identify two fundamental problems, which are present no matter how precisely we can process these 10 nm transistors:

1. The atomic variance (only ~5 doping atoms in the channel) makes the spread of transistor thresholds larger than the allowed supply voltage, so not all of them can be turned on or off.

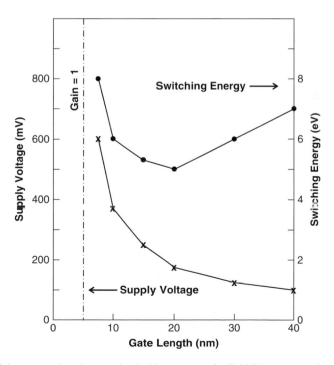

Fig. 1.1 Minimum supply voltage and switching energy of a CMOS inverter as a function of gate length for an equivalent gate-oxide thickness (EOT) of 1 nm

2. The available voltage gain (because of drain-induced barrier lowering) becomes
   <1, and without amplification, circuits lose their signals in noise (at least in all
   electronics operating in natural environments).

Because of these fundamental limits, we actually find that the most important
basic amplifier in modern chip electronics, the CMOS (complementary metal–
oxide–semicondiuctor) inverter, has its minimum energy per operation at a gate
length of 20 nm (Fig. 1.1), because, moving towards 10 nm, we have to raise the
supply voltage in order to compensate for the facts listed above. Nevertheless, the
switching energy of this fundamental amplifier is just $5\,eV = 10^{-18}$ J, which means
that we have to move just 25 electrons through a potential difference of 200 mV.

A 10 nm transistor and, for that matter, even a 20 nm transistor, performs this by
moving just one electron at a time through the channel at speeds (Fig. 1.2) depending
on the instantaneous voltages. That is, practically *all minimum transistors with gate
lengths <20* nm *are single-electron transistors.* We present these fundamental
details at the outset of this book in order to show that, at the end of the nano-
roadmap, we are putting (at least digital) *information on single electrons.*

Our emphasis when discussing the International Technology Roadmap for
Semiconductors (ITRS) in Chap. 7 is the projection for many of the other components

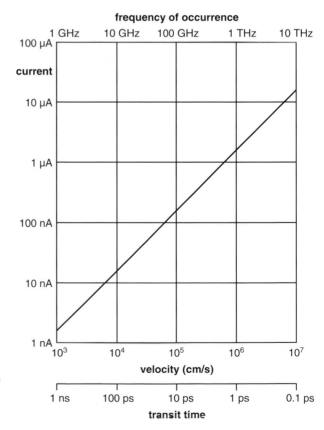

Fig. 1.2 Current carried by a
single electron at a given
velocity over a 10 nm
distance

on nanochips, such as capacitors and multilevel interconnects, which are important in an assessment of the processing, memory, and communication functions on a chip.

Any of the aggressive projections for down-sizing the features on future chips assume that we can image these features onto the Si wafers, and everyone agrees that nanolithography is the absolute key technology in realizing these incredibly complex structures effectively for many successive layers per wafer and millions of wafers per year. Burn Lin, as vice president of lithography at the world's largest contract manufacturer of wafers, TSMC, recognized as one of the three semiconductor companies at the forefront of technology, addresses both physical and economic issues in Chap. 8. He analyzes the two major contending techniques, EUV (extreme ultraviolet) and MEB (multiple-electron-beam direct write), and he tells us that EUV systems impose extreme requirements on the primary laser source, on the precision of mirrors and masks, and that the overall energy efficiency from plug to wafer is so low that the necessary primary electrical power might be bigger than the total power available for a gigafab (giant semiconductor fabrication plant). On the other hand, an e-beam system, attractive because it is maskless, would need 130,000 parallel, steered beams in order to be competitive, and such a system has not been built as of 2010 owing to a lack of product development. These enormous problems are a third, practical reason besides the fundamental ones listed above for a 10 nm limit on chip features.

1.3   Which Femtoelectronic Solutions Can We Find for Critical Functions?

Regarding the question of femtoelectronic solutions, we take the standard position that innovations in technology drive progress (technology push), but we focus this fundamental force on the energy question, combined with operating speed and chip area (footprint/function), in order to meet the great expectations made of future chips. The topics covered and the chapters in which they can be found are listed in Table 1.1.

Table 1.1 Femtoelectronic topics in this book	Topic	Chapter or section
	Logic and computing	3, 10
	Analog/digital interfaces	4
	Interconnects and transceivers	5
	Memories	11
	3D integration	3.5, 12
	MEMS sensors	13
	Vision sensors	14
	Retinal implants	17
	Power-efficient design	9
	Energy harvesting and chip autonomy	19

Fig. 1.3 The cross-coupled inverter pair (*quad*) as a differential signal regenerator, the fundamental building block of ultra-low-voltage digital femtoelectronics

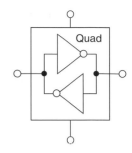

We start out with digital logic in Sect. 3.3, with the conclusion from the preceding section that the variance of 10 nm transistors forces the application of a new type of robust regenerator, the cross-coupled inverter pair, offering a differential, high-speed, full-swing regenerator with minimum dynamic energy. This block of four properly connected transistors (two NMOS and two PMOS) becomes the fundamental circuit unit to be optimized regarding chip footprint and process complexity, presented as a device-level 3D-integrated process in Sects. 3.4 and 3.5. The message here is that, in 10 nm CMOS, the target of making a transistor falls short of the necessary fundamental building block, which is the four-transistor cross-coupled inverter pair with minimum footprint, interconnects, and process steps (Fig. 1.3).

Out of the multitude of applications of the differential regenerator in digital logic, we select the toughest component, the high-speed, minimum-energy, and minimum-footprint multiplier, the largest macro in arithmetic and, especially, in all signal-processing units (images and video). As an example of possible development directions, we describe the HIPERLOGIC multiplier in Sect. 3.6, which has the potential to perform 600 M 16 $\times$ 16 bit multiplications with a power-efficiency of one TOps mW^{-1} = 1 fJ, which would exceed the 2010 state-of-the-art by a factor of 1,000 (Fig. 1.4).

On the higher levels of digital chip functionality, high-performance processors and their future are described by Peter Roth, Christian Jacobi, and Kai Weber of IBM in Chap. 10. Owing to the clock speed limits of <10 GHz towards the end of the nano-roadmap, multicore processors have become a must, and they discuss the associated challenges for operating systems and application development, needing new levels of hardware–software co-design.

This complex design scenario is analyzed systematically by Barry Pangrle of Mentor Graphics in Chap. 9, with a focus on power efficiency on all levels: system, chip, process technology, test, and packaging. He comes to the conclusion that "it won't be the technology that's the limit but the cost of implementation."

Processors are the much-quoted hearts of advanced integrated chips, but much more functionality is integrated today, namely true *systems-on-a chip* (SOCs). These SOCs incorporate

– Sensors, which sense the environment and mostly produce analog output signals;
– Analog–digital interfaces converting analog sensor or receiver outputs to digital signals for processing;

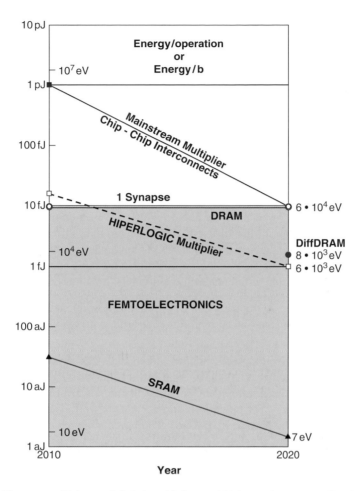

Fig. 1.4 The power-efficiency of digital multipliers and high-speed memory cells advancing to femtojoule and attojoule levels

- Receivers, which receive radio signals or signals on wires as well as optical fibers;
- Transmitters, which send output signals on lines or into space;
- High-speed interconnects on-chip or chip-to-chip;
- Memories, which store digital data and instructions for processing these;
- Energy sources on-chip to produce or/and harvest a part or all of the chip energy necessary for its operation;
- 3D integration as the key technology to merge these functionalities.

The state-of-the-art and the future of the highly sophisticated disciplines treating these specialties are presented in dedicated chapters.

Boris Murmann of Stanford University with Yiannos Manoli and Matthias Keller of the University of Freiburg, Germany, are recognized leaders in analog–digital converters, and they show in Chap. 4 that these converters have proceeded along their own, remarkable roadmap of energy per conversion and that there is great potential for future improvements in energy and speed using advanced, high-density CMOS technology nodes for digital-processing-based iteration and correction schemes at low voltages for precision conversion rather than traditional high-voltage, operational-amplifier-based converters.

Interconnects, transmitters, and receivers are studied in Chap. 5, and we find that these communication functions are the most difficult to reduce in their energy requirements. Even today, they make up ~1/3 of the total energy of chip systems, and even with a rigorous transition to low-voltage differential signaling as well as with 3D integration for shorter communication distances, the energy per bit will only be reduced by <100 times in a decade (Fig. 1.4).

Memory energy is responsible for about another 1/3 of a chip-system's energy, and we will see in Chap. 11 that different memory types need different strategies for their advancement. The high-speed static CMOS random-access memory (SRAM) is best prepared for down-sizing both in footprint and in dynamic energy because of its differential *active* cell and differential sense amplifiers. It will also benefit fully from the 3D device-level integration of the fundamental four-transistor differential regenerator, so that it can reach a density of 16 Gbit/cm^2 by 2020 with a dynamic energy per bit of just 7 eV (Fig. 1.4). The density of a dynamic RAM (DRAM) is 5 times higher in 2010. However, DRAM can only be scaled to 22 nm and 13 Gbit/cm^2. A differential DRAM is proposed, which would enable a 2D density of 50 Gbit/cm^2 by 2020 and a read energy of 8,000 eV, eight times better than the best DRAM.

Figure 1.4 shows the development potential of the power efficiency of digital multipliers (their energy per operation) and SRAM as well as DRAM (dynamic energy per bit) over the decade 2010–2020. The remarkable advances are detailed in the corresponding chapters. All these functions reach or surpass the femtoelectronic mark or that of a neuron's synapse at 10 fJ $= 6 \times 10^4$ eV per operation (per bit).

Multilevel flash memories, the large-scale leaders among nonvolatile memories, have made impressive progress since their first realization 1985. They are heading towards a minimum of 500 eV/bit, but again with a scaling limit at 22 nm (3 bits per cell) or 16 nm (2 bits per cell). These nonvolatile memories have seen the largest development effort on 3D integration into chip stacks of up to 128 chips, producing densities of many terabits per square centimeter. These *solid-state drives* (SSDs) are displacing magnetic-disk memories. Among the alternative styles of nonvolatile RAMs, phase-change (PCM) and resistive (RRAM) memories are progressing towards 20 Gbit/cm^2. More aggressive tokens for a nonvolatile bit, for example the spin of an electron, continue to be burdened by energy- and space-hungry write-and-read transistor circuitry. Single-electron wells with Coulomb confinement may become the elements for a quantum-CCD (charge-coupled device) memory.

Regarding the variety of sensors, Jiri Marek and Udo-Martin Gómez of Robert Bosch GmbH, Germany, world leaders in micro-electro-mechanical systems (MEMS) for automotive applications, present, in Chap. 14, a comprehensive and quantitative overview of how much it takes in creative and extreme quality-conscious development to put >1 billion safety-critical electromechanical sensing systems on the road. Tiny 3D mechanical structures combined with calibration and processing chips are applied for the realization of new sensing systems as well as for optimizing energy and manufacturing costs. These sensing systems are now expanding into many, particularly portable, consumer products.

The other very important, highly integrated sensor systems are those for imaging and video. These are discussed in Chap. 15 with respect to selected topics such as sensitivity and pixel dark current, resolution and lens aperture, 3D integration or backside illumination for high sensitivity, eye-like high dynamic range, and 3D imaging. The performance of vision sensors is the first level in the complex and endlessly challenging system of intelligent vision and performance- and bandwidth-efficient video.

Very demanding vision specialties are retinal implants for blind people. Albrecht Rothermel, the chip specialist in one of the world-leading consortia for such implants, gives a balanced overview of the global status in 2010 in Chap. 17. He describes in a holistic spirit all aspects, including chip design, physiology, surgery, and training with patients; he further outlines what we can expect by 2020.

Implants are the exemplary type of chips for which energy is the dominant issue. How can we possibly produce energy on-chip or *harvest* energy on the chip or from its environment? One of the leading institutes in the field of energy harvesting is IMTEK, the Institute for Microsystems Technology in Freiburg, in cooperation with IMIT, Germany. Yiannos Manoli and his team present an exceptional tutorial, assessing the state-of-the art and highlighting future developments of such sources as motion, vibration, temperature differences, light, biofuel, and fuel cells. These sources and transducers operate in ultralow-voltage systems, and the authors show in Fig. 19.17 that, as of 2010, minimum supply voltages have been realized at $L = 130$ nm, while voltages had to be higher at shorter gate lengths, results confirming – as in our introductory Fig. 1.1 – that, energy-wise, *shorter is no longer better*.

1.4 What Are the Requirements for New Chip Products and How Can We Meet Them?

After the assessment of Si technology and circuit capabilities and limits, we reflect these against the requirements in the major market categories for nanoelectronics (Table 1.2).

Table 1.2 Areas of applications and the chapters in which they are treated

Topic	Chapter
Overview and markets	6
Superprocessors and supercomputers	10
Wireless and mobile companions	12, 13
Automotive applications	14
Medical implants	17

The overview in Chap. 6 tells us that the semiconductor industry, with aggregate growth rates of 4–7% per year, advancing toward >450 billion US$ by 2020, continues to be the technology driver for the information and communication economy. Communication, consumer, and automotive chip markets show growth rates above average. The perspective of

- Six billion mobile-phone subscribers,
- A computer tablet per student with free access to the world's libraries, and
- A carebot (personal robot) per family (in Asia)

explains our emphasis on wireless mobile. Georg Kimmich, the system-on-chip product manager at ST Ericsson, describes, in Chap. 12, the remarkable progress of the 3D integration of heterogeneous chips, processors, memories, and transceivers, in close interaction with wide-bandwidth architectures, toward high-quality, cost-effective multimedia. In Chap. 13 Greg Delagi, Senior Vice-President for Wireless Systems at Texas Instruments, takes on the full perspective of the personal mobile companion of 2020:

- 3D imaging and display,
- Gesture interface,
- Integrated projection,
- Object and face recognition,
- Context awareness,
- Internet of things
- Brain–machine interface,
- Body-area network connection.

His chapter is the most explicit catalog of requirements on functional through-put, transceiver sensitivity, power, and bandwidth, and, throughout, a 1,000 times less energy per function, ultralow-voltage circuitry being the first thing to be promoted.

On the other side of the energy equation, he stresses the needed progress on energy harvesting. Considering the magnitude of the challenges, he asks for a "strong collaboration in research and development from universities, government agencies, and corporations" (see Sect. 1.6 and Chap. 22).

1.5 Which Radical Femtoelectronic Solutions Should We Seek for Intelligent Computing?

As alternatives to traditional computing architectures with processor, instruction- and data-memory, artificial neural networks (ANNs) saw about a dozen years of intense research between 1988 and 2000, as often, a bit too early to have a broad, disruptive impact. Systems with a large number of inputs and with tasks of making decisions or recognizing patterns on the basis of rules, learning, and knowledge can be realized with these networks at speeds and energy levels unmatched by other architectures. Their final success is more likely now, because integrated electronic *neurons* are becoming effective on all counts at Si technology nodes <100 nm and because their fundamental 3D architecture can finally be realized with the large-scale introduction of 3D chip integration. Digital neural networks and silicon brains are treated in Chaps. 16 and 18, respectively.

Ben Spaanenburg of Lund University, Sweden, has had a distinguished career in high-level chip design and cellular neural networks (CNNs). He and his coworker Suleiman Malki present a concise tutorial on working CNN chips and chip systems, both analog and digital. They describe new digital neuroprocessors for *vision-in-the-loop* new media tasks, from software, through scalable architecture, to digital-neuron design.

These CNNs are based on the extensively studied multilayer perceptron, which has become a quasi standard. By contrast, really radical research programs were launched in 2008 on *silicon brains*; on the one hand because of the availability of ultralarge-scale nano-CMOS and 3D chip integration, but actually with the intention of building general-purpose, biomorphic chip systems with the brain's complexity of 10^{11} neurons and 10^{14} synapses (Chap. 18).

1.6 What Are the Challenges for the Femtoelectronics Ecosystem of Education, Research, and Business?

The broad front along which chips can be advanced raises questions concerning all kinds of resources: the energy crisis (Chap. 20), the extreme-technology industry (Chap. 21), and education and research (Chap. 22).

The globally installed electric power in 2010 is ~2 TW (2×10^{12} W). Information and communication technology (ICT), operated by electronic chips, is estimated to need >20% of the world's electric power. In Chap. 20, we look just at the needs of data centers, those *server farms* with 36 million servers in 2010, the processing backbone of the internet. They require ~36 GW, 10% of the total electric power installed in Europe. These data centers need and are expected to increase their performance 1,000-fold within a decade. This is obviously incompatible with the electric power available, unless this performance increase is achieved with, hopefully, a 1,000-fold improvement in power efficiency per function, which is synonymous with a 1,000-fold reduction in energy per function, the magic factor throughout this book. We also note in Chap. 20 that energy in a gigafab makes up >80% of the cost of a fully processed Si wafer. This leads us to Chap. 21 on the chip industry as "the extreme-technology industry", marked by an investment rate >15% as well as an R&D rate >15% of revenues, twice as high as the R&D rate of the top *pharmaceuticals* company in the Organisation for Economic Co-operation and Development (OECD) list of high-tech industries.

Naturally, these unique rates of progress can only be achieved with a unique interest in and unique investments in highly skilled manpower at all levels from kindergarten to retirement (Chap. 22). The concerns about the ecosystem of education, research, industry, government, and the public are global. However, no wave, like the one triggered by *Sputnik* in the 1960s or that in the 1980s is in sight as of 2010.

After the world financial crisis of 2008–2009, with limited global financial and energy resources, a clear, requirements-driven research strategy should be established. For chips, as consistently stated throughout this book, the strategy should be *sustainable nanoelectronics towards femtojoule electronics*, with the key applications *educational tools*, *health and care*, *communication*, and *safe mobility*.

1.7   How Does the 2020 World Benefit from Femtoelectronic Chips?

The refocusing from a nanometer to a femtojoule strategy for nanoelectronics makes possible functionally powerful, energy-conscious chip systems serving the global base of six billion potential customers. As stated in Chap. 23, many of these pervasive chip innovations support health and safety in public, in private, and on the shop floor, so they should be accompanied, from the outset, by qualification and certification programs. Again, meeting the concerns of twenty-first century people, minimum-energy nanoelectronics can attract broad interest in the public, well beyond just the nanometer experts.

Chapter 2
From Microelectronics to Nanoelectronics

Bernd Hoefflinger

Abstract We highlight key events in over 100 years of electronic amplifiers and their incorporation in computers and communication in order to appreciate the electron as man's most powerful token of information. We recognize that it has taken about 25 years or almost a generation for inventions to make it into new products, and that, within these periods, it still took major campaigns, like the Sputnik effect or what we shall call 10× programs, to achieve major technology steps. From Lilienfeld's invention 1926 of the solid-state field-effect triode to its realization 1959 in Kahng's MOS field-effect transistor, it took 33 years, and this pivotal year also saw the first planar integrated silicon circuit as patented by Noyce. This birth of the integrated microchip launched the unparalleled exponential growth of microelectronics with many great milestones. Among these, we point out the 3D integration of CMOS transistors by Gibbons in 1979 and the related Japanese program on Future Electron Devices (FED). The 3D domain has finally arrived as a broad development since 2005. Consecutively, we mark the neural networks on-chip of 1989 by Mead and others, now, 20 years later, a major project by DARPA. We highlight cooperatives like SRC and SEMATECH, their impact on progress and more recent nanoelectronic milestones until 2010.

2.1   1906: The Vacuum-Tube Amplifier

At the beginning of the twentieth century, the phenomenon of electricity (the charge and force of electrons) had received over 100 years of scientific and practical attention, and signals had been transmitted by electromagnetic waves, but their detection was as yet very limited, because signal levels were small and buried in noise. This changed forever when the vacuum-tube amplifier was invented in 1906

B. Hoefflinger (✉)
Leonberger Strasse 5, 71063 Sindelfingen, Germany
e-mail: bhoefflinger@t-online.de

B. Hoefflinger (ed.), *CHIPS 2020*, The Frontiers Collection,
DOI 10.1007/978-3-642-23096-7_2, © Springer-Verlag Berlin Heidelberg 2012

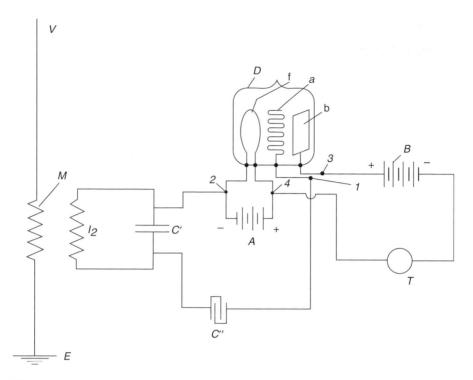

Fig. 2.1 The vacuum-triode amplifier after De Forest 1907. In this field-effect triode, the electrodes are the heated cathode on the left, which we would call the source today, the grid in the center, which would be the gate, and the anode on the right, which we would call the drain (© USPTO)

by Robert von Lieben in Austria [1] and Lee De Forest in the USA [2]. Its predecessor was the vacuum-discharge diode, a two-terminal device consisting of a heated cathode electrode emitting *thermionic electrons*, which are then collected through a high electric field by another electrode, the anode, biased at a high voltage against the cathode. This two-terminal device acts as a rectifier, offering a large conductance in the described case of the anode being at a higher potential than the cathode, and zero conductance in the reverse case of the anode being at a potential lower than the cathode. The invention was the insertion of a potential barrier in the path of the electrons by placing a metal grid inside the tube and biasing it at a low potential with respect to the cathode (Fig. 2.1). The resulting electric field between cathode and anode would literally turn the electrons around. Fewer or no electrons at all would arrive at the anode, and the conductance between cathode and anode would be much smaller. A variation of the grid potential would produce an analogous variation (modulation) of the cathode–anode conductance. This three-terminal device, the vacuum triode, consisting of cathode, anode, and control grid, became the first electronic amplifier: It had a certain voltage gain A_V, because the grid–cathode input control voltage could be made much smaller than the cathode–anode voltage, and it had infinite current gain A_I at low rates of input

Fig. 2.2 Equivalent circuit of an ideal amplifier (voltage-controlled current amplifier)

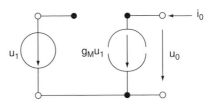

changes, because there was no current flow in the input between cathode and grid, while large currents and large current changes were effected in the output circuit between cathode and anode. As a consequence, the power gain $A_V A_I$ approaches infinity.

It is worthwhile drawing the abstraction of this amplifier as a circuit diagram (Fig. 2.2), because inventors have by now spent over 100 years improving this amplifier, and they will spend another 100, even if the signal is not electrons. We see that the input port is an open circuit, and the output port is represented by a current source $g_m V_{in}$ in parallel with an output resistance R_{out}.

The inventors of the vacuum-tube amplifiers were tinkerers. They based their patent applications on effects observed with their devices and achieved useful products within just a few years (1912).

10×: Long-range radio (World War I)

These amplifiers launched the radio age, and they triggered the speculative research on building controlled-conductance amplifying devices, which would replace the bulky vacuum tubes with their light-bulb-like lifetime problems.

2.2   1926: The Three-Electrode Semiconductor Amplifier

The best solid-state analogue to the vacuum tube would be a crystal bar whose conductance could be varied over orders of magnitude by a control electrode. This is what the Austro-Hungarian–American physicist Julius Lilienfeld proposed in his 1926 patent application "Method and apparatus for controlling electric currents" [3]. He proposed copper sulfide as the semiconducting material and a capacitive control electrode (Fig. 2.3). This is literally a parallel-plate capacitor, in which the field from the control electrode would have an effect on the conductance along the semiconducting plate.

He did not report any practical results. However, since the discovery of the rectifying characteristics of lead sulfide by K.F. Braun in 1874, semiconductors had received widespread attention. However, it was not until 1938 that Rudolf Hilsch and Richard Pohl published a paper, "Control of electron currents with a three-electrode crystal and a model of a blocking layer" [4], based on results obtained with potassium bromide. Shockley wrote, in his article for the issue of the *IEEE Transactions on Electron Devices* commemorating the bicentennial of the United

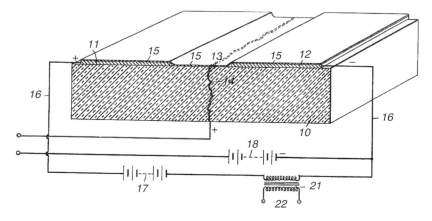

Fig. 2.3 The field-effect triode proposed by Lilienfeld in 1926 (© USPTO)

States in 1976 [5], that he had this idea in December 1939: "It has occurred to me today that an amplifier using semi conductors rather than vacuum is in principle possible". Research continued during World War II on the semiconductor amplifier [6], but a critical effort began only after the war. As we shall see, it was not until 1959 that the Lilienfeld concept was finally reduced to practice.

2.3   1947: The Transistor

One possible launch date of the Age of Microelectronics is certainly the invention of the transistor in 1947 [5]. Shockley himself described the events leading to the point-contact transistor as the *creative failure mechanism*, because the invention resulted from the failure to achieve the original goal, namely a field-effect transistor (FET) with an insulated gate in the style of the Lilienfeld patent. Nevertheless, this *failure*, implemented as Ge or Si bipolar junction transistors, dominated microelectronics into the 1980s, when it was finally overtaken by integrated circuits based on insulated-gate FETs, the realization of the Lilienfeld concept.

2.4   1959: The MOS Transistor and the Integrated Circuit

Shockley described the first working FET in 1952, which used a reverse-biased pn junction as the control gate, and junction FETs (JFETs) were then used in amplifiers, where a high input impedance was required. In fact, when I was charged in 1967 at Cornell University with converting the junior-year lab from vacuum tubes to transistors, I replaced the vacuum triodes in the General Radio bridges by junction FETs, and one of my students wrote in his lab report: "The field-effect transistor thinks that it is a tube".

Fig. 2.4 The MOS transistor in Kahng's patent filed in 1960 (© USPTO)

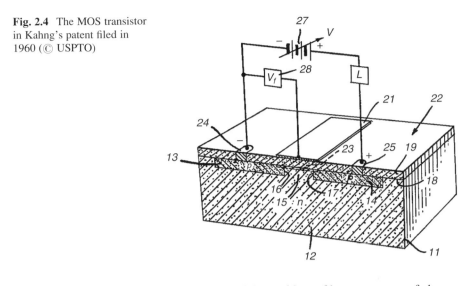

The insulated-gate FET presented the sticky problem of large amounts of charge at the dielectric–semiconductor interface masking any control effect by the gate electrode. It was the use of thermally grown SiO_2, the native oxide of Si, as the dielectric on Si as the semiconductor that produced the breakthrough in 1959 [7]. It enabled Atalla and Kahng to make the first working FET with oxide as the insulator and a metal gate (Fig. 2.4) [8]. Its name, MOSFET, metal-oxide field-effect transistor, was soon abbreviated to MOS transistor. It rarely appeared on the market as a discrete, separately packaged device with three terminals, because it was easily damaged by external voltages causing a breakdown of the insulator. MOS transistors really developed their full potential as soon as many of them were connected to perform functions inside an *integrated circuit.*

The coincidence with the invention of the integrated circuit at the same time can be linked to the victory of Si over Ge as the favored semiconductor material and to the rapid evolution of thermally grown SiO_2 as the insulator perfectly compatible with Si as the semiconductor. At Fairchild Semiconductor, founded in 1956 by Robert Noyce and the so-called *Treacherous Seven*, who had left Shockley Transistor, the Swiss-born Jean Hoerni made his historic notebook entry on December 1, 1957, that SiO_2 should be used to passivate the surface edges of p-n junctions, and he illustrated that in his later patent application [9] in May 1959 as shown in Fig. 2.5. The critical metallurgical junction between the bulk crystal and the diffused region is fully shielded from the outside, and the surface is a plane oxide insulator.

This planar structure led his boss Noyce to integrate various devices, such as diodes and transistors, avoiding their interference by isolating them from each other with reverse-biased pn junctions and by connecting their electrodes with thin Al lines etched from a thin Al film evaporated on the oxide surface (Fig. 2.6). This most famous integrated-circuit patent [10], filed by Noyce on July 30, 1959, is so explicit, as shown in Fig. 2.6, that even today, 50 years later and at dimensions

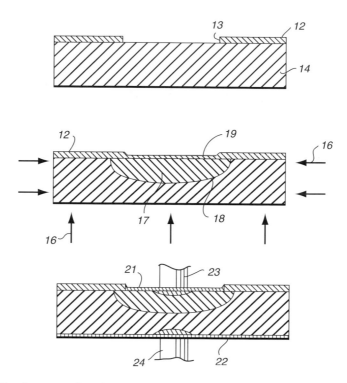

Fig. 2.5 The planar manufacturing process according to Hoerni (© USPTO)

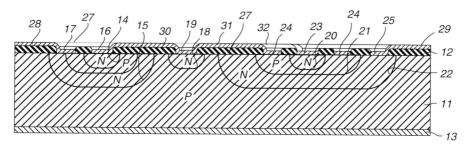

Fig. 2.6 The planar integrated circuit according to Noyce (patent filed 1959) [10] (© USPTO)

1,000 times smaller, the basic method of integrating devices side-by-side in Si is still the same, resulting in a two-dimensional arrangement of transistors and other devices, today at the scale of billions of these on the same chip.

The events of 1956–1959, pivotal for microelectronics, have been covered in great detail, for example in [6 and 11], and we will not elaborate here on the parallel patent by Kilby. It did not have any technical impact because of its Ge mesa technology and soldered flying wires as interconnects.

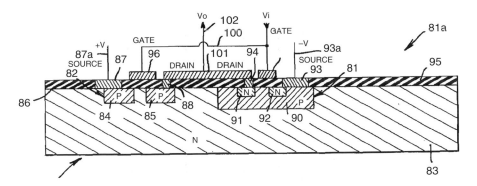

Fig. 2.7 Cross section through the chip surface showing a PMOS and an NMOS transistor side-by-side and isolated from each other, after the Wanlass patent filed in 1963 [12] (© USPTO)

Fig. 2.8 The complementary MOS inverter in the Wanlass patent. Note that the gates are connected and that the PMOS drain is connected to the NMOS drain, making this pair an important fundamental functional unit (© USPTO)

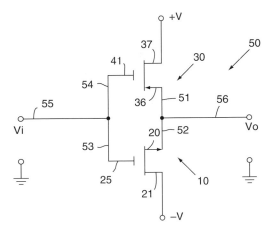

While the Noyce patent showed a bipolar transistor, others in his company concentrated on MOS and came up in 1963 with the ultimate solution for integrated circuits: *complementary MOS (CMOS) integrated circuits*. Frank Wanlass filed this famous patent in 1963 [12], and he presented a paper in the same year with C.T. Sah [13]. Nothing explains the power of this invention better than the figures in the patent (Figs. 2.7 and 2.8). The first one shows a PMOS transistor on the left with p-type doped source and drain, conducting positive carriers (holes) under the control of its gate, and an NMOS transistor on the right with its n-type source and drain conducting electrons under the control of its gate. The transistors are isolated from each other by a reverse-biased p-n junction consisting of a p-well and an n-type substrate. The construction and functionality are listed in Table 2.1.

The complementary transistor pair as connected in Fig. 2.8 is the world's most perfect inverting amplifier for the restoration and propagation of digital signals. It establishes a perfect HI and a perfect LOW at the output V_o (no. 56). There is zero current flow, which means zero power consumption except during a signal transition, when the PMOS would provide current to charge the output to HI and the

Table 2.1 Functionality of complementary MOS transistors

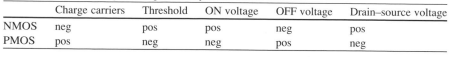

	Charge carriers	Threshold	ON voltage	OFF voltage	Drain–source voltage
NMOS	neg	pos	pos	neg	pos
PMOS	pos	neg	neg	pos	neg

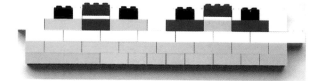

Fig. 2.9 Cross section of a CMOS technology structure on an insulator as the substrate. The Lego blocks indicate materials and aspect ratios. *Yellow*: n-Type Si (doped with donor-type atoms such as As and P), *Blue*: p-Type Si (doped with acceptor-type atoms such as B), *Red*: Metal (or as yet mostly highly As-doped polycrystalline Si), *White*: SiO$_2$ insulator, *Black*: Interconnect metal

NMOS would discharge the output to LOW. As an amplifier, it has an operating region with infinite voltage gain so that it is the best ever embodiment of the ideal amplifier as shown in Fig. 2.2.

Fairchild did not pursue this ingenious invention, and Wanlass eventually left to join the dedicated MOS companies General Instruments and AMI. His boss Noyce did not embrace CMOS until 1980, almost 20 years later. The real CMOS pushers were at RCA, where a pervasive CMOS culture was established through the first textbook on FETs [14]. They also went one step further towards the ideal CMOS structure by building the transistor layer in a thin Si film on insulator (later named SOI), Fig. 2.9.

This obviously optimizes the vertical isolation, it allows a higher transistor density, and it minimizes parasitic capacitances. In the mid-1960s at RCA, they chose sapphire as the insulating substrate, which was expensive and did not match with the Si lattice so that this approach was only used for space chips because of its immunity to radiation. As of about 2000, 35 years later, the SOI structure finally gained ground, based on massive process development over the last 20 years.

Back at Fairchild, Gordon Moore saw by 1965 the potential for doubling the number of transistors per chip with every new generation of planar IC technology, and in 1965 he had already sufficient data to show that this would happen every 18 months [15], as shown in Fig. 2.10, a reproduction of his 1975 update on this famous curve [16].

Another powerful and lasting driving force for integrating more and more functionality on a single chip is the minimum power and superior reliability achievable in the planar process. We highlight this here with the observation that, if we succeed in doubling the number of transistors or digital gates composed by these on a single piece of silicon, the number of contacts to the outside world will only increase by 40%, or the square root of 2. This is the famous Rent's rule. Because contacts mean cost and a reduction of reliability (lifetime), this means extra pressure on increasing the chip area beyond the force provided by Moore's

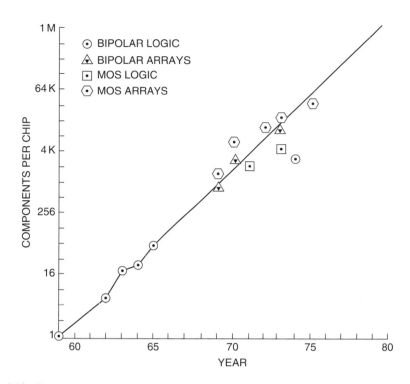

Fig. 2.10 Components per chip vs. year after Gordon Moore, 1965 and 1975 [16] (© 1975 IEEE)

law. These forces led to a new manufacturing paradigm, namely that of producing chips with a manufacturing percentage yield smaller than the high nineties customary in classical manufacturing. This was a fundamental debate in the 1960s between the classical camp of perfect-scale or right-scale integration and the advocates of large-scale integration, who won and went through very-large-scale integration (VLSI) to the giga-scale or giant scale integration (GSI) of today.

As an important historical coincidence, the Sputnik shock provided the perfect scenario for a 10× push.

10×: The computer on a silicon wafer (1966)

This enormous endeavor was launched through the US Air Force Office by giving three contracts to Philco-Microelectronics, RCA, and Texas Instruments. The speaker for the companies was Richard Petritz [17], then with TI and later a co-founder of MOSTEK. This computer never flew, but these projects created design automation, floor planning for yield and testability, automated test and the necessary *discretionary wiring* for each wafer, the birth of *direct electron-beam writing on wafer*, the best technology for rapid prototyping then, today, and probably also in 2020 (see chap. 8).

Another important side-effect of this project was that MOS circuits, which were discredited as being slow and not suitable for computing as compared with bipolar circuits, gained momentum because of their density, their topological regularity (making them so suitable for design automation), and their much lower power density and overall power. Through this project, the USA advanced toward putting processing and memory functions with higher complexity on a larger scale on a single chip, eventually leading to the first *microprocessor* [6].

When, in the internal Japanese competition for the electronic desktop calculator, Sharp was looking for a competent source for their custom-designed calculator chips, they finally gave a much publicized contract in 1969 to the Autonetics division of Rockwell, because at home in Japan this capability did not exist. This shame became the origin of a big national R&D program funded and coordinated by MITI, the Ministry for International Trade and Industry:

10×: The Joint Very-Large-Scale Integration (VLSI) Laboratory in Japan (1972–1980)

This initiative caught the rest of the world unprepared: the USA was deeply entrenched in Vietnam, and the collapse of the dollar caused a long global recession. The VLSI Lab in Japan had researchers from all the large semiconductor companies united under one roof. The result was the biggest boost, on a relative scale, in microelectronic history for new, large-scale equipment and manufacturing. Memories were identified as lead devices, and by 1980 NEC had become the world's largest semiconductor company and the Japanese dominated the global memory market. Moore's law and the *scaling law* provided simple yardsticks.

The scaling law in its 1974 version became the persistent driving force for microelectronics. It takes us back to the fundamental structure of the FET, as shown schematically in Fig. 2.11.

It had been noted in 1962 (see the history of the 1974 milestone of *The Silicon Engine* [6]), that the FET was uniquely suited to miniaturization, by shrinking its lateral dimensions L and W. To first order, the transistor area would be halved if L and W shrank by a scaling factor of 0.7, providing a simple model for Moore's law. The scaling law declared that, in order to observe the maximum gate-field and source-drain-field limits, the gate and drain–source voltages would have to be

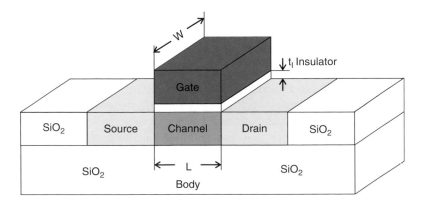

Fig. 2.11 Cross section of the MOS transistor annotated for scaling

reduced by the same amount, as would the insulator thickness t_I, in order to maintain the aspect ratio of the transistor.

The zero-order model for the drain current in the MOS transistor:

$$\text{current} = \frac{\text{channel charge}}{\text{transit time}}, \tag{2.1}$$

$$\text{channel charge} = \text{capacitance} \cdot \text{voltage} \propto \frac{W \cdot L}{tI} \cdot \text{field} \cdot L, \tag{2.2}$$

$$\text{transit time} \propto \text{field} \cdot L, \tag{2.3}$$

assumes the simple result in the constant-field case

$$\text{current} \propto \frac{W \cdot L}{tI}, \tag{2.4}$$

suggesting a constant maximum current per micrometer for a given technology. Many in-depth publications followed, analyzing the potential and limits of scaling-down transistors, and we will discuss some in Sect. 3.2, but the one in 1974 by Dennard et al. [18] triggered a strategy that is very much alive today, more than 30 years later, as we shall see in the International Technology Roadmap for Semiconductors (ITRS) in Chap. 7.

Ever since the 1960s, we have seen an extreme rate of technology improvements marked by the fact that the leaders rapidly increased their R&D budgets dispropor-tionately to well over 20% of their total revenue. We have selected eight technologies to analyze their history and future in Chap. 3. However, for the grand picture, which innovations stand out since 1959 and 1963? The scaling principle took it for granted that devices sat side-by-side in a 2D arrangement and progress would come from making them smaller. Obviously there was the potential of going into the third dimension.

2.5   1979: The Third Dimension

As we saw in Fig. 2.8, the complementary MOS transistor pair shares the input gate contact. This pair was the target of Gibbons and Lee at Stanford, and they produced the *CMOS hamburger* in 1979 [19]. In Fig. 2.12, we see the cross section of their PMOS/NMOS transistor pair.

This can be wired into the basic inverting amplifier of Fig. 2.8 on a footprint of close to one transistor, which would double the circuit density. Their invention is indicative of the process innovation at the end of the 1970s: The upper Si film, in which their NMOS transistor was formed, was deposited as a second polycrystalline-silicon film after the first film for the gate. That upper layer was then recrystallized with the energy from a laser. Recrystallizing Si deposited on oxide was one way to

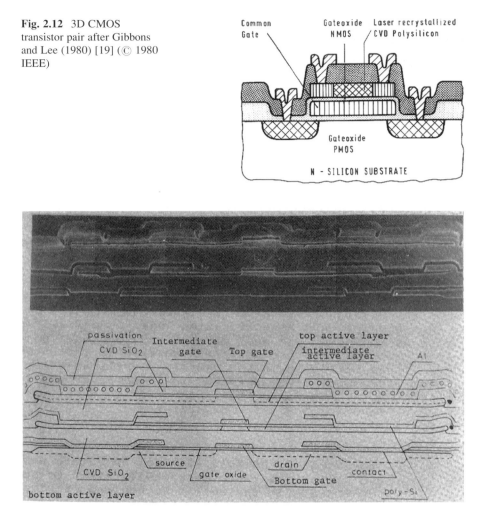

Fig. 2.12 3D CMOS transistor pair after Gibbons and Lee (1980) [19] (© 1980 IEEE)

Fig. 2.13 Cross section with three transistor layers, after Kataoka (1986) [20] (© 1986 IEEE)

obtain multiple layers of silicon on top of each other separated by insulating layers of SiO_2. High-energy implantation of oxygen and proper high-temperature annealing would produce buried layers of SiO_2, thereby realizing a Si–SiO_2–Si sequence of films potentially suitable for multiple device layers. In fact, after the heavily product-oriented VLSI Laboratory, the Japanese launched another, more fundamental effort:

10×: The program on Future Electron Devices, FED (1980)

Here we will only focus on the part *3D integration* and recapitulate some results, because it took until about 2005, roughly 25 years, for this strategic direction to take center-stage. In Fig. 2.13, we see the cross section of a Si chip surface with two crystallized Si films, allowing three transistor layers on top of each other [20]. 3D integration became part of the research efforts that evolved to be the greatest global renaissance in microelectronics history.

<hr>

10×: VHSIC, SRC in the USA, ESPRIT in Europe (1981)

<hr>

At the end of the 1970s, the Vietnam decade and a long global recession, entrepreneurial and intellectual forces stepped forward – and were heard: Simon Ramo, Ray Stata, and Robert Noyce in the USA, François Mitterand and Jean-Jacques Servan-Schreiber in Europe were quoted everywhere in association with new, substantial initiatives. The strategic program on very-high-speed integrated circuits (VHSIC) made possible a set of very advanced, highest quality CMOS manufacturing lines, certified by the RADC (Rome Air Development Center in upstate New York) as qualified manufacturing lines (QMLs), which also provided foundry services to several totally new or refurbished university departments. Hardware, software, and tools for automated chip design were donated on a large scale to US universities. What started as a summer course, given by Carver Mead of Caltech and Lynn Conway of Xerox Palo Alto Research Center (Xerox PARC) at MIT, *VLSI Design*, became a bible for tens of thousands of students and industrial engineers [21]. The Defense Advanced Research Projects Agency (DARPA) launched the foundry service MOSIS (MOS Implementation Service) for universities under the leadership of Paul Losleben and a special university equipment program of hundreds of millions of dollars, incredible dimensions those days. The VHSIC Program produced the first standard hardware-description language VHDL (originally VHSIC Hardware Description Language), which is still the standard today. A uniquely successful university–industry research cooperative, the SRC (Semiconductor Research Cooperative), was founded in 1982 under the chairmanship of Robert Noyce, which quickly grew to a capacity of 500 Ph.D. students working on all aspects of microelectronics together with assignees from industry. A leader from industry, Eric Bloch from IBM, became president of the National Science Foundation (NSF). He created the university Engineering Research Centers (ERCs) on the premise that at least three faculties and more than 15 professors would work together with over 50 Ph.D. students in such a center. The program started with ten centers and now has over 50 in the USA.

The European Commission in Brussels, until then focused on coal, steel, and agriculture, prepared a technology program. As a historical note, as founders and leaders of two of the existing university pilot lines for large-scale integrated circuits in Europe, in Leuven, Belgium, and Dortmund, Germany, Roger Van Overstraeten and I were invited in 1979 to advise Giulio Grata, the responsible person in Brussels, on the elements of a European Microelectronics Research Program, which was launched eventually as ESPRIT (European Strategic Programme for Research in Information Technology).

The 1980s were marked by numerous *megaprojects*, multi-hundred million dollar investments per new facility, many with public money, one typical target being the first 1 Mbit DRAM memory chip. The communist German Democratic Republic went bankrupt on its megaproject, because they had to pay outrageous amounts of money to acquire US equipment and computers through dark channels in the communist bloc. New regions and states competed for microelectronics industry investments, and today's map of chip manufacturing centers was pretty much established then. That period was also the origin of the global-foundry business model exemplified by TSMC (Taiwan Semiconductor Manufacturing Co.) established by Morris Chang after a distinguished career at Texas Instruments (TI).

One special result of the VHSIC program was high-speed electron-beam direct-write-on-wafer lithography for rapid prototyping and small-volume custom integrated circuits. The variable-shape beam, vector-scan systems Perkin-Elmer AEBLE 150, used by ES2 (European Silicon Structures) in France, and Hitachi HL700, used in Japan, Taiwan, and in my institute in Germany, established electron-beam lithography as a viable technology to help the scaling-driven industry to realize one new technology generation after the other. Line widths of 400 nm or less were considered the limit of optical lithography so that sizeable centers for X-ray lithography were set up at large synchrotrons in the USA at Brookhaven and in Germany at BESSY in Berlin and COSY in Karlsruhe. Wafer shuttles were planned to transport wafers there in volume, plans which never materialized because optical lithography with deep-UV laser sources was able to do the job.

With a capability of millions of transistors on a single chip and supported by a foundry service for up-to-date prototyping, the research community was encouraged to take on big questions, such as information processing not with the von Neumann architecture of processor, program memory, and data memory, but closer to nature, also named biomorphic.

2.6 1989: Neural Networks on Chips

The integration of large numbers of electronic functions on a single microchip was exploited from the early 1960s for array-type tasks, one particular direction being imager arrays made up of optical sensor elements and circuits for reading out the signals from the picture elements (pixels) [14]. The Japanese FED Program produced a 3D technology in which a Si photodiode layer was fabricated on top of two CMOS layers [22], where the CMOS layers would be used for reading and processing the pixel signals (Fig. 2.14).

This can be viewed as an early embodiment of a layered Si retina. A specific retina was then proposed by Mead and Mahowald (Fig. 2.15) in Mead's book *Analog VLSI and Neural Systems* [23]. This book again became a bible and the opener for worldwide activity on building neural networks on Si microchips. Figure 2.15 shows an analog implementation with resistors as synapses. Even in such a rigid setup, powerful functions such as the detection of shapes could be performed. Many other researchers chose general, programmable digital two-layer perceptrons. An example is shown in Fig. 2.16 [24], which can help to explain why these networks have significant potential for intelligent information processing.

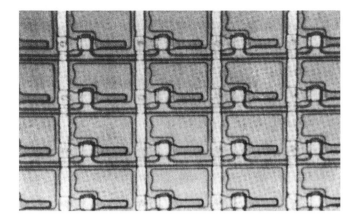

Fig. 2.14 Micrograph of an array of phototransistors with vias to the underlying CMOS layers for read-out electronics (1986) [22] (© 1986 IEEE)

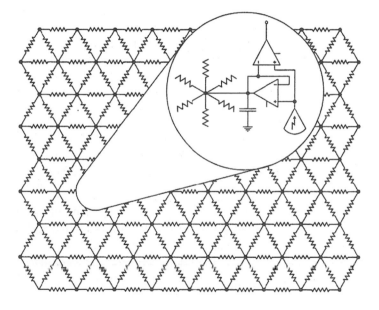

Fig. 2.15 The analog neural network of Mead and Mahowald for vision tasks such as shape detection consisting of active *pixels* and resistors as synapses (1989) [23] (© Springer 1989)

For the neural controller for automatic steering of a vehicle, driving data obtained on a 2 km section of a normal road with a human driver in the car were enough to train the neurocontroller. This is just one example of the prolific research that was launched through the investments in the 1980s.

The 1990s were marked by the worldwide re-engineering of large companies, enhanced by structural consequences after the end of the Cold War. This could have

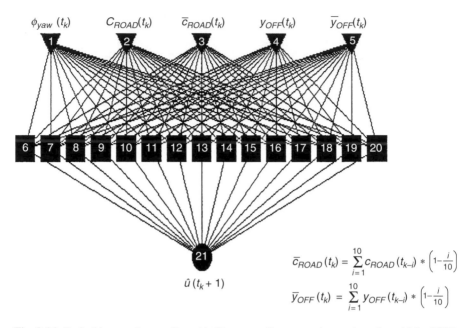

Fig. 2.16 Trainable neural controller with 21 neurons for automatic steering of a vehicle (1993) [24] (© 1993 IEEE)

hit the semiconductor industry much worse if there had not been two large-scale effects offsetting these problems:

- The force of Moore's law and of the scaling law.
- The tremendous push of the Tigers Korea, Taiwan, and Singapore, with just about 80 million people taking on the world of chip manufacturing, and dominating it today.

The effect of both of these factors has been the refocusing of the traditional players and an unparalleled strategic and global alliance of all major players.

10×: International SEMATECH and the Roadmap (1995)

Originally, SEMATECH started in 1988 as a cooperation of the US semiconductor and equipment industry, including government funding, in order to strengthen the US position in the field. In Europe, the JESSI (Joint European Submicron Silicon) project was initiated under the umbrella of EUREKA, the less bureaucratic agency that manages European cooperative strategy, while the funding runs on a per-country basis, which means that national governments fund their constituency. The focus on manufacturing, which relies much on global resources and partnering, soon let non-US companies join so that International SEMATECH was formed in 1995, first concentrating on developing the capability to manufacture on the basis of newly available 300 mm wafers. The International Technology Roadmap for Semiconductors (ITRS) was started with worldwide

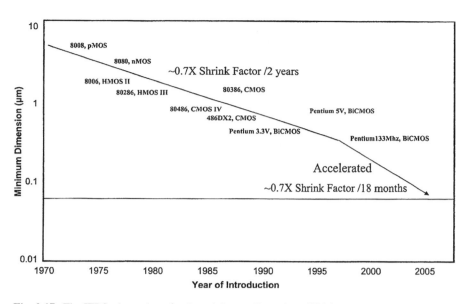

Fig. 2.17 The ITRS, shown here for the minimum dimension: 1999 issue

expert committees, and it became the single most powerful document defining the global strategy of the semiconductor industry. SEMATECH has expanded its charter since 2000 by operating large cooperative R&D facilities in Austin, TX, and Albany, NY, including major funding by the respective states. The Europeans concentrated their most significant efforts in Grenoble, France, and Leuven, Belgium, and IMEC (the Inter-University Microelectronics Center) in Leuven, founded by Roger Van Overstraeten in 1983, is today the world's largest independent R&D center for nanoelectronics. No other industry has developed such a joint global strategy and infrastructure to master the progress manifested by a cumulative R&D budget above 20% of revenue. It is interesting to overlay an early Roadmap for the minimum required lateral dimension on chips with more recent data (Fig. 2.17). It shows that the rate of progress on the nanometer scale has been pushed repeatedly beyond earlier targets [25].

Towards 100 nm, the limits of scaling down MOS transistors were seen as being near, triggering speculative research on that limit and how to go beyond it. A very short channel of well under 100 nm between the metallurgical source and drain junctions was producing a situation in which electrons would tunnel through these junctions and might get trapped in the channel island if the electrostatic "binding" energy exerted by the gate were large compared with the kinetic energy $k_B T$ at the environmental temperature T (k_B is the Boltzmann constant). Obviously, cooling would help to create this situation, as well as a very small channel island, where this electron would be locked up and only released for certain gate voltages and source–drain voltages. This tiny island containing a single electron (or none) was called a *quantum dot*, and it became very attractive as a memory element. A three-terminal *field-effect triode* (Fig. 2.18) consisting of a gate, a quantum dot and two

Fig. 2.18 Concept of a SET (1987) (© Wiley-VCH)

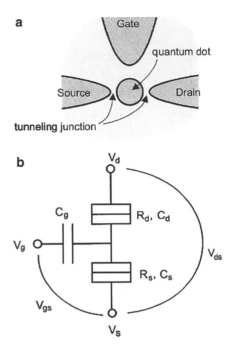

tunnel junctions was proposed as a single-electron transistor (SET) in 1987 [26]. The first experimental SETs were reported a few years later, and, clearly, one electron per bit of information would be a major quantum step of progress.

For comparison, at the time, a memory transistor stored a ONE as 50,000 electrons on its floating gate, and a ZERO was about 5,000 electrons in a non-volatile memory. The programming pulse had to place these amounts of charge on the floating gate with narrow margins, and the read-out amplifier had to distinguish this difference. Research began on techniques to place three distinguishable amounts of charge on the floating gate, which would establish three threshold voltages and to refine the readout so that it could differentiate between these voltages [27]. Figure 2.19 shows measured threshold-voltage distributions. There was no overlap, and the means were about 1 V apart. Only 10,000 electrons were needed now to distinguish the bits [27]. This capability provided a true quantum jump in bit density. This storage capability development advanced with remarkable speed to about 250 electrons in 2010 as the length of a memory transistor decreased to 34 nm.

2.7   2000: The Age of Nanoelectronics Begins

The Y2K effect in microelectronics was that volume chip production reached the 100 nm lithography level, and compatible overall processing was achieved. Total capital expenditure for a factory based on 300 mm wafers exceeded US$ 1 billion.

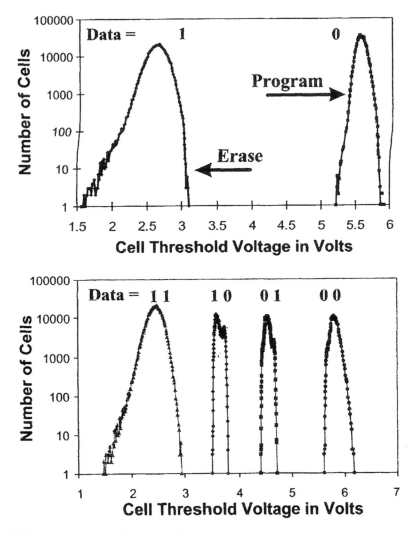

Fig. 2.19 Distribution of multiple threshold voltages achieved in 1995 [27] (© 1995 IEEE Press – Wiley)

10×: Next-Generation Lithography, NGL (2000)

The performance of optical lithography was seen to be at its limits, and this, in fact can be considered as another attribute of the new age of nanoelectronics, namely that, the generation of these sub-100 nm lateral structures and zillions of these on a wafer would require a lithography beyond short-wavelength refractive-optics-based patterning. The largest project in the history of International SEMATECH now became NGL (next-generation lithography), aimed at providing a non-optical alternative for anything smaller than 45 nm to be available for prototyping in 2005. The contenders were ion-beam lithography (IBL) and extreme ultraviolet (EUV) lithography.

IBL was favored in Europe. A prototype was built in Vienna, Austria, based on a hydrogen-ion beam going through silicon stencil masks produced in Stuttgart, Germany. It was completed and demonstrated 45 nm capability in 2004. However, the international SEMATECH lithography experts group decided to support solely the EUV project, because IBL was assessed to not have enough throughput and less potential for down-scaling. For EUV lithography, a sufficiently effective and powerful source for 13 nm radiation, as well as the reflective optics and masks, had not become available by 2010, and EUV lithography has been rescheduled for introduction in 2012.

On the evolutionary path, optical lithography survived once more as it did against X-rays in the 1980s. At the level of deep UV (DUV), 193 nm, a major breakthrough was achieved by immersing the lens into a liquid on top of the silicon wafer. The liquid has a refractive index several times higher than air or vacuum significantly increasing the effective aperture. This immersion lithography saved the progress on the roadmap for one or two generations out to 22 nm. Thereafter, the future beyond optical is an exciting topic to be covered in Chap. 9.

It is natural that the scaling law cannot be applied linearly. Although lithography provided almost the factor 0.7 per generation, the transistor size could not follow for reasons of manufacturing tolerance, and its switching speed did not follow for physical reasons. Therefore, the industry as of 2000 had to embrace a more diversified and sophisticated strategy to produce miniature, low-power, high-performance chip-size products. It finally embraced the third dimension, 20 years after the FED Program in Japan, which we discussed in Sect. 2.6. Of course, this renaissance now occurred at much reduced dimensions, both lateral and vertical. While in the mid-1980s it was a touchy art to drill a hole (from now on called a *via*) through a 200 μm thick silicon/silicon dioxide substrate, wafers were now thinned to 50 μm or less, and 20 years of additional process development had produced a formidable repertory for filling and etching fine structures with high aspect ratios of height vs. diameter. Test structures on an extremely large scale emerged since 2006 with wafers stacked and fused on top of each other and thin W or Cu vias going through at high density to form highly parallel interconnects between the wafer planes. One early example is shown in Fig. 2.20 [28].

The mobilization in this direction of technology has been remarkable. Stacked wafers with through-silicon vias (TSVs) obviously provided a quantum jump in bit and transistor density per unit area. Enthusiastic announcements were made by industry leaders proclaiming new laws of progress beyond Moore. And it is true that, besides the gain in density, interconnect lengths are much reduced, partly solving the problem of exploding wiring lengths in 2D designs. The added manufacturing cost is accrued at dimensions that are more relaxed, and pre-testing each wafer plane before stacking provides more options for handling the testability of the ever more complex units. Stacking and fusing processor planes and memory planes offers a divide-and-conquer strategy for the diverging roadmaps for

Fig. 2.20 Cross sections through a 3D memory produced by stacking eight wafers with TSVs (2006) [28] (© 2006 IEEE)

processor and memory technologies and for the partitioning of the total system. We will come back to the significance of this type of 3D evolution in Chap. 3.

The first decade of the new millennium has seen further tremendous progress in CMOS technology in nanoelectronics, best described by the recent consensus that NMOS and PMOS transistors have reasonable and *classical or conventional* characteristics down to channel lengths of 5 nm, so that the worldwide design know-how and routines can be leveraged for new and improved products to an extent that is only limited by our creative and engineering capacity to develop these.

For the grand long-term picture, we address two recent achievements that have great potential or that indicate the direction in which we might perform speculative research on how to achieve entirely new levels of electronic functionalities.

2.8   2007: Graphene and the Memristor

The ultimately thin conducting film to which one could apply control, would have a thickness of one atomic layer. It would be a 2D crystal. It is not surprising that, in the context of widespread carbon research, this 2D crystal was eventually realized, observed, and characterized in carbon, where it is called graphene (Fig. 2.21). In 2007, Geim and Novoselov pulled off this single-atom-thick film of carbon from graphite with Scotch tape and transferred it to a SiO_2 layer on top of silicon [29]. The graphene layer fit on the oxide layer so well that the measurements confirmed theories on 2D carbon crystals going back to 1947, and high electron mobilities were observed. The already large carbon research community converged and expanded on graphene. A high-frequency transistor and an inverter with complementary transistors were reported soon after. The film deposition techniques for producing these graphene layers appear to be compatible with large-scale Si manufacturing so that graphene has high potential for future nanoelectronics.

Fig. 2.21 Real graphene
single-atom layer with the C
atoms clearly visible (From
Wikipedia)

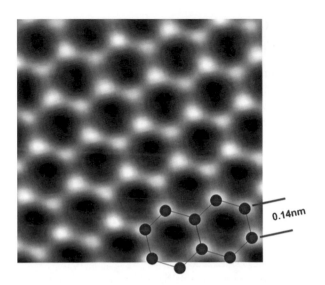

Fig. 2.22 Micrograph of 17
memristors (J.J. Yang, HP
Labs)

Another recent achievement resulting from nanometer-scale electronics research
is a new two-terminal device, which has an analog memory of its past with high
endurance. It was reported in 2007 by Williams and members of his Laboratory for
Information and Quantum Systems at Hewlett-Packard [30]. The device consists of
two titanium dioxide layers connected to wires (Fig. 2.22). As the researchers
characterized their devices, they arrived at a model that corresponded to the
memristor, a two-terminal device postulated and named by Leon Chua in 1971
[31] on theoretical grounds. The memristor would complement the other three
devices resistor, capacitor, and inductor.

A two-terminal device that could be programmed or taught on the go would be very powerful in systems with distributed memory. For example, the resistor synapses in Mead's retina (Fig. 2.15) could be replaced by these *intelligent resistors* to build very powerful neural networks for future silicon brains.

With graphene as a new material and the memristor as a new device, we conclude our grand overview of the technological arsenal that has been developed over more than 60 years and which forms the basis of our 2020 perspective in the following chapter.

References

1. von Lieben R.: Kathodenstrahlenrelais. German Patent No. 179807. Issued 4 March 1906
2. De Forest L.: Device for amplifying feeble electrical currents. US Patent No. 841387, filed 25 Oct 1906. Issued 15 Jan 1907
3. Lilienfeld J.E.: Method and apparatus for controlling electric currents. US Patent No. 1745175, filed 8 Oct 1926. Issued 18 Jan 1930
4. Hilsch R, Pohl R.W.: Steuerung von Elektronenströmen mit einem Dreielektrodenkristall und ein Modell einer Sperrschicht [Control of electron currents with a three-electrode crystal and a model of a blocking layer]. Z. Phys. **111**, 399 (1938)
5. Shockley W.: The path to the conception of the junction transistor. IEEE Trans. Electron Dev. **23**, 597 (1976)
6. See "The Silicon Engine" at www.computerhistory.org/semiconductor/. Accessed Feb 2011
7. Atalla M.M.: Stabilisation of silicon surfaces by thermally grown oxides. Bell Syst. Tech. J. **38**, 749 (1959)
8. Kahng D.: Electric field controlled semiconductor device. US Patent No. 3102230, filed 31 May 1960. Issued 27 Aug 1963
9. Hoerni J.A.: Method of manufacturing semiconductor devices. US Patent No. 3025589, filed 1 May 1959. Issued 20 March 1962
10. Noyce R.N.: Semiconductor device-and-lead structure. US Patent No. 2981877, filed 30 July 1959. Issued 25 April 1961
11. Saxena A.N.: Invention of Integrated Circuits – Untold Important Facts. World Scientific, Singapore (2009)
12. Wanlass F.M.: Low stand-by power complementary field effect circuitry. US Patent No. 3356858, filed 18 June 1963. Issued 5 Dec 1967
13. Wanlass F.M, Sah C.T.: Nanowatt logic using field-effect metal-oxide semiconductor triodes. IEEE ISSCC (International Solid-State Circuits Conference) 1963, Dig. Tech. Papers, pp 32–33
14. Wallmark J.T, Johnson H (eds.): Field-Effect Transistors. Prentice-Hall, Englewood Cliffs (1966)
15. Moore G.: Cramming more components onto integrated circuits. Electron Mag. **38**(8), 114–117 (1965)
16. Moore G.: Progress in digital integrated electronics. IEEE IEDM (International Electron Devices Meeting) 1975, Tech. Dig., pp. 11–13
17. Petritz R.L.: Current status of large-scale integration technology. In: Proceedings of AFIPS Fall Joint Computer Conference, Vyssotsky, Nov 1967, pp. 65–85
18. Dennard R.H, Gaensslen F.H, Yu H.N, Rideout V.L, Bassous E, LeBlanc A.R.: Design of ion-implanted MOSFET's with very small physical dimensions. IEEE J. Solid-State Circuits **9**, 256 (1974)
19. Gibbons J.F, Lee K.F.: One-gate-wide CMOS inverter on laser-recrystallised polysilicon. IEEE Electron Dev. Lett. **1**, 117 (1980)

20. Kataoka S.: Three-dimensional integrated sensors. IEEE IEDM (International Electron Devices Meeting) 1986, Dig. Tech. Papers, pp. 361–364
21. Mead C, Conway L.: Introduction to VLSI Systems. Addison-Wesley, Reading (1979)
22. Senda K, et al.: Smear-less SOI image sensor. IEEE IEDM (International Electron Devices Meeting) 1986, Dig. Tech. Papers, pp. 369–372
23. Mead C, Ismail M.: Analog VLSI Implementation of Neural Systems, ISBN 978-0-7923-9040-4, Springer (1989).
24. Neusser S, Nijhuis J, Spaanenburg L, Hoefflinger B.: Neurocontrol for lateral vehicle guidance. IEEE Micro. **13**(1), 57 (1993)
25. www.ITRS.net/. Accessed Feb 2011
26. Likharev K.K.: IEEE Trans. Magn. **23**, 1142 (1987)
27. Bauer M, et al.: A multilevel-cell 32 Mb flash memory. IEEE ISSCC (International Solid-State Circuits Conference), Dig. Tech. Papers, 1995, pp. 132–133
28. Lee K et al.: Conference on 3D Architectures for Semiconductor Integration and Packaging, San Francisco, Oct–Nov 2006
29. Geim A.K, Novoselov K.S.: The rise of graphene. Nat. Mater. **6**, 183 (2007)
30. Wang Q, Shang D.S, Wu Z.H, Chen L.D, Li X.M.: "Positive" and "negative" electric-pulse-induced reversible resistance switching effect in $Pr_{0.7}Ca_{0.3}MnO_3$ films. Appl. Phys. **A 86**, 357 (2007)
31. Chua L.O.: Memristor – the missing circuit element. IEEE Trans. Circuit Theory **18**, 507 (1971)

Chapter 3
The Future of Eight Chip Technologies

Bernd Hoefflinger

Abstract We select eight silicon chip technologies, which play significant roles in the decade 2010–2020 for the development of high-performance, low-energy chips 2020 and beyond. In the spirit of the 25-years rule, all of these technologies have been demonstrated, and some, in fact, are very mature and yet are worth to be revisited at the nano-scale.

The bipolar transistor remains superior in transconductance and bandwidth, and the complementary cross-coupled nano-pair can become the best ultra-low-energy signal-regenerator.

MOS and CMOS circuits continue to be the most effective solutions for giant-scale integration in a silicon-on-insulator technology. However, the end of progress with just scaling down transistor dimensions is near, and this is not a matter of technology capability, but one of atomic variance in 10 nm transistors. Once only ~6 doping atoms are responsible for their threshold and voltage gain, 95% of these transistors would have between 1 and 9 such atoms. Their threshold would vary more than their supply voltage. We show that, at these dimensions, not a transistor, but the cross-coupled pair is CMOS inverters is the elementary and necessary signal regenerator at the heart of ultra-low-voltage differential logic. This assembly of four transistors is the perfect target for 3D integration at the transistor-level on-chip, and selective Si epitaxy is shown as an exemplary solution, including self-assembly eliminating certain lithography steps. This optimized 4T building block enables a 6T SRAM memory cell scalable to a 2020 density competitive with a DRAM cell, which cannot be scaled because of capacitor size and transistor leakage. The 4T block is also the key accelerator in the differential logic HIPERLOGIC, exemplified by an n-bit by n-bit multiplier, which also illustrates the importance of new circuit architectures. DIGILOG, a most-significant-bit-first multiplier has a complexity $O(3n)$ versus $O(n^2/2)$ in commonly used Booth multipliers. With HIPERLOGIC, a

B. Hoefflinger (✉)
Leonberger Strasse 5, 71063 Sindelfingen, Germany
e-mail: bhoefflinger@t-online.de

B. Hoefflinger (ed.), *CHIPS 2020*, The Frontiers Collection,
DOI 10.1007/978-3-642-23096-7_3, © Springer-Verlag Berlin Heidelberg 2012

16×16 bit multiplier is projected to require only 1fJ per multiplication in 2020, about 1,000-times less than the status in 2010 and 10-times less than a synapse in the human brain. Therefore, processing has the potential to become 1,000-times more energy efficient within a decade.

The 3D integration of chips has become another key contributor to improve the overall energy efficiency of systems-of-chips incorporating sensing, transceiving, and computing.

In the previous chapter, we highlighted those technologies that significantly advanced microelectronics to the present level of nanoelectronics. In this chapter, we focus on eight chip technologies, the history of their evolution, and their future role and potential. We highlight their S-curves of innovation rate and market penetration with four milestones:

- First prototype
- First product
- First *killer* product
- Market dominance

   Our eight candidates are:

- Bipolar transistors
- Metal–oxide–semiconductor integrated circuits (MOS ICs, PMOS or NMOS)
- Complementary MOS (CMOS) and bipolar CMOS (BiCMOS) ICs
- Silicon-on-insulator (SOI) CMOS ICs
- 3D CMOS ICs
- Dynamic and differential MOS logic
- Chip stacks
- The single-electron transistor and other concepts

   The evolution of all eight technologies and their future significance are displayed in Fig. 3.1, and for each one, we will comment on their past and present role as well as describe the challenges and opportunities looking at 2020 and beyond.

1. The *bipolar transistor*, first prototyped in 1950, was the transistor type of choice for the first integrated circuits in 1960. Its killer products, the operational amplifier and transistor–transistor logic (TTL), brought an IC market share for this type of IC of over 95% in 1970. It had lost its lead by 1980, and bipolar ICs are today, on the basis of heterojunction bipolar transistors (HBTs) with maximum frequencies of 600 GHz, a high-value niche specialty for terahertz (THz) transceivers and photonic modulators.
2. *MOS field-effect-transistor* ICs appeared as prototypes in 1960, earned some limited recognition as controllers and random-access memories (RAMs) by 1968, and began their March to the top in 1972 with microprocessors and dynamic RAMs (DRAMs) as killer products, which led MOS ICs to pass bipolar ICs in 1980.
3. *CMOS (complementary MOS) ICs* were first presented in 1965 and became a success for those western and Japanese companies specializing in watch

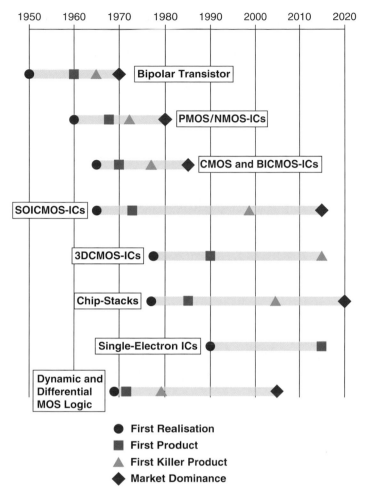

Fig. 3.1 The life cycles of eight chip technologies

circuitry and high speed static RAM (SRAM). Killer products such as these SRAMs, high-speed, low-power microprocessors, and application-specific, mixed-signal circuits and gate arrays forced all semiconductor companies by 1981 to quickly establish a CMOS capability, often enhanced by on-chip CMOS-bipolar drivers and amplifiers, a specialty that had been proposed in 1969. As early as 1985, CMOS had taken over the lead from single-transistor-type MOS ICs, and it has kept this lead ever since.

4. A fundamental improvement on CMOS, namely putting the complementary NMOS and PMOS transistors into a thin Si film on an insulator (*SOI*), had been realized as early as 1964 on sapphire and began to gain ground on silicon dioxide as of 1978, but it took until 2000 for major players to introduce volume products, and it may take until 2012 for SOI-CMOS to become the technology

type with the leading market share. It is the only type that is scalable to transistor channels <10 nm long and will be the workhorse for the 2020s.

5. *3D CMOS* technology, folding PMOS and NMOS transistors on top of each other with the gate sandwiched in between, was first demonstrated in 1979 and was brought to maturity with selective epitaxial overgrowth (SEG) by 1990, but it may take until about 2012 for its net reduction of processing steps and of area and wire savings to catch on. As an exemplary goal, a 3D six-transistor SRAM cell has been presented for the 10 nm node, requiring just ten electrons per bit and providing a density of 10 Gbit/cm^2.

6. *Dynamic MOS logic* was first realized in 1965 to exploit the fundamental synergy of charge storage, switching, and amplification, that is, signal restoration, provided in MOS ICs. It helped NMOS ICs to offer speed, to save power, and to survive for a long time against CMOS. Standard CMOS required two complementary transistors per logic input variable. In order to improve the transistor efficiency, various styles of dynamic CMOS logic were invented, where logic functions are mapped on arrays of NMOS switches, and precharge, sampling, and drivers are built in CMOS (examples are Domino and cascode voltage switch logic, CVSL). The persistent drive towards lower voltages V (to save energy CV^2) was advanced significantly with *differential CMOS logic* such as HIPERLOGIC in 1999, and this type of logic is seen as a must for robust logic, when the fundamental variance of transistor parameters such as their "threshold" becomes as large as the supply voltage for transistor lengths <20 nm.

7. We date today's drive towards *3D integration* of chip technologies back to 1968, when IBM introduced large-scale flip-chip ball bonding to ceramic carriers as the C4 (controlled-collapse chip connect) process. It was introduced industry-wide 25 years later. The next significant contributions came from the Japanese Future Electron Devices (FED) program, which promoted wafer thinning and contacts going fully through the wafer, so-called through-silicon vias (TSVs), in the late 1980s. It took until 2005 for these techniques to be finally adopted mainstream, and they have now become a massive industry drive to keep pace with market requirements and to benefit from the economic quantum jump that TSV chip stacking provides for memory density.

8. *The single-electron transistor* (SET) was first proposed in 1987 as a nanometer-size MOS transistor, in which the source and drain PN junctions have been replaced by tunnel junctions such that the transistor *channel* becomes a quantum box that can hold just one electron. In the search for the most advanced nonvolatile memory, this would take us to the impressive milestone of just one electron per bit. Among the read-out options for such a memory, one may be reminded of the charge-coupled device (CCD) serial memories in the 1970s. However, no practical prototype had been reported by 2010, so circuits of this type will only become products in the 2020s, a horizon where they will have to compete with practical nano-CMOS RAMs needing just a few electrons per bit.

Overall, we see that it took major innovations 25–35 years to make it onto the market. With a horizon of 2020, we see that all candidate technologies should have been implemented in convincing prototypes by now and that it may pay off to revisit innovations of 25 years ago that appeared then too early for their time.

3.1   The Bipolar Transistor

Our starting line for the bipolar transistor is the Noyce patent [1] and the planar Si bipolar transistor shown there (Fig. 3.2). It displays a junction-isolated NPN transistor, for which we would have to imagine a top contact to the deep collector to make it suitable for integration.

The first product milestone in 1960 was planar bipolar ICs, both analog and digital, and the progress on Moore's curve started with this technology. Digital logic exploded with three families: diode–transistor logic (DTL), transistor–transistor logic (TTL), and emitter-coupled logic (ECL). With well-organized customer support and multi-volume textbooks, Texas Instruments (TI) and Motorola dominated the market with TTL and ECL products. Digital products from controllers to mainframes were the high-volume killer-apps around 1970, and TI quickly became the world's largest semiconductor manufacturer. The progress in bipolar technology, as in any electronic device, depended on the critical transit time of the electrons, here the time to traverse the neutral base of width W_B, and on the *parasitics,* in rank order: collector resistance, collector capacitance, base resistance, and collector–base Miller capacitance. Decades of hard work have gone into this, and a status of the mid-1980s is shown in Fig. 3.3 with the buried collector layer and lateral oxide isolation.

With its emitter-last process flow, bipolar has been a fundamental headache for integration. So many circuits are common-emitter type, which means that the emitter is connected to ground, and since there is no better ground than the silicon bulk crystal, the emitter should be first. It is worth remembering that IBM and Philips once chose this disruptive route with integrated injection logic (I^2L), a low-power logic intended to compete with CMOS. But the deep emitter had a small emitter efficiency, and switching was slow because of a deep, thick base. At that time, bipolar technology had already lost the battle for large-scale integration against MOS ICs, which surpassed the bipolar IC revenue in 1980.

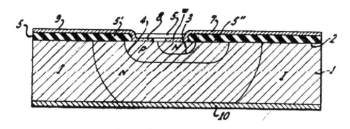

Fig. 3.2 Cross section of a bipolar transistor after Noyce (1959) [1]. © USPTO

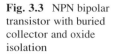

Fig. 3.3 NPN bipolar transistor with buried collector and oxide isolation

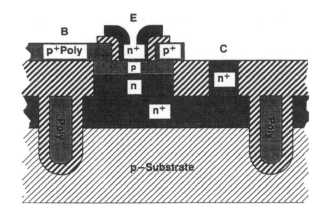

In the standard bipolar process flow, a lateral PNP transistor is also available and used in analog circuits. Because its base width W_B is determined by lithography, its switching speed is low, and a complementary bipolar technology did not evolve. However, in the vertical NPN transistor, the critical base width is determined by the difference between the depth of the base diffusion and that of the emitter diffusion, and this configuration has been tailored continuously towards higher frequencies. The maximum intrinsic frequency f_i is the inverse of the base transit time t_B:

$$f_i = \frac{2D_{nB}}{W_B^2}. \tag{3.1}$$

The diffusion constant D_{nB} of the electrons in the base is related to their mobility μ via the Einstein relation

$$D_{nB} = \mu_n k_B T / q. \tag{3.2}$$

A high electron mobility and a short base have been the target for decades, and heterojunction bipolar transistors are the optimal solution. In these transistors, the base is a SiGe compound, in which the electron mobility is two to three times that in Si. We can describe the state of the art as $\mu = 1{,}000$ cm^2 V^{-1} s^{-1} and a neutral, effective base width of 50 nm, so that we obtain an intrinsic frequency limit of 2 THz.

Parasitic capacitances and resistances result in maximum reported frequencies of over 500 GHz [2]. These have been achieved with 0.13 μm technologies, for which the typical parameters are:

Emitter area: 0.13 μm $\times$ 0.13 μm,
Max. current (HBT): 1.3 mA,
Collector area: 5 $\times$ 0.13 μm $\times$ 8 $\times$ 0.13 μm,
Collector capacitance: 2.8 fF.

This results in a minimum intrinsic switching time of 2.8 fF/1.3 mA = 2.15 ps, which does not yet consider the collector series resistance and load resistance as well as capacitance and inductance.

3.1.1 Scaling Laws for Bipolar Transistors

Scaling laws or rules for bipolar transistors are not so well known. However, there is a body of modeling relations mostly derived from the Gummel–Poon transistor model [3], and in the intrinsic frequency limit, (3.1), we have the first driving force towards higher speed, namely reducing the base width W_B. If we pursue this, we have to observe a number of scaling rules in order to maintain the transistor performance as measured by its current, voltage, and power gain.

The forward current gain

$$\beta_F = \frac{D_{nB}}{D_{pE}} \frac{N_E L_E}{N_B W_B} \tag{3.3}$$

is determined first by the ratio of the diffusion constants for electrons and holes in Si, which is 3. Then we see the product of emitter dopant (typically As) concentration and diffusion length of the minorities (holes). This product is practically a materials constant for a Si technology, because the larger the majority doping the smaller the minority diffusion length. For reference, we take this product as $10^{14}/cm^2$. Now we see that, in order to maintain current gain, we have to raise the base doping level as we shorten the base width. There are more reasons to watch the number of acceptors in the neutral base $N_B W_B$, which is called the Gummel number.

The *intrinsic* voltage gain of the transistor in the common-emitter mode, determined by the output conductance, is characterized by the ratio of collector current to Early voltage V_A (an extrapolated voltage at which the collector current for a given base current would be zero)

$$V_A = q N_B W_B W_C / \varepsilon_{Si}, \tag{3.4}$$

where ε_{Si} is the dielectric constant of silicon. We obtain the available voltage gain as

$$A_V = g_m / g_{CE} = \frac{I_C}{V_t} \frac{V_A}{I_C} = \frac{V_A}{V_t}, \quad where \quad V_t = k_B T / q. \tag{3.5}$$

High voltage gain requires a large Gummel number and a large collector depletion width, which means a high-resistivity collector. This is associated with a large collector series resistance, causing slower speed. We also notice that the larger Gummel number means a lower current gain (3.3) as a direct trade-off situation. In fact, the available power gain appears as the figure of merit of a Si bipolar technology:

$$A_P = \beta_F A_V = \frac{q}{\varepsilon_{Si}} \frac{D_{nB}}{D_{pE}} N_E L_E W_C / V_t. \tag{3.6}$$

We see that the Gummel number has disappeared, and all other quantities are material or basic technology parameters, the collector depletion depth being the only variable with a choice. Inserting numbers, we find

$$A_P = \frac{1.6 \times 10^{-19}}{10^{-12}} 3 \times 10^{14} \cdot 3 \times 10^{-4}/0.025 = 6 \times 10^6 \qquad (3.7)$$

as the figure of merit for our reference technology with a 3 μm collector depletion typical for a 5 V transistor. Voltage scaling towards a shorter collector depletion would mean a direct reduction of available voltage and power gain.

Another important technology and design parameter critical for scaling is the current density range for high transconductance and current gain. It is determined by the *knee current*. At that current level, the collector current density is so high that the density of minority electrons entering the base reaches the acceptor density N_B in the base, and the transconductance drops by 50% (Fig. 3.4). This critical collector current density is reached as the peak electron density rises to N_B:

$$J_K = qD_{nB} \frac{N_B}{W_B}. \qquad (3.8)$$

At current densities above this level, the transconductance drops to $g_{mK} = I_C/2V_t$, which occurs for scaled-down emitter areas at fairly small currents so that such a transistor loses its edge as a high-speed load driver. Even for a heterojunction transistor, where we may reach a diffusion constant of 25 cm^2 s^{-1}, an effective acceptor level of 10^{19} cm^{-3}, and a base width of 50 nm, the knee current density

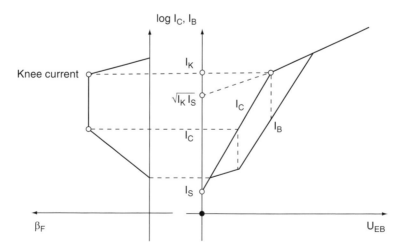

Fig. 3.4 Semilog plots of base and collector currents as well as current gain showing the base-current knee at low levels due to recombination and the collector-current knee at high levels due to strong injection

occurs at 8×10^6 A cm^{-2}, which for our 0.13 μm × 0.13 μm emitter means a current of 1.36 mA.

Nevertheless, SiGe bipolar transistors offer terahertz performance and are the active devices filling the *terahertz gap*. They provide the needed performance in the specialty areas of broadband transceivers, millimeter radar and terahertz imaging. As to the compatibility with mainstream IC processing: They require extra processing steps and thicker epitaxial Si films at extra cost. More cost-effective, high-speed CMOS technologies are infringing on the territory held by bipolar today. We will address that again in Chap. 5 on interconnects and transceivers.

3.2   MOS Integrated Circuits

Our first milestone for MOS integrated circuits is the realization of an MOS transistor by Kahng and Atalla in 1959 [4, 5]. The first products integrating hundreds of these into digital circuit functions on a single chip appeared in the mid-1960s, and the first killer products were the microprocessor and the random-access memory (RAM). Among the many incremental improvements of the MOS technologies, an early outstanding one was the replacement of the Al gate late in the process by a deposited polycrystalline gate earlier in the process, where it served as a mask for the diffusion of the source and drain regions, thereby self-aligning these electrodes with the gate. Because this same polysilicon layer could also serve as a first interconnect layer for local wiring, only four masks and a minimum number of processing steps were needed for this technology. We illustrate this process with Lego blocks in Fig. 3.5 on the basis of a more recent SOI structure.

The LegoMOS model presented here serves very well to demonstrate the technology properties and constraints:

– The 2 × 2 brick marks the smallest feature possible with lithography.
– The 1 × brick marks a thin layer, whose thickness is controlled by thin-film growth or deposition.
– Each button will stand for roughly one acceptor (yellow or red) or one donor impurity (blue) in our ultimate NMOS nanotransistor.

The starting material is a Si wafer, which has a buried-oxide insulating layer (white) and a thin Si film with P-type doping on top (blue). The intended transistor sites (active areas), of which we just show one, are masked against the thermal oxidation of the rest of the surface, which is thereby converted to Si dioxide (Fig. 3.5, top). The masking is then removed from the Si islands, and the thin gate dielectric is grown on the wafer. The polycrystalline Si film is deposited next (red), and the gates are defined by lithography and etched. These gate structures (red) serve as masks for the implantation of N-type dopants (As), which produce the source and drain electrodes (yellow) of the transistor (Fig. 3.5, middle).

An intermediate oxide is deposited next. On this, the contact holes are defined by lithography and etched. The following vapor deposition of Al will also fill the

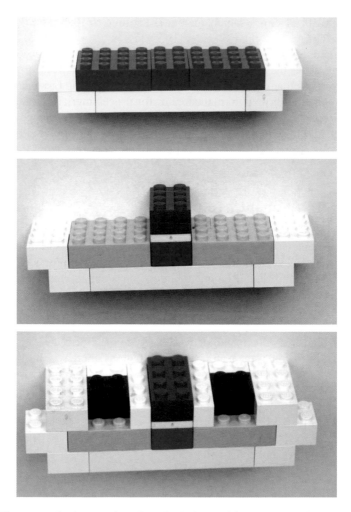

Fig. 3.5 Three steps in the manufacturing of a basic MOS integrated circuit. *Top*: Active area (*blue*). *Middle*: Poly gate (*red*) and source/drain (*yellow*). *Bottom*: Contacts (*black*)

contact holes (black, Fig. 3.5, bottom), and this layer is structured by lithography and etching into thin-film interconnect lines. The resulting topography, where the n-type Si (yellow) allows interconnects as well as the poly layer and the Al layer, is sufficient to wire any circuit function.

This four-mask process, active areas, gates, contacts, metal, tells the MOS success story of low cost and high production yield, and it led to the race for miniaturization: With higher-resolution lithography and etching, transistors could be made smaller and faster and be packed more densely. In about 1970, a second metal interconnect layer, separated by another deposited insulating layer, was added. This required two extra masks, one for defining contact holes, called vias, between the first and second metal layers, and a second mask for etching the second metal layer. Today more than six layers of metal are used. Table 3.1 and Fig. 3.6 present a grand overview of the evolution of the critical dimensions of MOS ICs, including those of the International Technology Roadmap for Semiconductors (ITRS), which has many more details to be covered in a later chapter.

We continue here by analyzing the key performance criteria of MOS transistors: *switching speed* and *energy*. We consider the ideal cross section of an NMOS transistor on top of an insulating layer as shown in Fig. 3.7. As a parallel-plate capacitor, it has a capacitance

$$C_0 = \frac{\varepsilon_I}{t_I} LW, \tag{3.9}$$

Table 3.1 Evolution of the critical dimensions of MOS ICs

Year	1970	1980	1990	2000	2010	2020
L [nm]	20,000	2,000	500	120	32	8
t_I [nm]	120	15	10	5.0	2.5	1.0
L/t_I	167	133	50	24	13	8
Gain $\left(\frac{\varepsilon_I L}{\varepsilon_{Si} t_I} - 2\right)$	54	42	15	6	2	0.7

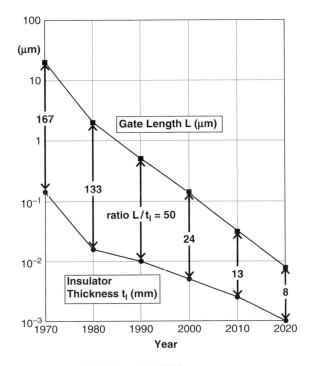

Fig. 3.6 Critical dimensions of MOS ICs 1970–2020

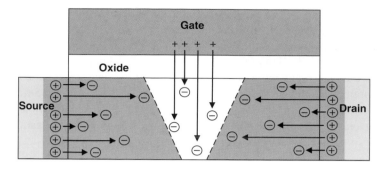

Fig. 3.7 Cross section of an NMOS transistor including charges

with the dielectric constant of the insulator ε_I, its thickness t_I, the length L between the metallurgical junctions of source and drain, and its width W. If we apply a positive voltage V_G, we find Q_0 positive charges on the gate and a negative charge $-Q_0$ in the silicon:

$$Q_0 = C_0 V_G.$$

With a voltage V_{DS} applied between drain and source, we establish an electric field $E = V_{DS}/L$ pointing in the negative x direction. If all the negative charges were mobile electrons, they would now move towards the drain with a certain velocity v. Within the channel transit time L/v, the entire charge Q_0 would be collected at the drain, so that we would obtain the drain current

$$I_D = Q_0 v/L = C_0 V_G v/L. \tag{3.10}$$

This is only correct for very small drain–source voltages. We have to consider that the voltage difference between gate and channel at the drain end is only $V_{GD} = V_{GS} - V_{DS}$, so that the charge density decreases as we move along the channel

$$Q_0'(x = 0) = C_0' V_{GS},$$
$$Q_0'(x = L) = C_0'(V_{GS} - V_{DS}).$$

We introduce the average charge

$$\langle Q \rangle = [Q_0'(0) + Q_0'(L)]LW/2,$$
$$\langle Q \rangle = \frac{\varepsilon_I}{t_I} WL(V_{GS} - V_{DS}/2). \tag{3.11}$$

Under the force of a moderate electric field E, mobile charges in semiconductors move with a velocity proportional to this field $v = \mu_0 E$, where the mobility μ is about 250 cm^2 V^{-1} s^{-1} for electrons in a Si MOS transistor. We take an effective transit time for the charge $\langle Q \rangle$ as

$$1/\tau = v/L = \mu_0 V_{DS}/L^2,$$

so that we obtain a more accurate drain current as

$$I_D = \frac{\langle Q \rangle}{\tau} = \mu_0 \frac{\varepsilon_I}{t_I} \frac{1}{L} (V_{GS} - V_{DS}/2) V_{DS}. \tag{3.12}$$

This is the drain current according to the so-called Shockley gradual-channel approximation. It reaches a maximum current as the drain voltage reaches the gate voltage:

$$V_{DS} \rightarrow V_{GS},$$
$$I_{D0} = \frac{\mu_0 \varepsilon_I}{t_I} \frac{W}{L} \frac{V_{GS}^2}{2}. \tag{3.13}$$

It was already evident in 1970 that this current was larger than what was measured in NMOS transistors. The reason is carrier-velocity saturation. The maximum electron velocity in most semiconductors at room temperature is

$$v_L = 10^7 \text{cm s}^{-1},$$
$$v_L = \mu_0 E_C,$$
$$v = \frac{\mu_0}{1 + E/E_C} E. \tag{3.14}$$

For a mobility of 250 cm^2 V^{-1} s^{-1}, the critical field is 40,000 V cm^{-1}. Here the mobility is already reduced by 50%, and actual fields are much higher. Not only is the maximum current reduced, but so is the drain voltage at which it is reached. For a modeling of this effect, see [6] and [7]. We settle here for an optimistic estimate of the maximum current. The most aggressive assumption is that all electrons in the channel travel at the velocity limit:

$$I_{D\infty} = Q_0 v_L/L,$$
$$I_{D\infty} = \frac{\varepsilon_I}{t_I} W v_L V_{GS}. \tag{3.15}$$

This optimistic extreme has a number of significant implications. The channel length does not appear in this equation, removing one basic argument for scaling. The outer limit for the transconductance would be

$$G'_\infty = \frac{\varepsilon_I}{t_I} v_L = \frac{0.4 \times 10^{-12}}{10^{-7}} \times 10^7 \times 10^{-4} = 4 \text{ mS } \mu\text{m}^{-1}$$

and the maximum current

$$I'_{D\infty \text{ max}} = \varepsilon_I v_L E_{I\text{max}} = 0.4 \times 10^{-12} \times 10^7 \times 5 \times 10^6 \times 10^{-4} = 2 \text{ mA } \mu\text{m}^{-1}.$$

This simple assumption overestimates the velocity on the source side of the channel. We now consider this and still simplify the situation by letting the field at the drain end approach infinity in order to transport the vanishing charge there at pinch-off. The maximum current becomes [7]

$$I_{DS\,max} = k \cdot I_{D\infty},$$

$$k = 1 - \frac{E_C L}{V_{GS}}\left(\sqrt{1 + 2\frac{V_{GS}}{E_C L}} - 1\right). \tag{3.16}$$

This correction factor k is shown in Fig. 3.8.

So far, we have assumed that the NMOS transistor starts to conduct current as soon as the gate voltage is larger than zero, in other words that its threshold voltage V_T is zero. However, for most practical applications, a positive threshold voltage is required so that our model drain current only starts to flow when the gate voltage exceeds this threshold. We introduce this excess voltage as the effective gate voltage:

$$V_{Geff} = V_{GS} - V_T.$$

In Tables 3.2 and 3.3 we assess the maximum channel charge, the maximum current per micrometer width and the maximum current for a minimum transistor with $W/L = 2$ for all the past milestones and for the 2020 target, and we illustrate these in Fig. 3.9.

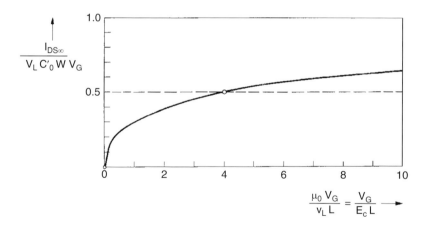

Fig. 3.8 Correction factor k for the maximum available current [7]

Table 3.2 Transistor parameters and NMOS transistor performance 1970–2020

Year	1970	1980	1990	2000	2010	2020
Length [nm]	20,000	2,000	500	120	32	8
Thickness [nm]	120	15	10	5.0	2.5	1.0
Supply voltage [V]	20	5.0	3.3	1.5	0.5	0.4
Threshold [V]	4.0	1.5	1.0	0.7	0.2	0.1

Table 3.3 Maximum current per micrometer width and intrinsic switching energy 1970–2010

Year	1970	1980	1990	2000	2010	2020
Channel electrons	2.7×10^7	5×10^5	3×10^4	1.2×10^3	60	15
Current [μA μm^{-1}]	80	200	300	270	240	190
Switching energy [eV]	7×10^8	3×10^6	1.2×10^5	1870	50	3

Our assessment of the intrinsic maximum current and transconductance shows that scaled-down Si NMOS transistors approach a limit dictated by the dielectric constant of the gate insulator and the maximum electron velocity fairly independent of the channel length, contrary to the decades up to 2000, where they improved almost in proportion to the reduction of the channel length.

The minimum intrinsic switching time is

$$\tau_S - C_0 V_{DD}/I_{DS\,max},$$

$$\tau_S = \frac{L}{k \cdot v_L} \cdot \frac{V_{DD}}{V_{DD} - V_T},$$

offering a linear improvement as L is reduced. The intrinsic bandwidth is

$$f_{3dB} = \frac{k \cdot v_L}{2\pi L},$$

which, for $k = 0.6$, $V_T = 0$ and a 10 nm channel, results in 200 GHz. The evolution of bandwidth and switching times of NMOS transistors from 1970 with projections for 2020 can be found in Fig. 3.10 and Table 3.4.

After this generic assessment of maximum available current at a maximum gate voltage given by the breakdown field of the dielectric, typically 5 MV cm^{-1}, we now take a closer look at the transistor channel. Figure 3.7 already tells us that there is a more complex balance of charges in the transistor channel, in our case the dominant normally-off NMOS transistor with a positive threshold voltage. This requires an acceptor doping density N_C [cm^{-3}], which we recognize as fixed negative charges in the channel. On its source and drain side, we see the PN junction depletion layers against source and drain set up as pairs of positively charged donors in the source and drain layer and negatively charged acceptors in the channel. In equilibrium (drain–source voltage zero), these depletion layers set up potential barriers each 0.7 V high. The width of each depletion layer is

$$W_D = \sqrt{\frac{2 \times 0.7 \times \varepsilon_{Si}}{qN_C}},$$

which is 3 nm wide for a doping density of 10^{20}/cm^3. For a metallurgical channel length of 8 nm, this leaves us with an effective channel length of 2 nm. In this region, for a Si film 2 nm thick and a channel width of 10 nm, we find just

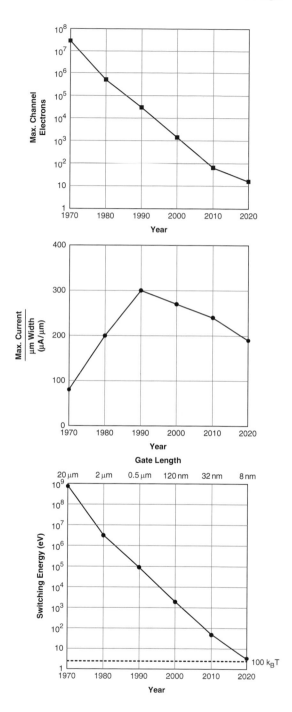

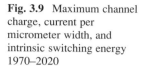

Fig. 3.9 Maximum channel charge, current per micrometer width, and intrinsic switching energy 1970–2020

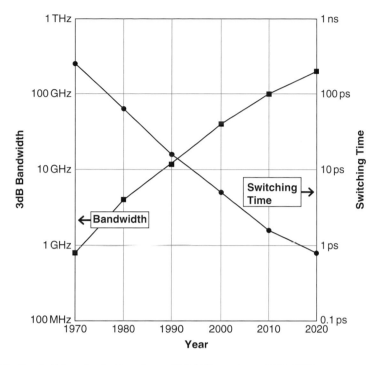

Fig. 3.10 Bandwidth and switching times of NMOS transistors 1970–2020

Table 3.4 Bandwidth and switching times of NMOS transistors 1970–2020

Year	1970	1980	1990	2000	2010	2020
Bandwidth [GHz]	0.7	4.0	10	40	100	200
Switching time [ps]	250	60	17	5	1.6	0.8

four acceptors. We show in Fig. 3.7 that it takes an equal number of charges on the gate to compensate these acceptors, and the necessary gate voltage would be the threshold voltage. Any higher voltage would now induce free electrons in the channel. This would be the electrostatic model, which we have considered so far. However, there are more forces present than we have identified as yet. As soon as we apply any voltage between the gate and the source, we lower the potential barrier between source and channel below the equilibrium of 0.7 V. As a consequence, the electric field in the source depletion layer, which confined electrons to the source so far, is reduced, and electrons can traverse the barrier and will be collected at the drain under any applied drain bias. This current component is a diffusion current, as in bipolar transistors controlled by the gate as an insulated base. It is alternatively called leakage or subthreshold current. We will call it the barrier-control mode of transistor operation, because not only the gate but also the drain has an effect on this barrier.

In Fig. 3.11, we show a short-channel NMOS transistor in which the potential barrier Φ_B is controlled by applied voltages $V_G - V_S$ and $V_D - V_S$. Impurities are

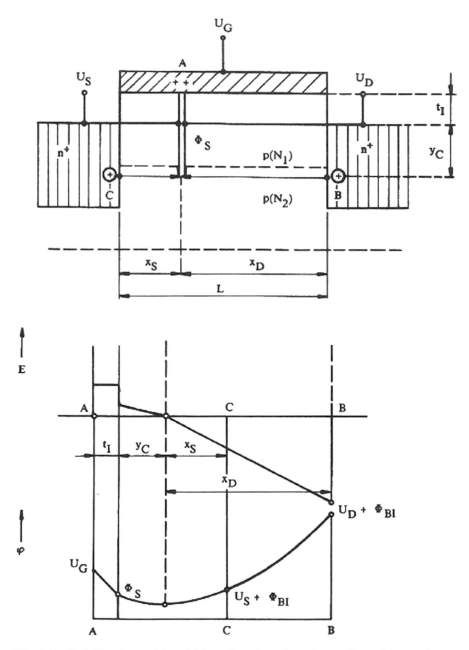

Fig. 3.11 2D field and potential model for a short-channel transistor under barrier control

engineered in the channel such that there is a density N_1 near the surface and N_2 in a *buried* layer. The height of the potential barrier can be modeled as [6]

$$\Phi_B = A(V_{GS} - V_T) + BV_{DS} - CV_{DS}(V_{GS} - V_T). \tag{3.17}$$

As we expect, the threshold voltage here is dependent on the drain voltage:

$$V_T = V_{T0} - \frac{BV_{DS}}{A - CV_{DS}}.$$

We introduce

$$n = \frac{C_0 + C_d}{C_0},$$
$$\eta = \frac{\varepsilon_{Si} t_I}{\varepsilon_I L} \sqrt{\frac{N_1}{N_2}}, \tag{3.18}$$

and we obtain simple and powerful expressions for the barrier dependence on gate and drain voltage:

$$A = \frac{1 - 2\eta}{n},$$
$$B = \frac{\eta}{n}, \tag{3.19}$$
$$C = \frac{4\varepsilon_{Si}}{qN_2L^2} \cdot \frac{\eta}{n}.$$

At the threshold voltage, $\Phi_B = 0$, and we define a maximum subthreshold current I_c such that

$$I_{D1} = I_C \exp(\Phi_B / V_t) \tag{3.20}$$

is the subthreshold or barrier-controlled current. The log of this current is plotted in Fig. 3.12 vs the input gate voltage and the drain voltage. As we see, both voltages exert control over the current, and a quantitative assessment of these effects becomes essential.

In order to connect this current to the drift current (3.16) above the threshold voltage, we set

$$I_C = \frac{1}{e} \cdot I_{DS\,max}(V_{Geff} = V_t),$$

and we obtain a continuous current from subthreshold diffusion type to above-threshold drift type as [8]

$$I_{tot} = I_C \cdot \ln\left[\exp\left(\frac{I_{DSmax}}{I_C}\right) + \exp\left(\frac{\Phi_B}{V_t}\right)\right] \tag{3.21}$$

for $\Phi_B < 0$ and noting that $\ln(1 + x) = x$ for $x < 1$.

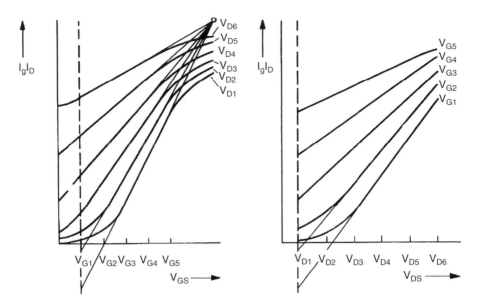

Fig. 3.12 The drain current of a short-channel MOS transistor in its subthreshold or barrier-control mode

We now find that, for transistors shorter than 30 nm, the operating range of <0.5 V means mostly the barrier-control mode. We derive the *transconductance* from (3.17) and (3.20) as

$$g_m = I_{D1}/AV_t,$$

the output conductance as

$$g_{DS} = I_{D1}/BV_t,$$

and the *available voltage gain* as

$$A_V = g_m/g_{DS} = A/B = (1 - 2\eta)/\eta = \frac{\varepsilon_I L}{\varepsilon_{Si} t_I} \sqrt{\frac{N_2}{N_1}} - 2. \qquad (3.22)$$

We see immediately that a high dielectric constant of the insulator together with its scaling directly with the gate length are essential to maintain a gain >1, that the channel should be doped inversely, which means that the deep density N_2 should be higher than N_1 at the Si surface, and that we keep the depletion capacitance C_d in (3.18) small compared to the gate capacitance.

We build a 10 nm reference NMOS *nanotransistor* with the following parameters:

$L = 10$ nm

$t_I = 3$ nm

$\varepsilon_I/\varepsilon_{Si} = 1$
$N_2/N_1 = 4$
$C_d/C_0 = 3/4$
We obtain the following results:

Transconductance:	150 mV per decade
Output conductance:	700 mV per decade
Voltage gain:	4.7
ON/OFF current ratio:	27:1 for 0.4 V supply voltage
Max. current at 0.4 V:	20 µA, $V_T = 0$ V
Min. current at 0 V:	18 nA for $V_T = 400$ mV
Max. current at $V_{GS} - V_T = 25$ mV:	0.5 µA

For this critical current at the top of the subthreshold region and the onset of the drift-current region and any current smaller than this, there is on the average just one electron in the channel: *Our 10 nm NMOS nanotransistor is a single-electron transistor*.

For a comparison with published results [9], we study the characteristics in Fig. 3.13. We recognize that these are mostly subthreshold characteristics with transitions at the high-current sections into drift-current modes. All characteristics show a strong dependence on the drain–source voltage as we expect from our study

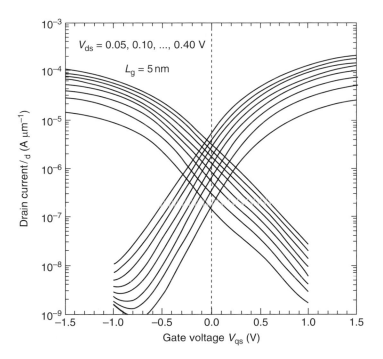

Fig. 3.13 Current–voltage characteristics of 5 nm NMOS and PMOS transistors after Wakabayashi et al. (2006) [9]. © 2006 IEEE

of the barrier-control mode. The analysis of the characteristics shows for the two transistor types:

NMOS Transistor:

Transconductance:	320 mV/decade
Output conductance:	320 mV/decade
Voltage gain:	1

PMOS Transistor:

Transconductance:	270 mV/decade
Output conductance:	320 mV/decade
Voltage gain:	1.18.

These are very informative and instructive results: *At L = 5 nm, it becomes very difficult to build useful MOS transistors.*

Our models as described by (3.18) and (3.22) tell us that a higher dielectric constant and a thinner gate dielectric would be needed at the cost of further reducing the operating voltage, and careful channel engineering would be needed. On top of this squeeze, we have to realize that anywhere below 20 nm we operate not only with serious manufacturing tolerances, but with the fundamental variance of numerical quantities: Any quantity with a mean value N has a variance N and a standard deviation σ equal to the square root of N.

3.2.1 The Fundamental Variance of Nanotransistors

We have noticed already in Fig. 3.7 that, in the MOS transistor channel, we deal with a very sensitive balance of charges, which becomes a matter of small numbers at nanometer dimensions. In our reference nanotransistor, we have carefully doped the channel with a density of 10^{20} acceptors/cm^3 near the source and drain metallurgical junctions.

In equilibrium, at this density, 36 acceptors set up the source and drain depletion layers, each 3 nm wide. In the remaining 4 nm channel region, we place five acceptors, corresponding to 2×10^{19} acceptor volume concentration, to control the threshold voltage and the gain of the transistor: N_1: one acceptor near the Si surface; N_2: four acceptors underneath in the 3 nm Si film. No matter how precisely we have processed, these five acceptors have a fundamental statistical variance of 5 and a standard deviation σ of two acceptors. In this statistical distribution, about 90% of all values lie within $+2\sigma$ and -2σ of the mean value 5. If we settle for this range as an acceptable design space, the low threshold voltage would be given by one acceptor and the high threshold voltage by 9 acceptors. Our gate capacitor has a capacitance of 8×10^{-18} F, which requires 20 mV per elementary charge. This results in a fundamental threshold range of our transistor of threshold voltage high $(9q)$ − threshold voltage low $(1q)$ = 160 mV.

Can we build useful circuit functions with this? The answer is yes, if we can allow currents per gate to vary by orders of magnitude. However, the situation gets even tougher fundamentally when we consider the variance of the transconductance.

We have used channel engineering, Fig. 3.11 and (3.18), to optimize the subthreshold slope (on/off current ratio) and the voltage gain, and we implemented that in our reference transistor with $N_2/N_1 = 4$. Counting acceptors, we allocated 4 deep and 1 near the surface out of our mean total of 5 ions. The spread of $N_1 + N_2$ from 1 to 9 can result in a maximum $N_2/N_1 = 8$ and a minimum $N_2/N_1 = 1$. Using our model, we obtain the following ranges:

Transconductance high: 136 mV/decade, gain high: 7.2
Transconductance low: 300 mV/decade, gain low: 1.3.

The resulting range of transistor characteristics is shown in Fig. 3.14.

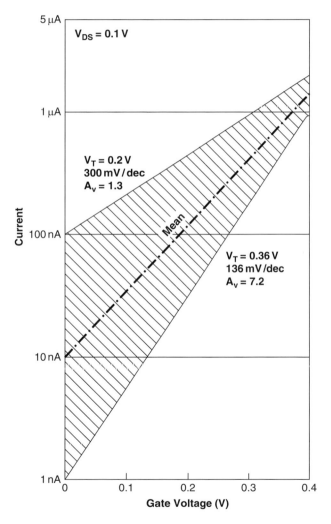

Fig. 3.14 The 90% corridor of current vs gate-voltage characteristics at $V_{DS} = 0.1$ V for our 10 nm reference nanotransistor

This technology design space with a fundamentally large spread of threshold voltages, a transconductance range of 2:1, and gains as low as 1.3 requires robust circuit techniques, which we will cover in the next section, where we shall use exclusively the availability of both NMOS and PMOS transistors.

We mention here that all of the above modeling equations, which we developed for NMOS transistors, can be used to describe PMOS transistors provided we remember that the mobility of holes is about 1/3 that of electrons normally and that it can almost reach that of electrons in Si strained by the introduction of Ge.

Historically, PMOS ICs were first, because it was easier to engineer the (negative) threshold voltage for the necessary normally-off transistors. In order to benefit from the higher mobility of electrons, a lot of energy went into controlling the channel impurities in NMOS transistors so that they could be made normally-off. The biggest boost for this impurity control came with ion implantation [10] of the dopants, which was demonstrated for MOS first by Mostek in 1970. NMOS became the basis for killer products such as microprocessors and memories with such impressive growth that MOS ICs passed bipolar in world-market share in 1980, and Nippon Electric Corporation (NEC) passed TI to become the world's largest semiconductor company. The investment in NMOS computer-aided design, circuit libraries, and process technology became so large that the major players kept their emphasis on this technology well past 1980, while the more radical complementary MOS (CMOS) technology was gaining ground steadily for reasons that we will assess in the following section.

3.3   CMOS and BiCMOS Integrated Circuits

The CMOS process and circuit as invented in 1963 by Wanlass, [11, 12] and shown in Fig. 2.7, derives its success from the features of the pair of complementary transistors as illustrated in Figs. 2.8 and 3.15.

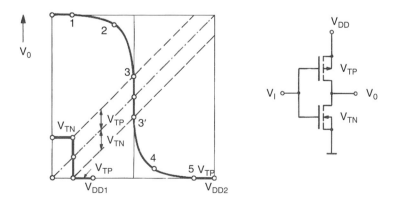

Fig. 3.15 Circuit diagram (*right*) and transfer characteristic (*left*) of the CMOS inverter

Table 3.5 Properties of PMOS and NMOS transistors

	Charge carriers	Threshold	ON voltage	OFF voltage	Drain–source voltage
NMOS	Neg	Pos	Pos	Neg	Pos
PMOS	Pos	Neg	Neg	Pos	Neg

The complementary properties and parameters of NMOS and PMOS transistors are listed in Table 3.5, and from these we obtain, somewhat simplified, the transfer characteristic in Fig. 3.15 for the inverter. It has three sections.

1. For input voltages V_I between zero and the threshold of the NMOS transistor, the output voltage is at the positive supply voltage, because the PMOS transistor is turned on and charges the output to the supply voltage, uncompromised by the NMOS transistor, which is turned off below threshold.
2. For high input voltages beginning at a threshold voltage below the supply voltage, the output voltage is zero, because now the NMOS transistor is turned on and discharges the output to zero, while the PMOS transistor is turned off and does not compete.
3. Between these perfect HI and LO levels, we have a transition region, which we have simplified here with a constant slope, namely the voltage gain of the inverter. If indices N and P denote NMOS and PMOS respectively, this gain is

$$A_V = \frac{g_{mN} + g_{mP}}{g_{DSN} + g_{DSP}}. \tag{3.23}$$

Equation 3.23 shows that the combination of the transistor types establishes gain safely, although one type may even have an internal gain <1. This will be important when we deal with variances at the nanometer scale.

Historically, the magic of CMOS arose from the attributes that we see in Fig. 3.15.

- *CMOS provides the perfect regeneration amplifier*: Although the input may be less than a perfect HI or LO, the inverter output provides a perfect LO or HI, respectively.
- *A CMOS circuit ideally only draws a current and dissipates energy while it is switching*. As soon as signal levels are stable, there is no current flow.
- *CMOS circuits in subthreshold mode operate robustly and dissipate very little power*. Supply voltages can be lower than the sum of the threshold voltages for a very low energy CV^2.
- *CMOS is tolerant to large differences in on-resistance between PMOS pull-up and NMOS pull-down transistor networks*: PMOS and NMOS transistors can have *unit size* irrespective of their individual function. This allows very-large-scale design automation.

In spite of these significant advantages over single-transistor-type PMOS or NMOS circuits, which require tailored *ratio design* per gate type, and which have marginal gain, signal swings less than the supply voltage, and permanent power

Fig. 3.16 Transistor diagram
of a CMOS/bipolar inverter
after [13]

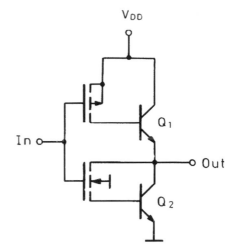

dissipation, CMOS started out in two niches only: digital watches, immediately controlled by Japan, and radiation-hard circuits (small-Si-volume SOI, large noise margin).

Because channel lengths were determined by lithography and significantly longer than the base widths of bipolar transistors, switching speeds were slow (Fig. 3.10), NPN bipolar drivers were proposed in 1969, [13] and Fig. 3.16, to improve the switching speed, particularly for large capacitive loads:

$$I_L = \beta_F \cdot I_{DS},$$

where β_F is the current gain (3.3) of the bipolar transistor. However, such technologies initially were very expensive, and it proved to be difficult to make good NPN transistors with optimum parameters such as given by (3.3–3.6) together with optimized MOS transistors so that progress was slow. About 10 years later, BiCMOS drivers had become economical with all-ion-implanted technologies, further assisted by lateral oxide isolation and, finally, by SOI since about 2000. We should still bear in mind that the bipolar transistor first has to handle its own collector RC load before handling the external load and that a properly designed pair of CMOS inverters is often competitive: If we measure the load as F times that of the internal capacitance of the first inverter with its channel width W, then the second inverter designed with a width of $WF^{1/2}$ delivers the optimum speed.

Moving back to digital CMOS logic, we highlight another unique feature: *The regular topography of CMOS*, which led to *design automation, gate arrays*, and *sea-of-gates*. A pioneering effort was made in 1974 by a group at Philips who recognized [14] that CMOS logic could be designed very effectively with

- Unit-size PMOS and NMOS transistors,
- Regular standard placement of rows of PMOS and NMOS transistors,
- Wiring channels between these rows and

- Highly automated generation of *standard logic cells,* as well as
- Cell placement and routing for customized functions.

This local oxidation CMOS (LOCMOS) gate array technology of 1974 became the prototype of the CMOS gate array strategies, which offered a *quantum jump in IC design productivity* and the *economic realization of large-scale application (customer)-specific ICs (ASICs).*

As wiring became the number-one cost factor compared with the cost of transistor sites on-chip, an endless sea-of-transistors, also called a sea-of-gates, was put on the chip with many degrees of freedom for dense wiring on top.

Another important step was the simplification and optimization of CMOS and CMOS/bipolar manufacturing with the extensive application of *ion implantation* as a *precision process for a wide range of impurity profiles, a room-temperature process with a low-thermal-budget anneal.* This led to an all-ion-implanted CMOS/bipolar process meeting all the specific requirements with a minimum of processing steps [15]. This process, offering optimum collector-diffusion-isolated (CDI) bipolar transistors together with CMOS needed only the following steps:

1. N-type implant for CMOS N-well and N-collectors,
2. active areas for transistors,
3. boron implant for NMOS-, PMOS-, and field threshold,
   (a) Boron implant for active base (Gummel number, (3.3, 3.4),
4. Gate,
5. NMOS source/drain,
6. PMOS source/drain,
7. Contact,
8. Metal 1.

At the time, the second half of the 1970s, CMOS had also made its entry into low-power analog circuits, and mixed-signal analog-digital circuits including analog–digital converters and filters meant the start of digital telephony and the *dawn of mobile communication.*

The CMOS-compatible realization of lateral high-voltage MOS transistors (double-diffused MOS (DMOS)) with operating voltages of hundreds of volts [7, 16] as well as vertical high-voltage, high-current transistors (power MOS) led to the first *systems-on-chip (SOC).*

Regarding the CMOS innovation process in the 1970s, university labs with a pilot line for prototyping new circuits, for example

- Stanford University, Palo Alto, CA: Prof. James D. Meindl and Jaques Baudouin,
- University of California, Berkeley, CA: Prof. Paul R. Gray and Prof. David A. Hodges,
- Katholieke Universiteit Leuven, Belgium: Prof. Roger Van Overstraeten,
- University of Dortmund, Germany: Prof. Bernd Hoefflinger played a key role.

In today's language, at the 10- to 5-μm technology nodes, these labs were able to prove new processes and circuit concepts, providing a critical arsenal of CMOS technology [7], which was widely publicized and exploited in Japan in the context of the Japanese

10×: VLSI (Very-Large-Scale Integration) program.

The first low-power, high-performance microprocessors and digital signal processors were announced by Japanese companies in 1979/1980 at about the 3 μm node, including all high-speed static random-access memories (SRAMs).

The US had missed large-scale entry into CMOS because of its emphasis on NMOS dynamic memories (DRAMs) and high-speed (high-power) bipolar and NMOS processors. The race to catch up started in 1981. TI, Digital Equipment Corporation (DEC), HP, AT&T and Motorola were the first to push CMOS, and CMOS passed NMOS in world-market share by 1985 (Fig. 3.1).

Logic-circuit styles have seen a lot of differentiation, and we will cover some important ones in Sect. 3.6 on dynamic and differential logic. Here we now concentrate on the future potential of complementary MOS circuitry at the nanoscale, and we assess this guided by the *fundamental variance of nano CMOS circuits*.

In Sect. 3.2.1 we assessed the fundamental variance of transistor characteristics at the nanoscale as demonstrated by the 90% corridor of current vs gate voltage for our 10 nm nanotransistor in Fig. 3.14. We now study the consequences of this variance at the transistor level on our fundamental key circuit, the complementary inverter and its transfer characteristic, which would look ideally like Fig. 3.15. We learned there the requirements for this robust amplifier-regenerator: finite threshold voltages and an internal voltage gain >1, (3.23). We found for our reference transistor

- 90% range of threshold voltages: 160 mV,
- 90% range of voltage gain: 1.3–7.2.

Let us assume that we can succeed in making PMOS transistors symmetrical to NMOS transistors, i.e., the same threshold range, and an inverter gain of 2:

$$g_{mN} = g_{mP} = 150 \text{mV/decade},$$

$$g_{DSN} = g_{DSP} = 300 \text{mV/decade}.$$

The corridor of transfer characteristics would then be as shown in Fig. 3.17, where the envelope is given on the left by a zero threshold of the NMOS and 160 mV of the PMOS, and on the right by 160 mV for the NMOS and zero for the PMOS transistor.

The envelopes stand for those pairs where we would just achieve a full swing from the positive to the negative supply voltage. These extreme cases are

Fig. 3.17 Voltage-transfer characteristics of the left (*black*) and right (*red*) branches of the cross-coupled inverter pair and the two stable states. (**a**) Nominally symmetric. (**b**) Variance causing worst-case offsets. *NM*: Noise margin

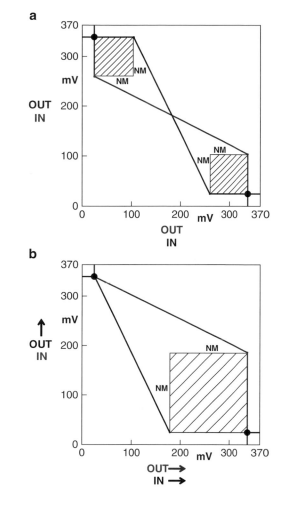

$$V_{TN1} = V_{TN2} = 0 \quad \leftrightarrow \quad V_{TP1} = V_{TP2} = high,$$

$$V_{TN1} - V_{TN2} = high \quad (\ ) \quad V_{TP1} = V_{TP2} = 0.$$

As a function of the threshold variance and the average gain, the minimum supply voltage for a full signal swing from within a V_t of the positive and negative supply voltages, often called a rail-to-rail swing, is

$$V_{DD\,min} = 2V_t + \Delta V_T(1 + 2/A_V). \qquad (3.24)$$

In our case, this supply voltage would be 370 mV.

In Fig. 3.17, we have also entered the static noise margin *NM*, the largest distortion on a logical ONE or ZERO that would be relaxed by the cross-coupled inverter pair, the quad, to a perfect ONE or ZERO. If the inverters are symmetrical (a), this noise margin, relative to V_{DDmin}, is

$$\frac{NM}{V_{DD\,min}} = 0.5 - \frac{1}{2A_V}.$$ (3.25)

In the symmetrical case, this noise margin would be 25% for both ONE and ZERO. In the marginal case (b), one noise margin is 0, and the other is 50%.

The lesson of this exercise is that properly engineered symmetrical 10 nm CMOS nano-inverters can serve as amplifying signal regenerators and drivers. However, it becomes clear as well that the single nano-inverter will be plagued by asymmetry both in its transfer characteristic and in its drive capability and that at this level of miniaturization

– The cross-coupled CMOS pair (the quad) becomes the most robust and the fastest voltage regenerator, and
– Fully differential signaling/logic can result in double the capacitance, but safe operation at half the operating voltage means a 50% net reduction of the switching energy CV^2.

The quad is key to many practical and robust nanoelectronic functions:

– Fully differential signal regeneration,
– Differential amplification,
– Low-voltage, high-speed memory.

The quad has long been the heart of the memory cells in CMOS static random-access memories (SRAMs). The six-transistor cell in Fig. 3.18 shows this cell with fully differential read and write capability.

This cell is suitable for low-voltage standby because of the immunity inherent in the storage of both the true and inverted data. How low can this standby voltage be? The true and inverted voltage levels have to be separated by at least $2V_t = 50\,mV$ at

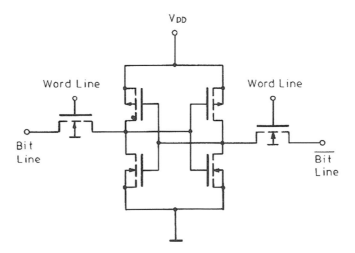

Fig. 3.18 Six-transistor CMOS SRAM cell for differential, nondestructive write and read

room temperature. To this we have to add the variance of the threshold voltages, giving a standby voltage of 210 mV for our 10 nm target technology.

The voltage difference between ONE and ZERO is about 50 mV, and the switching energy to flip the content of this cell would correspond to moving *2.5 electrons* from one pair of gates to the other: *The 10 nm CMOS SRAM cell in standby is a two-electrons perbit memory* for random-access, high-speed write and nondestructive read.

The price for this effective type of memory is the standby current. It is, to first order, the geometric mean of the maximum and minimum transistor current for a given supply voltage:

$$I_{standby} = \sqrt{I_{max} \cdot I_{min}}.$$

This amounts to 3 nA for our cell, or 3 A and 600 mW for a 1 Gbit 10 nm SRAM in standby. We treat the trade-offs between DC and dynamic-switching power in SRAMs further in Chap. 11.

We will discuss CMOS logic-options at the 10 nm node in Sects. 3.5 and 3.6. Here we summarize our assessment of CMOS technology and circuits on their move towards 10 nm:

- 10 nm CMOS transistors are dominated, like all semiconductor devices with functional volumes $<1{,}000$ nm^3, by the fundamental variance of the numbers of charge carriers and doping atoms involved in the electronic functions.
- MOS field-effect transistors <50 nm operate no longer in the familiar (Shockley) drift mode but in the diffusion or subthreshold mode, where the barrier for mobile charges is controlled by both the gate and the drain, resulting in small and sensitive intrinsic voltage gains of transistors.
- Variance and intrinsic gains marginally >1 require complementary NMOS/ PMOS circuitry, in particular robust signal regenerators and accelerators.
- Towards 10 nm, regenerators will have to contain CMOS quads, cross-coupled complementary pairs, and fully differential signaling and logic will be needed.
- The push for 10 nm transistors means supply voltages <400 mV, heading towards 200 mV, attractive for high-density memory and mobile systems. Allowances for speed and signal integrity can mean diminishing returns.

In spite of all these concerns, CMOS is the only robust, large-scale, and dominant chip technology for the next 20 years.

3.4 Silicon-on-Insulator (SOI) CMOS Technology

The ideal planar integration of CMOS integrated circuits was proposed and realized more than 35 years ago, as we saw in Sect. 2.4, and as illustrated in Fig. 2.5. Lateral and vertical oxide isolation of the complementary PMOS and NMOS transistors allows the highest integration density and the smallest parasitic capacitances in order to reduce the switching energy CV^2. We follow the lengthy evolution of this SOI technology. Although its implementation was shown virtually overnight in 1964 [17, 18] by Si epitaxy on sapphire, it acquired the label of being expensive and difficult, and it remained a very small niche technology for military and space applications, albeit meeting extreme reliability requirements.

It took 15 years for SOI to receive renewed attention. A lab in Japan and the Stanford Electronics Laboratory [20] deserve the credit for launching the SOI wave of the 1980s. Izumi and coworkers [19] realized the film sequence silicon–insulator–silicon substrate by high-dose, high-energy implantation of oxygen into silicon to produce a buried oxide layer, of course with a subsequent very-high-temperature anneal to create a stoichiometric buried silicon dioxide layer and to recrystallize the top Si film amorphized by the implant. Further milestones are listed in Table 3.6.

Table 3.6 Selected SOI milestones

Year	Milestone	Reference
1964	Silicon-on-sapphire (SOS)	[17, 18]
1978	SIMOX: Silicon isolation by implanting oxygen	[19]
1979	Laser recrystallization of polysilicon deposited on silicon dioxide	[20]
1981	FIPOS: Full isolation by porous silicon	[21]
	ELO: Epitaxial lateral overgrowth	[22]
1985	Wafer bonding	[23]
1994	ELTRAN: Epitaxial-layer transfer	[24]
1995	Smart cut: H implantation, bonding, and splitting	[25]
2005	SOI foil [53]	

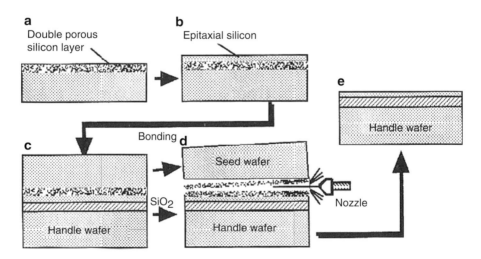

Fig. 3.19 SOI wafer obtained by the transfer of an epitaxial Si layer (after [26]). A: Formation of a two-level porous Si layer on the seed wafer. B: Growth of the epitaxial Si layer on the seed wafer. C: Wafer bonding of the seed wafer to the handle (product) wafer. D: Splitting the wafers. Polishing, etching, and possibly a smoothing epitaxy step for the handle (product) wafer. E: Surface cleaning of the seed wafer

A more detailed description of these processes can be found in [26]. We focus here on the quality of the Si films and on their economics for large-scale manufacturing. As a rule, the surviving SOI technologies for ICs have a high-quality crystalline film grown by Si epitaxy. That feature is available in epitaxial lateral overgrowth. We cover that in the following section on 3D CMOS. Epitaxial film is also available in processes following the ELTRAN scheme, which is shown in Fig. 3.19.

3.5 3D CMOS Technologies and Circuits

For 50 years, integrated circuits have been 2D arrangements of transistors and other components, such as resistors and capacitors, in a thin surface layer of a crystalline Si wafer. Recently, 3D integration has become popular in the sense of stacking chips on top of each other. We address that in Sect. 3.6. Here, we take a dedicated look at the history and the future potential of stacking transistors on top of each other with high density and functionality within the same Si substrate.

Early efforts in this direction were made with regard to bipolar transistors and using PN junction isolation, but without success. Gibbons and Lee deserve the credit for launching true 3D functional integration for CMOS in 1979. We marked that as a significant milestone in Chap. 2 (Fig. 2.12, [20]) and we highlighted the Japanese FED program that followed shortly thereafter. All these related efforts suffered from the fact that only the bottom crystalline transistor layer offered high-quality transistors; the higher layers consisted of poorer, recrystallized Si, producing an inferior transistor quality. The sole high-quality crystalline films on silicon dioxide as an insulator were those obtained by seeded, selective epitaxial lateral overgrowth (ELO) [22]. Because the lateral growth distance meant about as much useless vertical growth and the seed meant a short to the underlying Si layer, this technique did not receive much attention.

However, if one uses the seed as a vertical contact (a via) between devices stacked on top of each other, it turns into an advantage rather than being a shortcoming. This was proposed as a concept at one of the FED meetings [27] with a highly compact full-adder as an example (Fig. 3.20). The real consequence, namely to use as many seeds for selective epitaxial growth as there are CMOS transistor pairs, was expressed in 1984 [28], and a practical process patent application was filed, as illustrated in Fig. 3.21 [29].

Coordinated research between Purdue University and the Institute for Micro-electronics Stuttgart, Germany, in the following years produced a number of remarkable results based on localized epitaxial overgrowth for 3D CMOS:

– A top gate above the thin-film PMOS transistor provided a mask for the PMOS source and drain, and it made this transistor a dual-gate transistor with a top and a bottom channel in parallel. It was shown that this could more than double the

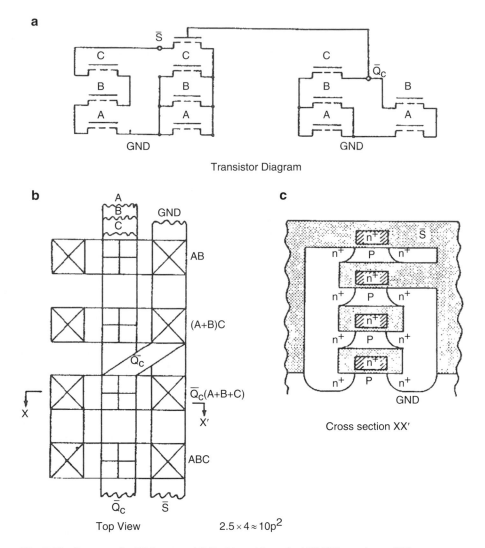

Fig. 3.20 Concept of a 3D-integrated full adder with stacked NMOS transistors [27]

PMOS transistor current to levels almost equal to the underlying NMOS transistor with the same width (Fig. 3.22) [30].

– Independent operation of the two PMOS gates provided a logic OR function with a perfect, very-low-leakage off state (Fig. 3.23)

– The PMOS OR function together with an additional NMOS transistor at the lower level provided a two-input NAND gate with a very small footprint (Figs. 3.24 and 3.25).

The topography of the stacked complementary transistor pair led to a primitive cell that, beyond the inverter, offers the function of a selector (Fig. 3.26), which can

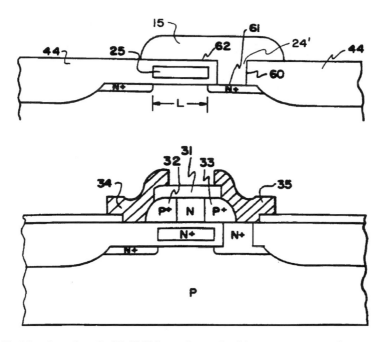

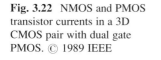

Fig. 3.21 Manufacturing of a 3D CMOS transistor pair with a common gate and common drains [29]. Top: After formation of the bottom NMOS transistor and localized epitaxial overgrowth (LOG) with the transistor drain as the seed. Bottom: After formation of the top thinned epitaxial-film PMOS transistor with its drain connected to the bottom NMOS transistor. The selective epitaxy acts here as self-assembly

Fig. 3.22 NMOS and PMOS transistor currents in a 3D CMOS pair with dual gate PMOS. © 1989 IEEE

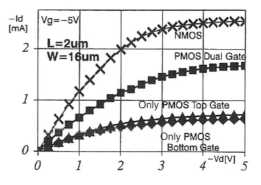

serve as a universal basic element for automatic logic synthesis with the benefit of a very dense, highly regular physical layout [31].

A part of the 3D strategy was to provide the lower supply voltage from the Si bulk, where a chip-size PN junction serves as a large charge store to reduce ground bounce and to provide a ground contact through a "sinker" where needed, for example, with the NMOS source for the inverter. Numerous 3D test circuits, both full-custom and sea-of-gates type, were designed. The comparison of these and of

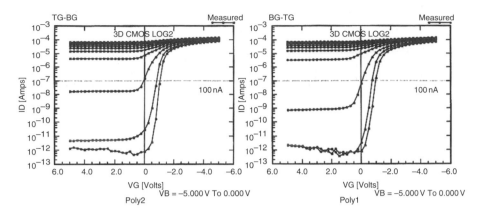

Fig. 3.23 Currents for different poly-gate voltages on the dual-gate PMOS transistor. © 1989 IEEE

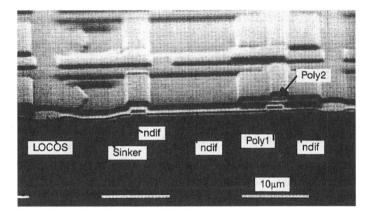

Fig. 3.24 Scanning electron microscopy (SEM) image of the cross section of a two-NAND gate. The stacked dual-gate PMOS on top of the NMOS transistor can be seen on the right. LOCOS: Local oxidation of silicon. © 1992 IEEE

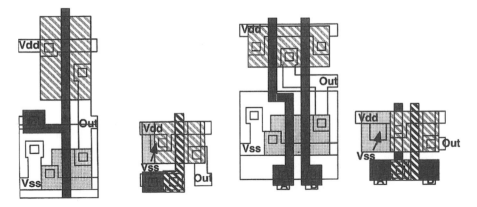

Fig. 3.25 Layout comparison of 2D and 3D CMOS inverters and two-NANDs for equal design rules. © VDI-Verlag 1993

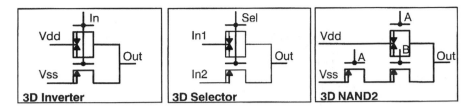

Fig. 3.26 Functions of the 3D complementary pair with dual-gate PMOS transistor. © VDI-Verlag 1993

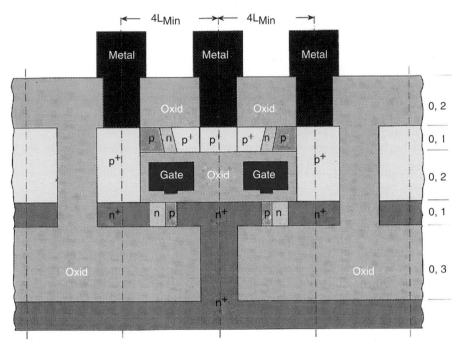

Fig. 3.27 Cross section of a 3D CMOS quad in SOI, the fundamental building block for high-performance nano-CMOS. © WEKA-Medien 2000

34 standard cells for the same design rules provided the average result [32] that *3D CMOS with localized epitaxial overgrowth requires only 50% of the area as 2D CMOS with the same mask count.*

3D CMOS is even more effective in the merger of two complementary pairs, the CMOS quad mentioned in the previous section as the universal core of CMOS differential amplifiers, regenerators (Fig. 3.27), and storage elements such as SRAM. In Fig. 3.28, we see such a memory cell with an additional LOG epitaxial thin-film transistor level for the NMOS word-select transistors [32].

The density potential of 3D CMOS logic and its compatibility with computer-aided design (CAD) automation were critically evaluated in the late 1990s and their significance clearly demonstrated [33]. However, the rapid progress in scaling 2D CMOS circuits met the density requirements until recently so that the 2D evolution

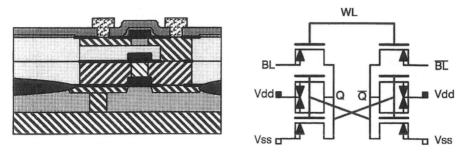

Fig. 3.28 Circuit diagram and cross section of a 3D CMOS SRAM cell with three levels of transistors, including dual-gate PMOS. © VDI-Verlag 1993

has ruled, and disruptive 3D structures remained on the sidelines. Other alternative introductions of the third dimension in vertical transistors, FinFETs, tri-gate transistors, and carbon nanotubes with wrapped-around control gates have been introduced in 2011 as 22 nm Tri-gate transistors by Intel. One can find surveys of these, for example, in [34–36].

Will the third dimension arrive, finally and in a disruptive fashion? The answer is

- 3D stacking of chips and wafers arrived, fairly disruptively, in 2005, to meet the density requirements of the market (more in the following section);
- 3D stacking of CMOS transistors on-chip will follow in view of the 10 nm barrier.

What are the requirements for future 3D CMOS?

- Compatibility with mainstream Si CMOS technology
- No increase in total process complexity
- No disruption of the physical-layer circuit design style
- Large-scale vertical Si interconnect by selective growth to replace contacts and metal
- High quality of the upper-level thin-crystalline-film transistors

There will certainly appear alternatives to the selective-epitaxy technology presented here. It merits, however, to be reconsidered: What required growth of 10 μm Si mountains and their thinning-back by chemo-mechanical polishing (CMP) almost 20 years ago, will now be <1 μm with a much smaller thermal budget, and selective molecular-beam epitaxy (MBE) could be considered [37], as well as atomic-layer epitaxy (ALE).

3.6 MOS Low-Voltage Differential Logic

MOS logic styles evolved rapidly in the 1960s and were mostly built on the availability of one transistor type, which was initially P-type. The gates functioned as voltage dividers, which required *ratio design* with narrow, long, high-resistance load transistors and wide, short switching transistors. Designs varied largely depending on designer skills and discipline. This situation improved when depletion-type load transistors became available in about 1970, after the introduction of ion implantation [10] allowed precise threshold control.

While this type of logic was pretty much a replica of bipolar logic with MOS transistors, a radical alternative exploited the MOS transistor as a switch with a very high off-resistance and a large on–off resistance ratio, and, at the same time, a high-quality capacitor to store the information as a charge. As a consequence, various styles of clocked MOS logic were introduced in the 1960s with distinct phases for precharging the output nodes of a gate, followed by a sampling phase, in which the status of the gate was evaluated. Unit-size transistors could be used, and only the standard enhancement-type was needed. Complex series–parallel logic switching networks could be used, generally with reduced total network capacitance and higher speed.

In our chart of milestones (Fig. 3.1), we note 1980 as the year of killer products based on dynamic MOS logic. Microprocessors and signal processors used high-density, high-speed complex dynamic logic to achieve performance. The basis was domino logic [38], followed by several enhancements to deal with speed problems caused by the resistance of high-fan-in, series-connected transistors and the asymmetry of pull-up and pull-down times of gates.

This led to the use of fully differential input trees, which were then connected to differential amplifiers, which

– Restore full signal swing,
– Deliver perfect TRUE and COMPLEMENT outputs,
– Suppress common-mode asymmetry and offset noise,
– Can drive large-fan-out loads.

In some versions, the construction appeared in the name, such as DCVSL (dynamic cascode voltage switched logic) [39].

The no.1 field for complex transmission-gate logic is adders and multipliers. In particular, long carry chains with short overall delays have seen continuous inventions and improvements, many dependent on (low-voltage) fully differential signaling (LVDS) for safe operation with sufficient noise margin and fast detection with differential amplifiers, eliminating common-mode voltages, crosstalk and other noise, and generating full-swing complementary outputs with high speed.

One implementation of a "Manchester carry chain" is shown in Fig. 3.29. In the center we see the cross-coupled pair, the *quad*, as the heart of the amplifier/ accelerator. It assures the safe and high-speed propagation of the carry signals in spite of a supply voltage of only 0.5 V, as shown in Fig. 3.30 [40–42].

Arithmetic units are those logic circuits where density, speed, and energy in the form of the number of operations per unit of energy (watts $\times$ seconds $=$ joules) or operations per second per watt (OPS/W) matter most. It is the show-stopper for media-rich mobile electronics. Watching the energy CV^2, we have to minimize the circuitry ($=C$) – and the voltage V by any means because of its quadratic effect. For the extremely low voltages of $300-500$ mV, differential logic has become a must.

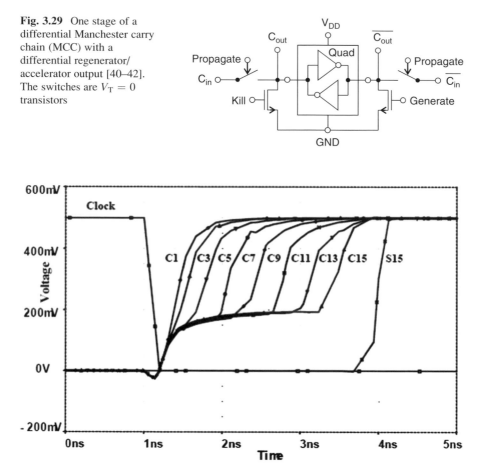

Fig. 3.29 One stage of a differential Manchester carry chain (MCC) with a differential regenerator/accelerator output [40–42]. The switches are $V_T = 0$ transistors

Fig. 3.30 Generation of the most-significant-bit carry and sum of a 16-bit HIPERLOGIC adder in a 100 nm SOI CMOS technology [40–42]. © 2000 IEEE

What can be done about the required circuitry? Beyond the optimization of adders, the key problem is the multiplier. The raw estimate says that the task of multiplying two N-bit words has a complexity of the order $O(N^2)$. A well-known improvement on this is Booth encoding, which reduces the complexity to $O(N^2/2)$. The result is normally accumulated to a length of 2 N bits, which is overkill in all signal-processing applications, because, *if the inputs have N bit accuracy, the accuracy of the multiplication result is only N bit.* One radical alternative, particularly considering signal processing, is to add the binary logarithms and to do this by getting the most significant bit of the result first, [40–42]. This is the basis of the DIGILOG (digital logarithmic) multiplier.

For our two binary numbers A and B, we determine the leading ONEs at positions j and k, respectively, and then proceed as follows:

$$A = 2^j + A_R,$$
$$B = 2^k + B_R,$$
$$A \cdot B = (2^j + A_R)(2^k + B_R),$$
$$A \cdot B = 2^j \cdot B + 2^k \cdot A_R + A_R \cdot B_R.$$

The first part can be obtained easily by shifting and adding, producing a first approximate result. Iterative procedures on the remainder $A_R B_R$ deliver more accurate results: After one iteration, the worst-case error is -6%, after two iterations it is -1.6%, and the probability that it is $<1\%$ is 99.8%. Operating with two iterations, we can estimate the complexity of this DIGILOG multiplier as $O(3\ N)$, which means an improvement of 24/32 for an 8×8 bit multiplier and of 48/128 for a 16×16 bit multiplier. This is just to show that arithmetic units have great potential for continuing improvement.

High-performance logic (HIPERLOGIC) based on the differential logic accelerator of Fig. 3.29 and the DIGILOG multiplier can produce a 16×16 bit multiplier performing 14 TOPS/W (14 GOPS/mW) at the 100 nm node and an operating voltage of 500 mV, corresponding to70 fJ per operation [40–42]. Moving to 20 nm and 300 mV, a performance of 700 GOPS/mW can be achieved for a 16×16 bit multiplier with the technique described here or other ongoing developments.

The pressure for better arithmetic performance in mobile multimedia applications is very high, as expressed in Chap. 13, with the demand for a 300-fold improvement compared to 2010. The state-of-the art, although impressive, underlines the 300-fold improvement required in that chapter. We have entered several results from 2010 publications in Figs. 3.31 and 3.32.

The first plot, Fig. 3.31, can serve as the master graph, because it plots the speed against the energy per operation. Certainly in mobiles, the issue is: At which energy can you realize speed? We have entered the HIPERLOGIC data of above for 16×16 bit multipliers and some recently reported results from ultra-large-scale media processors. These contain hundreds to thousands of arithmetic processing elements (PEs), mostly organized as 4×4 bit adder/multipliers.

The best reported achieves 27 MOPS with 12 μW at a supply voltage of 340 mV in a 32 nm technology, which gives a power efficiency of 2.8 TOPS/W for a 4×4 multiplier [43]. The adder inside this processor element is built in static CMOS logic, which naturally becomes very slow at 340 mV in spite of a very advanced technology node.

2048 four-bit arithmetic logic units (ALUs) were reported on an image processor performing 310 GOPS/W with a 200 MHz clock at 1.0 V [44]. The progress in multiplier performance between 2007 and 2010 is evident in [45], where 670 GOPS/W was reported. Similar efficiency levels were given in [46]. A trend line of 100×/decade has been drawn in Fig. 3.32, projecting 100 TOPS/W or 100 GOPS/mW for 4×4 multipliers in 2020. This may not be good enough considering the unabated push towards HD and 3D video as outlined in Chaps. 6 and 13, where the demand is for a 1,000-fold improvement in power efficiency

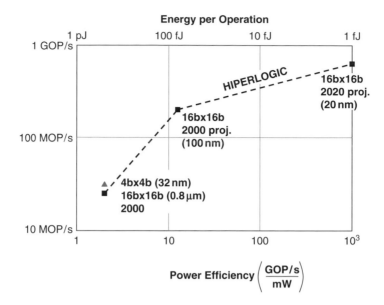

Fig. 3.31 This master chart of digital multipliers shows the operating speed vs the energy per operation (*top axis*) and its inverse, the operations per second per watt, also called the power efficiency (*bottom axis*)

between 2010 and 2020. The same rate is required for supercomputers (see Fig. 6.4). Some more radical innovations like HIPERLOGIC will be needed to meet the requirements of 2020. The HIPERLOGIC example not only offers a quantum step of an order of magnitude, but it does so for a 16 × 16 multiplier, and the PEs of 2020 will not be 4 × x4 but 8 × 8 as a minimum. *In 2020 a CMOS 16 × 16 bit digital multiplier will perform 600 MOPS with a power efficiency of 1 TOPS/mW (1fJ)* (500 times the 2010, much shorter 4 × 4 bit multiplier at 2.8 GOPS/mW).

3.7 3D Chip Stacks

2005 can be seen as the year of the breakthrough for a 3D integration strategy that has had a long history of sophisticated technology development in the shadow of scaling down Si transistors. It is the stacking of prefabricated Si crystals at the chip or wafer level. A number of milestones on this path can be identified as significant advances and pointers. We have marked 1980 in Fig. 3.1, when the

10×: Future Electron Devices (FED) program in Japan (1980–1990)

began to produce refined techniques for wafer thinning, for producing through-silicon vias (TSV's) and for wafer stacking. The

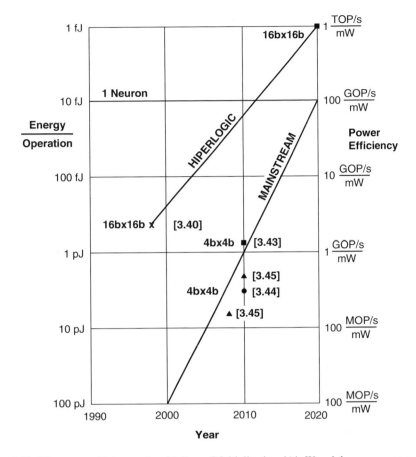

Fig. 3.32 The power efficiency of multipliers: (Multiplications/s)/mW and the energy per multiplication 1990–2020

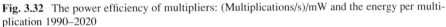

10×: Electronic Eye (Das Elektronische Auge) program in Germany (1994–1998)

led to the realization of a state-of-the-art intelligent vision system with the *biomorphic* structure of high-dynamic-range pixels in a top Si layer being merged with an underlying Si wafer containing first-level neural signal processing [47]. An exploded view of this imager is shown in Fig. 3.33.

Another development has been pivotal in creating multilayer Si systems, namely the use of Si as an *active* carrier for Si chips forming a Si multichip module (MCM) as shown in Fig. 3.34 [48]. In this approach, the chips are flipped upside-down and mounted face-down in the C4 (controlled-collapse chip connect) technique via μbumps, patented by IBM in 1968. Not only is the Si carrier compatible in thermal expansion, but it can be preprocessed to contain the proper blocking capacitance for the chips, Fig. 3.34, eliminating ground bounce, but it can actually be *smart,* providing electrostatic discharge (ESD) protection, temperature sensing, self-test

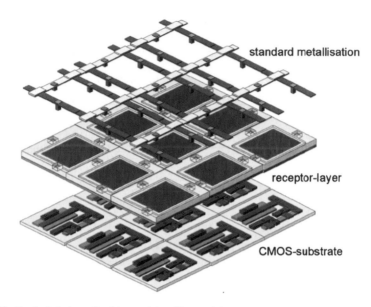

standard metallisation

receptor-layer

CMOS-substrate

Fig. 3.33 Exploded view of a chip-stack intelligent-vision system from 1997. The top layer is the back-end-of-line (BEOL) interconnect structure. The second layer shows the photoreceptor chip, and the third layer represents the neural processing chip [47]. Source: IMS CHIPS 1998

Fig. 3.34 Flipped chips on a Si carrier (1987) [48]. Courtesy IBM Boeblingen

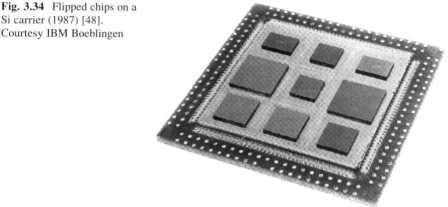

[49], and energy harvesting, like photovoltaic cells. When the IBM C4 patent expired in 1983, a new industry was formed providing *bumping services* and chip-scale packaging (CSP), a value-adding industry that has grown into a critical role today.

A further domain of chip stacking developed quickly with the 3D integration of heterogeneous Si system functions into smart-power and smart-sensor products. PowerMOS and insulated-gate bipolar (IGB) transistors are large-area Si devices, as are most Si sensors for pressure, acceleration, etc., so that there is area on the surface available to place small *intelligent* chips for control and communication.

To take into account this early introduction of chip stacks, we have placed a 1985 marker in Fig. 3.1 for these systems-on-chips (SoCs). Products in this domain are now produced in volumes of hundreds of millions per year and are specially covered in Chap. 14.

At about this time, digital, lightweight mobile phones began to appear, and the stacking of different chips, such as transceivers, processors, memories, and cameras, produced a high-volume, low-cost market for chip stacks as SoCs, and these deserve a chapter of their own (Chap. 12).

The 2005 explosion of chip stacking happened because

– The pressure on more flash memory/cm^2 became big enough to introduce chip-stacking and
– Significant know-how for thinning wafers and forming TSVs had been developing in the background at the interface between the semiconductor and the equipment industry to become quickly available.

IBM, for example, has had a continuous micropackaging research program over decades [50], from which experts migrated to the Asian Tiger countries and set up an industry providing stacking technology as a packaging service.

3D chip stacks developed into a multifaceted field, which can be organized into two major areas:

– Wafer-level and chip-level stacking and
– Uniform- or heterogeneous-function stacking.

For the stacking of wafers, we have basically three options, as shown in Fig. 3.35 [51].

This cross section shows the challenges and opportunities for progress in performance and cost of new products. In Fig. 3.36, we show a concept involving the bonding of wafers that contain two levels of transistors each produced by the localized epitaxial growth presented in the previous section on 3D CMOS (e.g., Fig. 3.27). The bonded wafer contains four transistor layers, and further stacking is proposed with µbumps.

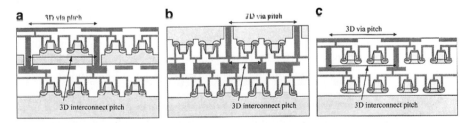

Fig. 3.35 Three options for stacking the upper wafer on the bottom one. (**a**) Face-up; fusing the back of the top wafer on the face of the bottom wafer requires full TSV on the top wafer. (**b**) Face-down; no TSVs needed, allowing higher density and less interconnect parasitics. (**c**) Face-up with oxide fusion, like wafer bonding, exploits strategies for vertical local and global routing to maximize speed and density [51]. © IBM Technical Journals

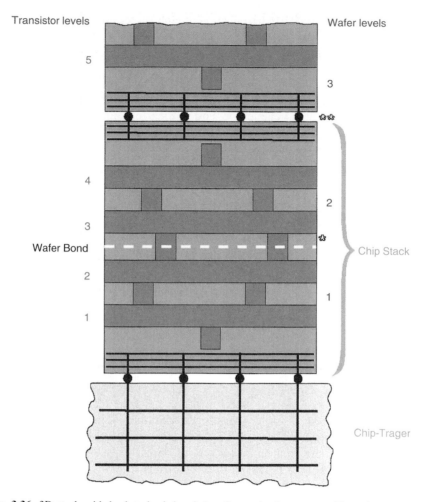

Fig. 3.36 3D stack with back-to-back bonded wafers and μ bumps to a Si carrier below and further stacks above. © WEKA Medien 2000

Uniform functions such as memories have the feature that vertical communication can be organized into a local area of each chip where the TSV signal and power *elevator* can be concentrated effectively over numerous levels of chips. Therefore, memories are candidates for integrating numerous layers of thinned wafers (Fig. 3.37 [52]).

The state-of-the art in 2010 for stacked DRAMs and Flash memories is four to eight wafers, and these devices certainly are the extreme test vehicles for managing yield and cost in the case of wafer-level stacking. Built-in redundancy will support a push towards more wafer levels for memories so that a trend curve as shown in Fig. 3.38 is possibly leading to stacks of 64 wafers by 2020.

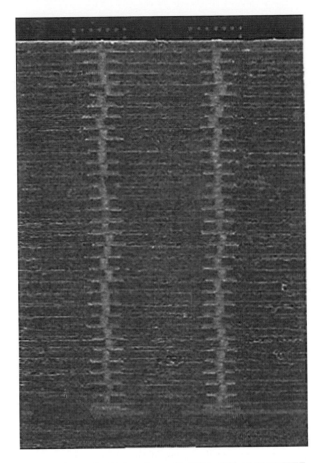

Fig. 3.37 Cross section of a stack of 32 wafers with TSVs [52]. © 2010 IEEE

When we talk about wafers, we mean wafers thinned to about 50 μm by grinding and polishing, or significantly thinner (10–20 μm) Si films, manipulated as handle wafers until they are fused to the next lower Si film and then detached from the handle with techniques similar to the ones shown in Sect. 3.4, for example Fig. 3.19.

The Si film thickness is, of course, essential for the TSV diameter, its parasitic resistance, capacitance, and inductance, and ultimately for the density achievable. We see in Fig. 3.39 that the trend curve for the density of TSVs takes 15 years for a 10-fold improvement. The TSV pitch of 40 μm in 2010 indicates a ratio of 1:1 between TSV pitch and wafer (film) thickness, so the keen interest in manufacturing technologies for Si films <20 μm will continue.

A future of its own opens up for chip-level stacking. It presents formidable challenges for handling these thin and miniature chips on the one hand, but it also opens up the opportunity of literally cracking the *complexity problems of large-scale system chips* containing different technologies and different functions with the strategy of *divide and conquer*.

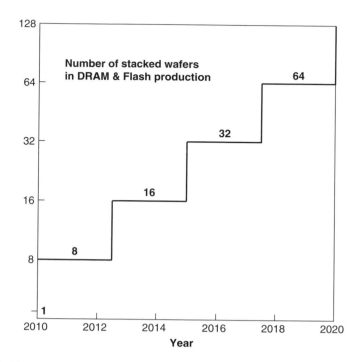

Fig. 3.38 The expected number of stacked wafers for DRAM or Flash products

One example for the production and handling of ultrathin chips is shown in Fig. 3.40 [53]. A double-layer porous Si surface is produced similar to the techniques described in Sect. 3.4 on SOI, Fig. 3.19, but now selectively and including buried cavities in just the areas of the chips. The resulting wafer is fully processed to implement the function of the chip, be it just a circuit or a sensor or actuator. And the chips are tested for functionality. A deep-trench etch into the chip boundaries predisposes the chips to be *cracked, picked, and placed* on an underlying substrate. This is one possible technique for face-up stacking of fully fabricated and tested chips.

We see that, in the world of stacking chips, we can partition the total system into chips optimized for function, design, technology, testability, yield, and manufacturing source, and pick up these individual chips as *known-good die* for 3D integration into a chip stack (or possibly several chip stacks on a Si carrier).

We can imagine how multifaceted the creation of new chip products for the best functionality at optimum cost is getting and that this will continually change the industry: the management of quality, the product reliability, and the product liability have to be assured and clarified in a truly global spirit and scenario, which will be both demanding and rewarding for those involved. This makes up another call for global education and research enabling growth and progress in our world with 3D-integrated chips.

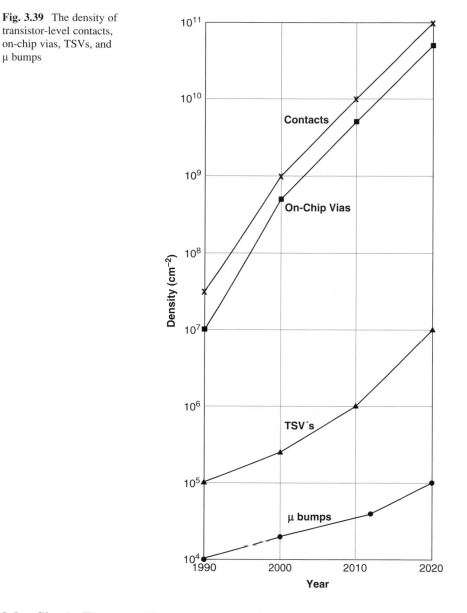

Fig. 3.39 The density of transistor-level contacts, on-chip vias, TSVs, and µ bumps

3.8   Single-Electron Transistor and Other Concepts

This is a book about nanoelectronic devices and products for everyone, and that makes it a book about room-temperature, natural-environment nanoelectronics. We have seen in Sect. 3.2 that our standard MOS transistor, at the end of its scaling towards 10 nm, has become a single-electron transistor (SET; Fig. 3.41) in the sense that we find at any given time just one electron in the channel moving across the

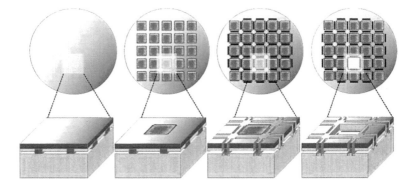

Fig. 3.40 Production of ultrathin-film chips, which have been fully processed and tested in a standard wafer process. Good die are *cracked, picked,* and *placed* on the underlying substrate [53]. © 2006 IEEE

potential barrier either faster, generating more current = high channel conductance, and an output ZERO or slower, generating less current = small channel conductance, and an output ONE.

As of 2010, no room-temperature circuits including signal regeneration, built out of SETs with Coulomb confinement, have been reported. Large-scale logic functions will probably not become competitive with ultralow-voltage differential nano-CMOS. An SE device chain with Coulomb confinement may be considered as a CCD-like serial memory (Chap. 23).

For 10 nm CMOS, we limited our assessment to single-gate transistor structures for processing and cost reasons, and we saw that the atomic variance of dopants in the channel had a large effect of 20 mV per doping ion on the threshold voltage and therefore on the minimum allowable supply voltage, amplified by the allowance for a worst-case voltage gain or noise margin. At this front, some improvement can be gained by dual-gate (Figs. 3.22 and 3.23) or surround gate structures [35], if these can be built with the EOT of 1 nm assumed in our quantitative study. The dual-gate fundamental threshold sensitivity might improve to 10 mV/charge. The surround-gate transistor topography should be vertical as in the following proposal for a complementary bipolar transistor stack (Fig. 3.43). If this could be realized with an equivalent oxide thickness EOT = 1 nm and a Si cylinder of 10 nm diameter, the sensitivity would decrease to 6.5 mV/charge.

We have to remember that all these 10 nm transistors operate in the subthreshold, barrier-control or carrier-diffusion mode. Since we operate in this barrier-control mode, analogous to the bipolar transistor, and since we can no longer turn the MOS transistor off anyway, it is worth revisiting the bipolar transistor, Sect. 3.1, and scaling it down to the <10 nm level.

Let us assume an NPN nanotransistor with a 10 nm × 10 nm emitter, a metallurgical base width of 10 nm and the base doped at a level of $N_A = 10^{20}$ cm^{-3}. This configuration is identical to the channel of our MOS nanotransistor in Sect. 3.2. In equilibrium, depletion widths $W_D = 3$ nm would extend from the emitter and the

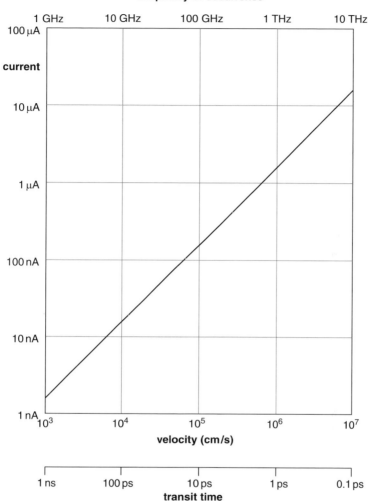

Fig. 3.41 Single-electron current across a channel of 10 nm length. For a channel cross section of 2 nm × 10 nm, the current density at 10 μA would be 5×10^7 A/cm^2, and the volume power density would be 20 GW/cm^3 for a drain voltage of 0.4 V

collector into the base, leaving a neutral base of 4 nm width. Within this neutral base, we would find 40 acceptors, the Gummel number of this nanotransistor with a standard deviation of $\sigma = (40)^{1/2} = 6$ acceptors. Our corridor with ±2 s would give us Gummel numbers between 28 and 52 or 40 ± 30%. Using (3.4), we find a mean Early voltage $V_A = 1.9$ V, giving us an intrinsic voltage gain according to (3.5) of

$$A_V = V_A/V_t = 1.9/0.025 = 76$$

Fig. 3.42 The cross-coupled
differential inverter pair

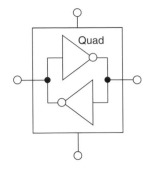

at room temperature. If we can realize an emitter doping $N_E = 10^{21}/cm^3$, we should be able to reach a current gain $\beta_F > 10$, see (3.3), resulting in a power gain $A_P > 760 \pm 60\%$, if we accept that variances accumulate. The maximum useful current for this nanotransistor, namely the knee current I_K, see (3.8), where the transconductance would be cut in half, would be $I_{max} = I_K = 1.2$ mA, much larger than any thermal limit.

This would make a reasonable device with several fundamental advantages over the MOS nanotransistor:

- No problems with gate dielectrics,
- Optimum transconductance of 60 mV/decade (>3 times that of the MOS transistor, Fig. 3.14),
- Perfect topology for vertical growth of transistor structures.

Following up on the last statement, we should pursue, as a vertical Si structure, the *pair of vertical complementary bipolar nanotransistors*: $(N^+PNN^+)(P^+PNP^+)$. The pair would form one half of the *quad,* the cross-coupled pair of inverters (Fig. 3.42), which is the necessary signal regenerator (Fig. 3.17) in low-voltage differential logic (Fig. 3.29) and the heart of the high-speed SRAM (Fig. 3.18), with the lowest operating voltage, nondestructive readout, and the smallest write and read energy.

A structure for the vertical complementary bipolar nanotransistor pair is proposed in Fig. 3.43. The realization of this transistor pair has associated with it the incorporation of three local interconnect layers for optimum surround contacts to the P and N bases and to the N^+P^+ output, and to provide three levels of local interconnects, for example, to form a high-density *quad.* The proposed process flow (Fig. 3.44) is

- SOI wafer with buried N^+ as ground supply and heat sink
- P^+-poly/salicide for P-base surround contact, INPUT, and interconnect
- Mask #1: P^+-poly/salicide etch
- Deposit SiO_2 and planarize
- Metal for N^+P^+ surround contact, OUTPUT, and interconnect
- Mask #2: Metal etch
- Deposit SiO_2 and planarize

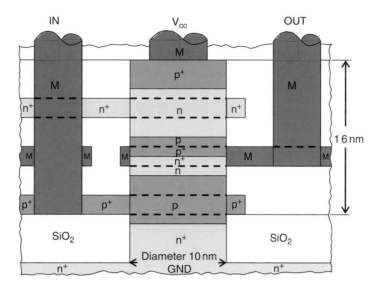

Fig. 3.43 Cross section of a vertical complementary bipolar nanotransistor pair as inverter

– N^+-poly/salicide surround contact of N base, INPUT, and interconnect
– Mask #3: N^+-poly/salicide etch
– Deposit SiO_2 and planarize
– Mask #4: Deep etch for transistor column
– Selectively doped Si crystal growth of complementary transistor column, matched with the poly and metal contact rings
– Mask #5: Deep etch of contact columns
– Selective deposition of metal columns, in situ contacted with the conductor rings and upwards-extendable as TSVs

The formation of complementary bipolar nanotransistors and three local contact layers with just five masks is indicative of the potential of

– Device-level 3D integration
– The replacement of masks, overlay, and etch by selective growth (self-assembly)
– Vertical contact columns as TSVs

The complementary bipolar structure together with lateral NMOS-transistor transmission gates ($V_T = 0$) and logic transistors ($V_T > 0$) would form a promising technology for low-voltage differential logic and high-speed SRAM memory.

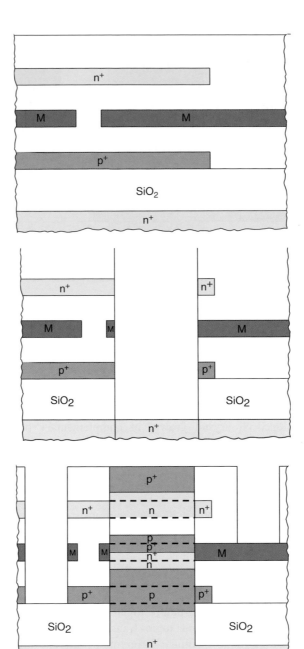

Fig. 3.44 Three steps in the formation of the complementary bipolar transistor pair. Top: After mask #3 and planar oxide. Middle: After mask #4 and deep etch for the growth of the transistor column. Bottom: After mask #5 and deep etch for contact columns

References

1. Noyce R.N.: Semiconductor device-and-lead structure. US Patent 2981877, filed 30 July 1959. Issued 25 Apr 1961
2. Pfeiffer U.R, Ojefors E, Zhao Y.: A SiGe quadrature transmitter and receiver chipset for emerging high-frequency applications. IEEE ISSCC (International Solid-State Circuits Conference), Digest of Technical Papers, pp.416–417, (2010). doi: 10.1109/ISSCC.2010.5433832
3. Gray P.R, Hurst P.J, Lewis S.H, Meyer R.G.: Analysis and Design of Analog Integrated Circuits. Wiley, New York (2001)
4. Atalla M.M.: Stabilisation of silicon surfaces by thermally grown oxides. Bell Syst. Tech. J. **38**, 749 (1959)
5. Kahng D.: Electric field controlled semiconductor device. US Patent 3102230, filed 31 May 1960. Issued 27 Aug 1963
6. Liu S, Hoefflinger B, Pederson D.O.: Interactive two-dimensional design of barrier-controlled MOS transistors. IEEE Trans. Electron Devices **27**, 1550 (1980)
7. Hoefflinger B.: "New CMOS technologies" In: Carroll J.E. (ed.) Solid-state devices 1980, Institute of Physics Conference Series, No. 57. pp. 85–139. Institute of Physics Publishing, Bristol (1981)
8. Grotjohn T, Hoefflinger B.: A parametric short-channel MOS transistor model for subthreshold and strong inversion current. IEEE J. Solid-State Circuits **19**, 100 (1984)
9. Wakabayashi H, et al.: Characteristics and modeling of sub-10-nm planar bulk CMOS devices fabricated by lateral source/drain junction control. IEEE Trans. Electron Devices **53**, 1961 (2006)
10. Hoefflinger B, Bigall K.D, Zimmer G, Krimmel E.F.: Siemens Forschungs- und Entwicklungsberichte **1**(4), 361 (1972)
11. Wanlass F.M.: Low stand-by power complementary field effect circuitry. US Patent 3356858, filed 18 June 1963. Issued 5 Dec 1967
12. Wanlass F.M, Sah C.T.: Nanowatt logic using field-effect metal-oxide semiconductor triodes. IEEE ISSCC (International Solid-State Circuits Conference), Digest of Technical Papers, pp. 32–33. (1963)
13. Lin H.C, Ho J.C, Iyer R, Kwong K.: Complementary MOS–bipolar transistor structure. IEEE Trans. Electron Devices **16**, 945 (1969)
14. Strachan A.J, Wagner K.: Local oxidation of silicon/CMOS: Technology/design system for LSI in CMOS. IEEE ISSCC (International Solid-State Circuits Conference), Digest of Technical Papers, pp. 60–61. (1974). doi: 10.1109/ISSCC.1974.1155337
15. Hoefflinger B, Schneider J, Zimmer G.: Advanced compatible LSI process for N-MOS, CMOS and bipolar transistors. IEEE IEDM (International Electron Devices Meeting), Technical Digest, pp. 261A–F. (1977). doi: 10.1109/IEDM.1977.189225
16. Plummer J, Meindl J.D, Maginness M G · An ultrasonic imaging system for real-time cardiac imaging. IEEE ISSCC (International Solid-State Circuits Conference), Digest of Technical Papers, pp. 162–163. (1974). doi: 10.1109/ISSCC.1974.1155347
17. Manasevit H.M, Simpson W.I.: Single-crystal silicon on a sapphire substrate. J. Appl. Phys. **35**, 1349 (1964)
18. Wallmark J.T, Johnson H (eds.): Field-Effect Transistors. Prentice-Hall, Englewood Cliffs (1966)
19. Izumi K, Doken H, Arioshi H.: CMOS devices fabricated on buried SiO_2 layers formed by oxygen implantation into silicon. Electron Lett. **14**, 593 (1978)
20. Gibbons J.F, Lee K.F.: One-gate-wide CMOS inverter on laser-recrystallised polysilicon. IEEE Electron Device Lett. **1**, 117 (1980)
21. Imai K.: A new dielectric isolation method using porous silicon. Solid-State Electron **24**, 159 (1981)

22. Ipri A.C, Jastrzebski L, Corboy J.F.: Device characterisation on monocrystalline silicon grown over SiO_2 by the ELO Process. IEEE IEDM (International Electron Devices Meeting), pp. 437–440 (1982)
23. Laskey J.B, Stiffler S.R, White F.R, Abernathey J.R.: Silicon on insulator (SOI) by bonding and etchback. IEEE IEDM (International Electron Devices Meeting), pp. 684 ff. (1985). doi: 10.1109/IEDM.1985.191067
24. Yonehara T, Sakaguchi K, Sato N.: Epitaxial layer transfer by bond and etch back of porous Si. Appl. Phys. Lett. **64**, 2108 (1994)
25. Bruel M.: Silicon on insulator material technology. Electron Lett. **31**, 1201 (1995)
26. Colinge J.P.: Silicon-on-insulator and porous silicon. In: Siffert P, Krimmel E (eds.) Silicon: Evolution and Future of a Technology, p. 139. Springer, Berlin/Heidelberg (2004)
27. Hoefflinger B.: Circuit considerations for future 3-dimensional integrated circuits. Proceedings of 2nd International Workshop on Future Electron Devices – SOI Technology and 3D Integration, Shujenzi, Japan, March 1985
28. Hoefflinger B, Liu S.T, Vajdic B.: Three-dimensional CMOS design methodology. IEEE J. Solid-State Circuits **19**, 37 (1984)
29. Liu M.S, Hoefflinger B.: Three-dimensional CMOS using selective epitaxial growth. US Patent 4686758, filed 2 Jan 1986. Issued 18 Aug 1987
30. Zingg R.P, Hoefflinger B, Neudeck G.: Stacked CMOS inverter with symmetric device performance. IEEE IEDM (International Electron Devices Meeting), Digest of Technical Papers, pp. 909–911 (1989)
31. Roos G, Hoefflinger B.: Complex 3D-CMOS circuits based on a triple-decker cell. IEEE J. Solid-State Circuits **27**, 1067 (1992)
32. Roos G.: Ph.D. Dissertation: *Zur drei-dimensionalen Integration von CMOS-Schaltungen*, Fortschrittsberichte VDI, Reihe 9: Elektronik, Nr.168, (VDI, Düsseldorf, 1993)
33. Abou-Samra S.J, Alsa, P.A, Guyot P.A, Courtois B.: 3D CMOS SOI for high performance computing. Proceedings of 1998 IEEE International Symposium Low-Power Electronics and Design (ISLPED). pp. 54–58 (1998)
34. Waser R (ed.): Nanoelectronics and Information Technology. Wiley-VCH, Weinheim (2003)
35. Risch L.: Silicon nanoelectronics: the next 20 years. In: Siffert P, Krimmel E (eds.) Silicon: Evolution and Future of a Technology. Springer, Berlin/Heidelberg (2004) (Chap. 18)
36. Wolf E.L.: Quantum Nanoelectronics. Wiley-VCH, Weinheim (2009)
37. Eisele I, Schulte J, Kaspar E.: Films by molecular-beam epitaxy. In: Siffert P, Krimmel E (eds.) Silicon: Evolution and Future of a Technology. Springer, Berlin/Heidelberg (2004) (Chap. 6)
38. Hossain R.: High Performance ASIC Design: Using Synthesizable Domino Logic in an ASIC Flow. Cambridge University Press, Cambridge (2008)
39. Ruiz G.A.: Evaluation of three 32-bit CMOS adders in DCVS logic for self-timed circuits. IEEE J. Solid-State Circuits **33**, 604 (1998)
40. Grube R, Dudek V, Hoefflinger B, Schau M.: 0.5 Volt CMOS logic delivering 25 Million 16×16 multiplications/s at 400 fJ on a 100 nm T-Gate SOI technology. Best-Paper Award. IEEE Computer Elements Workshop, Mesa, (2000)
41. Hoefflinger B, Selzer M, Warkowski F.: Digital logarithmic CMOS multiplier for very-high-speed signal processing. IEEE Custom-Integrated Circuits Conference, Digest, pp. 16.7.1–5 (1991)
42. cordis.europa.eu > EUROPA > CORDIS > Archive: esprit ltr project 20023
43. Agarwal A, Mathew S.K, Hsu S.K, Anders M.A, Kaul H, Sheikh F, Ramanarayanan R, Srinivasan S, Krishnamurthy R, Borkar S.: A 320 mV-to-1.2 V on-die fine-grained reconfigurable fabric for DSP/media accelerators in 32 nm CMOS. IEEE ISSCC (International Solid-State Circuits Conference), Digest of Technical Papers, pp. 328–329 (2010)
44. Kurafuji T, et al.: A scalable massively parallel processor for real-time image processing, IEEE ISSCC (International Solid-State Circuits Conference), Digest of Technical Papers, pp. 334–335 (2010)

45. Chen T.-W, Chen Y.-L, Cheng T.-Y, Tang C.-S, Tsung P.-K, Chuang T.-D, Chen L.-G, Chien S.-Y.: A multimedia semantic analysis SOC (SASoC) with machine-learning engine. IEEE ISSCC (International Solid-State Circuits Conference), Digest Technical Papers, pp. 3380150339 (2010)
46. Lee S, Oh J, Kim M, Park J, Kwon J, Yoo H.-J.: A 345 mW heterogeneous many-core processor with an intelligent inference engine for robust object recognition. IEEE ISSCC (International Solid-State Circuits Conference), pp. 332–333 (2010)
47. Graf H.G.: Institute for Microelectronics Stuttgart, Germany
48. Si carrier, courtesy IBM Laboratories, Böblingen, Germany
49. Werkmann H, Hoefflinger B.: Smart substrate MCM testability optimisation by means of chip design. IEEE 6th International Conference on Multichip Modules, pp. 150–155. (1997)
50. Knickerbocker J.U. (ed.): 3D chip technology, IBM J. Res. Dev. **52(6)**, November (2008)
51. Koester S.J, et al.: Wafer-level 3D integration technology. IBM J. Res. Dev. **52**, 583 (2008)
52. Kang U.: TSV technology and its application to DRAM. IEEE ISSCC (International Solid-State Circuits Conference), Forum 1: Silicon 3D Integration Technology and Systems (2010)
53. Zimmermann M, et al.: Seamless ultra-thin chip fabrication and assembly process, IEEE IEDM (International Electron Devices Meeting), Digest of Technical Papers, pp. 1010–1012. (2006). doi: 10.1109/IEDM.2006.346787

Chapter 4
Analog-Digital Interfaces

Matthias Keller, Boris Murmann, and Yiannos Manoli

Abstract This chapter discusses trends in the area of low-power, high-performance A/D conversion. Survey data collected over the past thirteen years are examined to show that the conversion energy of ADCs has halved every two years while the speed resolution product has doubled approximately only every four years. A closer inspection on the impact of technology scaling and developments in ADC design are then presented to explain the observed trends. Next, opportunities in minimalistic and digitally assisted design are reviewed for the most popular converter architectures. Finally, trends in Delta–Sigma ADCs are analyzed in detail.

4.1   Introduction

Analog-to-digital converters (ADCs) are important building blocks in modern electronic systems. In many cases, the efficiency and speed at which analog information can be converted into digital signals profoundly affects a system's architecture and its performance. Even though modern integrated circuit technology can provide very high conversion rates, the associated power dissipation is often incompatible with application constraints. For instance, the high-speed ADCs presented in [1] and [2] achieve sampling rates in excess of 20 GS/s, at power dissipations of 1.2 W and 10 W, respectively. Operating such blocks in a handheld application is impractical, as they would drain the device's battery within a short amount of time. Consequently, it is not uncommon to architect power constrained

M. Keller (✉) • Y. Manoli
Department of Microsystems Engineering –IMTEK, University of Freiburg, Georges-Köhler-Allee 102, Freiburg 79106, Germany
e-mail: mkeller@imtek.de

B. Murmann
Stanford University, 420 Via Palou, Allen-208, Stanford, CA 94305-4070, USA
e-mail: murmann@stanford.edu

B. Hoefflinger (ed.), *CHIPS 2020*, The Frontiers Collection,
DOI 10.1007/978-3-642-23096-7_4, © Springer-Verlag Berlin Heidelberg 2012

applications "bottom-up," by determining the analog/radio frequency (RF) front-end and ADC specifications based on the available power or energy budget. A discussion detailing such an approach for the specific example of a software-defined radio receiver is presented in [3].

With power dissipation being among the most important concerns in mixed-signal and RF applications, it is important to track trends and understand the relevant trajectories. The purpose of this chapter is to review the latest developments in low-power A/D conversion and to provide an outlook on future possibilities, thus extending and updating the results presented in [4]. Following this introduction, Sect. 4.2 provides survey data on ADCs published over the past 13 years (1997–2010). These data show that contrary to common perception, extraordinary progress has been made in lowering the conversion energy of ADCs. Among the factors that have influenced this trend are technology scaling, and the increasing use of simplified analog sub-circuits with digital correction. Therefore, Sect. 4.3 takes a closer look at the impact of feature-size scaling, while Sects. 4.4 and 4.5 discuss recent ideas in "minimalistic" and "digitally assisted" ADC architectures. Finally, the last section presents recently developed approaches for delta-sigma ADCs in order to face the limitations of aggressively scaled CMOS technologies.

4.2 General ADC Performance Trends

4.2.1 Survey Data and Figure of Merit Considerations

Several surveys on ADC performance are available in the literature [5–8]. In this section, recent data from designs presented at the IEEE International Solid-State Circuits Conference (ISSCC) and the VLSI Circuit Symposium will be reviewed. Figure 4.1 shows a scatter plot of results published at these venues over the past 13 years [9]. Figure 4.1a plots the conversion energy per Nyquist sample $P/f_{Nyquist}$, i.e., power divided by the Nyquist sampling rate with the latter being equal to two times the signal bandwidth f_b, against the achieved signal-to-noise-and-distortion ratio (SNDR).

This plot purposely avoids dividing the conversion energy by the effective number of quantization steps (2^{ENOB}), as done in the commonly used figure of merit [8]

$$FOM = \frac{P}{f_{Nyquist}2^{ENOB}} = \frac{P}{2f_b 2^{ENOB}} \qquad (4.1)$$

where

$$ENOB = \frac{SNDR[dB] - 1.76}{6.02}. \qquad (4.2)$$

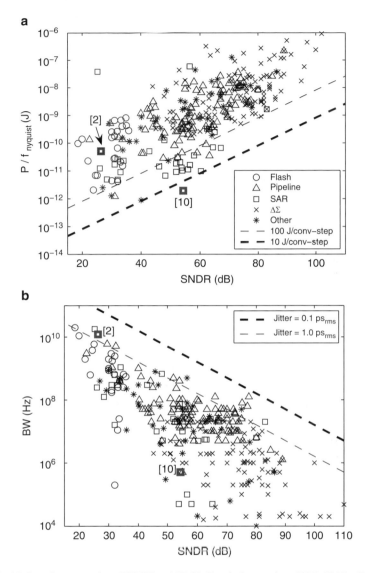

Fig. 4.1 ADC performance data (ISSCC and VLSI Circuit Symposium 1997–2010). Conversion energy (**a**) and conversion bandwidth (**b**) versus SNDR

Normalizing by the number of quantization steps assumes that doubling the precision of a converter will double its power dissipation, which finds only empirical justification [8]. Fundamentally, if a converter were purely limited by thermal noise, its power would actually quadruple per added bit (see Sect. 4.3). Nonetheless, since this assumption is somewhat pessimistic for real designs, it is preferable to avoid a fixed normalization between precision and energy altogether when plotting data from a large range of architectures and resolutions. The FOM given by (4.1) is useful mostly for comparing designs with similar resolution.

Figure 4.1a indicates that the lowest conversion energy is achieved by ADCs with low to moderate resolution, i.e., SNDR < 60 dB. In terms of the FOM given by (4.1), these designs lie within or near the range of 10–100 fJ/conversion-step, included as dashed lines in Fig. 4.1a.

In addition to an ADC's conversion energy, the available signal bandwidth is an important parameter. Figure 4.1b plots bandwidth against SNDR for the given data set. In this chart, the bandwidth plotted for Nyquist converters is equal to the input frequency used to obtain the stated SNDR (this frequency is not necessarily equal to half of the Nyquist frequency f_{Nyquist}). The first interesting observation from Fig. 4.1b is that, across all resolutions, the converters with the highest bandwidth achieve a performance that is approximately equivalent to an aperture uncertainty between 0.1 and 1 ps_{rms}. The dashed lines in Fig. 4.1b represent the performance of ideal samplers with a sinusoidal input and 0.1 and 1 ps_{rms} sampling clock jitter, respectively. Clearly, any of the ADC designs at this performance front rely on a significantly better clock, to allow for additional non-idealities that tend to reduce SNDR as well. Such non-idealities include quantization noise, thermal noise, differential nonlinearity and harmonic distortion. From the data in Fig. 4.1b, it is also clear that any new design aiming to push the speed–resolution envelope will require a sampling clock with a jitter of 0.1 ps_{rms} or better.

In order to assess the overall merit of ADCs (conversion energy and bandwidth), it is interesting to compare the locations of particular design points in the plots of Fig. 4.1a and 4.1b. For example, [2] achieves a bandwidth close to the best designs, while showing relatively high conversion energy. The opposite is true for [10]; this converter is the lowest energy ADC published to date, but was designed only for moderate bandwidth. These examples confirm the intuition that pushing a converter toward the speed limits of a given technology will sacrifice efficiency and increase the conversion energy. To show this more generally, the conversion energy of ADCs providing a bandwidth of 100 MHz or more is highlighted in red in Fig. 4.1a. As can be seen, there are only two designs that fall below the 100 fJ/conversion step line [11, 12]. The tradeoff between energy and conversion bandwidth is hard to analyze in general terms. There are however, architecture-specific closed form results, e.g., as presented in [13] for a pipeline ADC.

Yet another important (but often neglected) aspect in ADC performance comparisons is the converter's input capacitance (or resistance) and full-scale range. For most ADCs with a sampling-front-end, it is possible to improve the SNDR by increasing the circuit's input capacitance. Unfortunately, it is difficult to construct a fair single-number figure of merit that includes the power needed to drive the converter input. This is mostly because the actual drive energy will strongly depend on the driver circuit implementation and its full specifications. Therefore, an alternative approach is to calculate an approximation for the input drive energy based on fundamental tradeoffs. Such a metric was recently proposed in [14]

$$E_{Q,in} = \frac{C_{in}ELSB^2}{2^{ENOB}}. \qquad (4.3)$$

In this expression, ELSB is the effective size of the quantization step, i.e., the full-scale range (in volts) divided by the number of effective quantization steps (2^{ENOB}). For a converter that is limited by matching or thermal noise, the product in the numerator of (4.3) is independent of resolution, and captures how efficiently the capacitance is put to work. Normalizing by 2^{ENOB} then yields a metric similar to (4.1) in which the energy is distributed across all quantization levels.

The figure of merit defined by (4.3) is useful as a stand-alone metric to compare ADCs with otherwise similar performance specifications in terms of their input energy efficiency. Consider e.g. the 9–10 bit, 50 MS/s designs described in [15] (successive-approximation-register (SAR) ADC with $C_{in} = 5.12$ pF, FOM = 52 fJ/conversion-step) and [16] (pipeline ADC with $C_{in} = 90$ fF and FOM = 119 fJ/conversion-step). For the SAR ADC of [15], one finds $E_{Q,in} = 1.1 \times 10^{-19}$ J/step, while for the pipeline design of [16], one obtains $E_{Q,in} = 2 \times 10^{-21}$ J/step. This result indicates that the drive energy for the pipeline design is approximately two orders of magnitude lower compared to the SAR design. Whether this is a significant advantage depends on the particular application and system where the converter will be used.

4.2.2   Trends in Power Efficiency and Speed

Using the data set discussed above, it is interesting to extract trends over time. Figure 4.2a is a 3D representation of the conversion energy data shown in Fig. 4.1a with the year of publication included along the y-axis. The resulting slope in time corresponds to an average reduction in energy by a factor of two approximately every 1.9 years. A similar 3D fit could be constructed for bandwidth performance. However, such a fit would not convey interesting information, as the majority of designs published in recent years do not attempt to maximize bandwidth. This contrasts the situation with conversion energy, which is subject to optimization in most modern designs. In order to extract a trend on achievable bandwidth, Fig. 4.2b scatter-plots the speed–resolution products of the top three designs in each year. This metric is justified by the speed–resolution boundary observed from Fig. 4.1b, in which the straight lines obey a constant product of f_b and 2^{ENOB}. A fit to the data in Fig. 4.2b reveals that speed–resolution performance has doubled every 3.6 years, a rate that is significantly lower than the improvement in conversion energy. In addition, as evident from the data points, there is no pronounced trend as far as the top performance

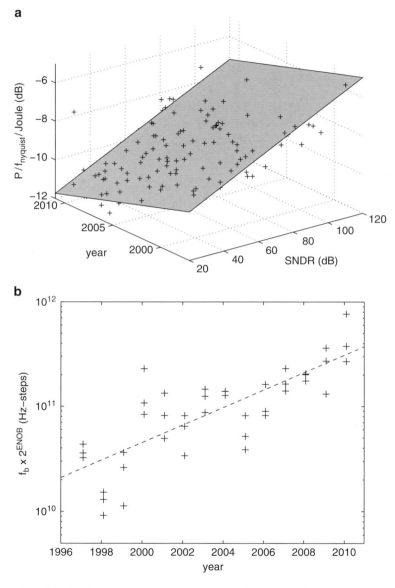

Fig. 4.2 Trends in ADC performance. (**a**) 3D fit to conversion energy using a fit plane with a slope of 0.5 ×/1.9 years along the time axis. (**b**) Fit to speed-resolution product of top three designs in each year using a fit line with a slope of 2×/3.6 years

point is concerned; designs of the early 2000s are almost on a par with some of the work published recently. Consequently, the extracted progress rate of speed–resolution performance should be viewed as a relatively weak and error-prone indicator.

4.3 Impact of Technology Scaling

As discussed in Sect. 4.2.2, the power dissipation of ADCs has halved approximately every 2 years over the past 13 years. Over the same period, CMOS technologies used to implement the surveyed ADCs have scaled from approximately 0.6 μm down to 45 nm. Today, the choice of technology in which an ADC is implemented strongly depends on the application context. For stand-alone parts, older technologies such as 0.18 μm CMOS are often preferred (see e.g., [17]). In contrast, most embedded ADCs usually must be implemented in the latest technologies used to realize large systems-on-chip (SoCs) [18]. Since the number of designs used in embedded SoC applications clearly outweighs the number of stand-alone parts, many ADC implementations in aggressively scaled technology have been presented over the past several years. Therefore, the role of technology scaling in the context of the trends summarized in the previous section is now investigated. Broader discussions on the general impact of scaling on analog circuits are presented in [5, 19, 20].

4.3.1 Supply Voltage and Thermal Noise Considerations

A well-known issue in designing ADCs in modern processes is the low voltage headroom. Since device scaling requires a reduction in supply voltage (V_{DD}), the noise in the analog signals must be reduced proportionally to maintain the desired signal-to-noise ratio. Since noise trades with power dissipation, this suggests that to first order power efficiency should worsen, rather than improve, for ADCs in modern technologies. One way to overcome supply voltage limitations is to utilize thick-oxide I/O devices [21], which are available in most standard CMOS processes. However, using I/O devices usually reduces speed. Indeed, a closer inspection of the survey data considered in this chapter reveals that most published state-of-the-art designs do not rely on thick oxide devices, and rather cope with supply voltages around 1 V. To investigate further, it is worthwhile examining the underlying equations that capture the tradeoff between supply voltage and energy via thermal noise constraints. In most analog sub-circuits used to build ADCs, noise is inversely proportional to capacitance

$$N \propto \frac{kT}{C},$$
(4.4)

where k is Boltzmann's constant and T stands for absolute temperature. For the specific case of a transconductance amplifier that operates linearly, it follows that

$$f_{Nyquist} \propto \frac{g_m}{C}.$$
(4.5)

Further assuming that the signal power is proportional to $(\alpha V_{DD})^2$ and that the circuit's power dissipation is V_{DD} multiplied by the transistor drain current, I_D, one finds

$$\frac{P}{f_{Nyquist}} \propto \frac{1}{\alpha^2} \frac{1}{V_{DD}} \frac{1}{(g_m/I_D)} kT \times SNR. \qquad (4.6)$$

The variable g_m/I_D in (4.6) is related to the gate overdrive $(V_{GS} - V_t)$ of the transistors that implement the transconductance. Assuming MOS square law, g_m/I_D equals $2/(V_{GS} - V_t)$, while in weak inversion, g_m/I_D equals $1/(nkT/q)$, with $n \cong 1.5$. Considering the fractional swing (α) and transistor bias point g_m/I_D as constant, it is clear from (4.6) that the energy in noise-limited transconductors should deteriorate at low V_{DD}. In addition, (4.6) indicates a very steep tradeoff between SNR and energy; increasing the SNR by 6 dB requires a fourfold increase in $P/f_{Nyquist}$.

Since neither of these results correlate with the observations of Sect. 4.2, it is instructive to examine the assumptions that lead to (4.6). The first assumption is that the circuit is purely limited by thermal noise. This assumption holds for ADCs with very high resolution, but typically few, if any, low-resolution converters are severely impaired by thermal noise. To get a feeling for typical SNR values at which today's converters become "purely" limited by noise, it is helpful to plot the data of Fig. 4.1a normalized to a $4 \times$ power increase per bit [22]. Figure 4.3 shows such a plot in which the $P/f_{Nyquist}$ values have been divided by

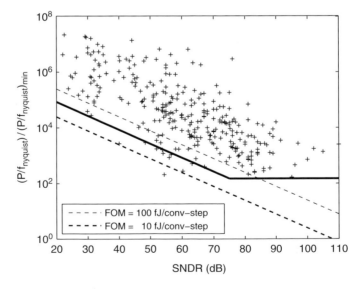

Fig. 4.3 Data of Fig. 4.1a normalized by $(P/f_{Nyquist})_{min}$ as given by (4.7). This illustration suggests the existence of an "SNR corner" (*bold line*). Only ADCs with SNR > 75 dB appear to be primarily limited by thermal noise

$$\left(\frac{P}{f_{Nyquist}}\right)_{min} = 8kT \times SNR \tag{4.7}$$

while assuming SNR $\cong$ SNDR. The pre-factor of 8 in this expression follows from the power dissipated by an ideal class-B amplifier that drives the capacitance C with a rail-to-rail tone at $f_{Nyquist}/2$ [23]. Therefore, (4.7) represents a fundamental bound on the energy required to process a charge sample at a given SNR.

The main observation from Fig. 4.3 is that the normalized data exhibit a visible "corner" beyond which $(P/f_{Nyquist})/(P/f_{Nyquist})_{min}$ approaches a constant value. This corner, approximately located at 75 dB, is an estimate for the SNDR at which a typical state-of-the-art design becomes truly limited by thermal noise. Since ADCs with lower SNDR do not achieve the same noise-limited conversion energy, it can be argued that these designs are at least partially limited by the underlying technology. This implies that, over time, technology scaling may have helped improve their energy significantly as opposed to the worsening predicted by (4.6).

To investigate further, the data from Fig. 4.1a were partitioned into two distinct sets: high resolution (SNDR > 75 dB) and low-to-moderate resolution (SNDR $\leq$ 75 dB). A 3D fit similar to that shown in Fig. 4.2a was then applied to each set and the progress rates over time were extracted. For the set with SNDR > 75 dB, it was found that the conversion energy has halved only every 4.4 years, while for SNDR $\leq$ 75 dB, energy halves every 1.7 years. The difference in these progress rates confirms the above speculation. For high-resolution designs, (4.6) applies and scaling technology over time, associated with lower supply voltages, cannot help improve power efficiency. As observed in [5], this has led to a general trend toward lower resolution designs: since it is more difficult to attain high SNDR at low supply voltages, many applications are steered away from using high-resolution ADCs in current fine-line processes. This is qualitatively confirmed in Fig. 4.4, which

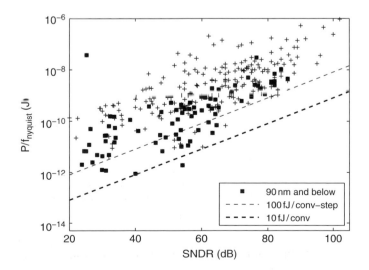

Fig. 4.4 Conversion energy for ADCs built in 90 nm and below

highlights the conversion energy data points of ADCs built in CMOS at 90 nm (and $V_{DD} \cong 1$ V) and below. As can be seen from this plot, only a limited number of ADCs built in 90 nm and below target SNDR > 75 dB.

Lastly, an additional and more general factor to consider is the tradeoff between the transconductor efficiency g_m/I_D and the transit frequency f_T of the active devices.

4.3.2 g_m/I_d and f_T Considerations

Switched capacitor circuits based on class-A operational transconductance amplifiers typically require transistors with $f_T \cong 50f_s$ [24]. Even for speeds of several tens of MS/s, it was necessary in older technologies to bias transistors far into strong inversion ($V_{GS} - V_t > 200$ mV) to satisfy this requirement. In more recent technologies, very large transit frequencies are available in moderate- and even weak-inversion. This is further illustrated in Fig. 4.5a, which compares typical minimum-length NMOS devices in 180 nm and 90 nm CMOS.

For a fixed sampling frequency, and hence fixed f_T requirement, newer technologies deliver higher g_m/I_D. This tradeoff is plotted directly, without the intermediate variable $V_{GS} - V_t$, in Fig. 4.5b. In order to achieve $f_T = 30$ GHz, a 180 nm device must be biased such that $g_m/I_D \cong 9$ S/A. In 90 nm technology, $f_T = 30$ GHz is achieved in weak inversion, at $g_m/I_D \cong 18$ S/A. From (4.6), it is clear that this improvement can fully counteract the reduction in V_{DD} when going to a newer process. Note, however, that this advantage can only materialize when the sampling speed is kept constant or at least not scaled proportional to the f_T improvement. This was also one of the observations drawn from Fig. 4.2b: a converter that pushes the speed envelope using a new technology typically will not simultaneously benefit from scaling in terms of conversion energy.

4.3.3 Architectural Impact

As discussed above, the high transistor speed available in new technologies can be leveraged to improve the energy efficiency of analog blocks (such as amplifiers). For similar reasons, it can also be argued that the high transistor speed has had a profound impact on architectural choices and developments. The very high-speed transistors in 90 nm CMOS and below have led to a renaissance or invention of architectures that were deemed either slow or inefficient in the past. As illustrated in Fig. 4.6, one example in this category is the SAR ADC [25], but also binary search architectures fall within this category (e.g. [26]). In addition, the high integration density of new processes makes massive time-interleaving (see e.g. [2]) with its associated benefits a possibility (see Sect. 4.5.2).

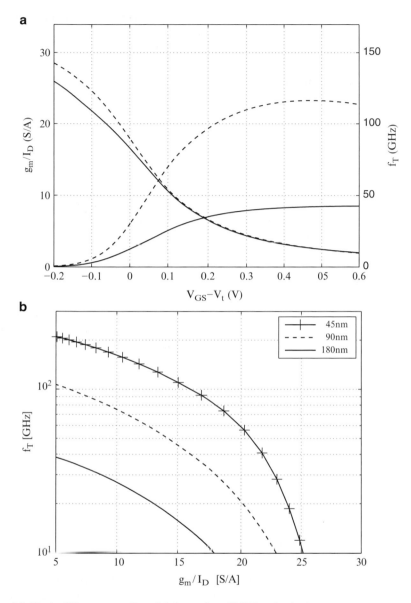

Fig. 4.5 Tradeoff between g_m/I_D and f_T in modern CMOS technologies

4.4 Minimalistic Design

In addition to technology scaling, the trend toward "minimalistic" and "digitally assisted" designs continues to impact the performance of ADCs. In this section, ideas in minimalistic design will be discussed, followed by a discussion on the importance of digitally assisted architectures in Sect. 4.5.

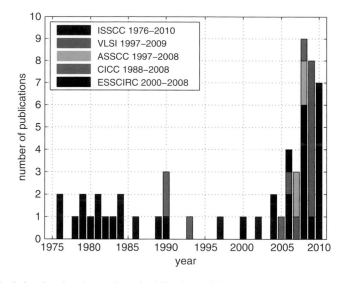

Fig. 4.6 Statistics showing the number of publications of SAR ADCs at various conferences from 1975 to 2010

Power dissipation in the analog portion of ADCs is strongly coupled to the complexity of the constituent sub-circuits. The goal of minimalistic design is to reduce power and potentially increase speed by utilizing simplified analog sub-circuits. In architectures that previously relied on precision op-amp-based signal processing, there exists a clear trend toward simplified amplifier structures. Examples include inverter-based delta-sigma modulators [27, 28] and various approaches emphasizing op-amp-less implementation of pipelined ADCs [16, 29–33]. Especially in switched capacitor circuits, eliminating class-A op-amps can dramatically improve power efficiency. This is for two reasons. First, operational amplifiers typically contribute more noise than simple gain stages, as for example resistively loaded open-loop amplifiers. Secondly, the charge transfer in class-A amplifier circuitry is inherently inefficient; the circuit draws a constant current, while delivering on average only a small fraction of this current to the load. In [34], it was found that the efficiency of a class-A amplifier in a switched capacitor circuit is inversely proportional to the number of settling time constants. For the typical case of settling for approximately ten or more time constants, the overall efficiency, i.e., charge drawn from the supply versus charge delivered to the load, is only a few percent.

As discussed further in [35], this inherent inefficiency of op-amps contributes to the power overhead relative to fundamental limits. Consider for instance the horizontal asymptote of Fig. 4.3, located at approximately 150 times the minimum possible P/f_{Nyquist}. The factor of 150 can be explained for op-amp based circuits as follows. First, the noise is typically given by $\beta \times kT/C$, where β can range from 5 to 10, depending on implementation details. Second, charge transfer using class-A circuits, as explained above, brings a penalty of approximately $20\times$. Third, op-amp circuits usually do not swing rail-to-rail as assumed in (4.7); this can contribute another factor of two. Finally, adding further power contributors beyond one dominant op-amp easily explains a penalty factor greater than 100–400.

A promising remedy to this problem is to utilize circuits that process charge more efficiently and at the same time contribute less thermal noise. A well-known example of an architecture that achieves very high efficiency is the charge-based successive approximation register (SAR) converter, see e.g. [2, 10, 25]. Such converters have been popular in recent years, primarily because the architecture is well-suited for leveraging the raw transistor speed of new technologies, while being insensitive to certain scaling implications, such as reduced intrinsic gain (g_m/g_{ds}). A problem with SAR architectures is that they cannot deliver the best possible performance when considering absolute speed, resolution, and input capacitance simultaneously (see Sect. 4.2.1). This is one reason why relatively inefficient architectures, such as op-amp-based pipelined ADCs, are still being used and investigated.

In order to make pipelined architectures as power efficient as competing SAR approaches, various ideas are being explored in research. Figure 4.7 shows an overview of amplification concepts that all pursue the same goal: improve the energy efficiency of the reside amplification.

In Fig. 4.7a, the traditional op-amp is replaced by a comparator [31], which shuts off the capacitor charging current when the final signal value is reached. In Fig. 4.7b, a "bucket brigade" pass transistor is used to move a sampled charge packet q from a large sampling capacitor C_S to a smaller load capacitor C_L, thereby achieving voltage gain without drawing a significant amount of energy from the supply [30, 36]. Lastly, in Fig. 4.7c, the gate capacitance of a transistor is used to acquire a charge sample (q_1, q_2). The transistor is then switched into a source-

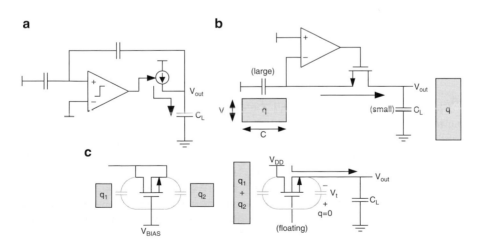

Fig. 4.7 Energy-efficient switched capacitor amplification concepts. (**a**) Comparator-based switched capacitor circuit [31]. (**b**) Bucket brigade circuit [30, 36]. (**c**) Dynamic source follower amplifier [16]

follower configuration, moving all signal charges $q_1 + q_2$ to the small capacitance from gate to drain, which also results in voltage amplification [16].

A general concern with most minimalistic design approaches is that they tend to sacrifice robustness, e.g., in terms of power supply rejection, common mode rejection, and temperature stability. It remains to be seen if these issues can be handled efficiently in practice. Improving supply rejection, for instance, could be achieved using voltage regulators. This is common practice in other areas of mixed-signal design, for example in clock generation circuits [37, 38]. Especially when the power of the ADC's critical core circuitry is lowered significantly, implementing supply regulation should be a manageable task.

A second issue with minimalistic designs is the achievable resolution and linearity. Op-amp circuits with large loop gain help linearize transfer functions; this feature is often removed when migrating to simplified circuits. For instance, the amplifier scheme of Fig. 4.7c is linear only to approximately 9-bit resolution. In cases where simplicity sacrifices precision, it is attractive to consider digital means for recovering conversion accuracy. Digitally assisted architectures are therefore the topic of the next section.

4.5   Digitally Assisted Architectures

Technology scaling has significantly reduced the energy per operation in CMOS logic circuits. As explained in [39], the typical $0.7 \times$ scaling of features along with aggressive reductions in supply voltage have led in the past to a 65% reduction in energy per logic transition for each technology generation.

As illustrated in Fig. 4.8a, a 2-input NAND gate dissipates approximately 1.3 pJ per logic operation in a 0.5 μm CMOS process. The same gate dissipates only 4 fJ in a more recent 65 nm process; this amounts to a 325-fold improvement in only 12 years. The corresponding reduction in ADC conversion energy considered in Sect. 4.2 amounts to a 64-fold reduction over the same time. This means that the relative "cost" of digital computation in terms of energy has reduced substantially in recent years.

To obtain a feel for how much logic can be used to "assist" a converter for the purpose of calibration and error correction, it is interesting to express the conversion energy of ADCs as a multiple of NAND-gate energy. This is illustrated in Fig. 4.8b assuming $E_{NAND} = 4$ fJ and $FOM = 100$ and 500 fJ/conversion, respectively. At low resolutions, e.g., $ENOB = 5$, a single A/D conversion consumes as much energy as toggling approximately 1,000 logic gates. On the other hand, at $ENOB = 16$, several million logic gates need to switch to consume the energy of a single A/D conversion at this level of precision.

The consequence of this observation is that in a low-resolution converter, it is unlikely that tens of thousands of gates can be used for digital error correction in the high-speed signal path without exceeding reasonable energy or power limits. A large number of gates may be affordable only if the involved gates operate at

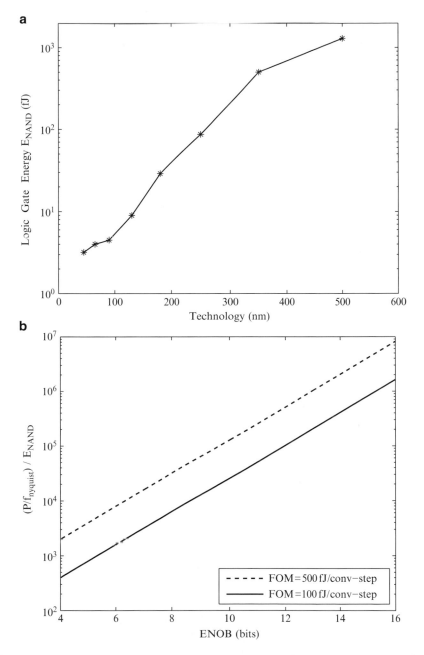

Fig. 4.8 Trends in ADC performance. (**a**) Typical energy per logic transition values (2-input NAND gate) for standard V_T CMOS technologies. (**b**) Ratio of ADC conversion energy ($P/f_{Nyquist}$) and energy per logic transition (2-input NAND gate in 65 nm CMOS)

a low activity factor, e.g., if they are outside the signal path, or if they can be shared within the system. Conversely, in high resolution ADCs, each analog operation is very energy consuming and even a large amount of digital processing may be accommodated in the overall power or energy budget.

In the following, a non-exhaustive discussion of opportunities for leveraging digital logic gates in the design of ADCs is provided.

4.5.1   Oversampling

The longest-standing example of an architecture that efficiently leverages digital signal processing abilities is the oversampling delta-sigma converter. This architecture uses noise-shaping to push the quantization error outside the signal band [22]. Subsequent digital filtering creates a high-fidelity output signal, while the constituent analog sub-circuits require only moderate precision. Even in fairly old technologies, it was reasonable to justify high gate counts in the converter's decimation filter, simply because the conversion energy for typical high-SNDR converters is very large.

A new paradigm that might gain significance in the future is the use of significant oversampling in traditional Nyquist converters. An example of such an ADC is described in [40]. As was noted from Fig. 4.5b, migrating a converter with a fixed sampling rate to technologies with higher f_T can help improve power efficiency. Ultimately, however, there is diminishing return in this trend due to the weak-inversion "knee" of MOS devices as shown in Fig. 4.5a. The value of g_m/I_D no longer improves beyond a certain minimum gate overdrive; it therefore makes no sense to target a transistor f_T below a certain value. This, in turn, implies that for optimum power efficiency, an ADC should not be operated below a certain clock rate. For example, consider the f_T versus g_m/I_D plot for 45 nm technology in Fig. 4.5b. For $g_m/I_D > 20$ S/A, f_T drops sharply without a significant increase in g_m/I_D. At this point, $f_T = 50$ GHz, implying that is still possible to build a switched capacitor circuit with $f_{clock} \cong 50$ GHz/50 $= 1$ GHz.

To date, there exists only a limited number of applications for moderate- to high-speed ADCs that require such high sampling rates, and there will clearly remain a number of systems in the future that demand primarily good power efficiency at only moderate speeds. A solution to this situation could be to oversample the input signal by a large factor and to remove out-of-band noise (thermal noise, quantization noise, and jitter) using a digital filter. Per octave of oversampling, this increases ADC resolution by 0.5 bit. In a situation where a converter is purely limited by noise, this improvement is in line with the fundamental thermal noise tradeoff expressed in (4.6).

4.5.2   Time-Interleaving

Time-interleaved architectures [41] exploit parallelism and trade hardware complexity for an increased aggregate sampling rate. Time-interleaving can be beneficial in several ways. First, it can help maximize the achievable sampling rate in a given technology. Second, time-interleaving can be used to assemble a very fast converter using sub-ADCs that do not need to operate at the limits of a given architecture or technology. As discussed above, this can help improve energy efficiency.

A well-known issue with time-interleaved architectures, however, is their sensitivity to mismatches between the sub-converters. The most basic issue is to properly match the offsets and gains in all channels. In addition, for the common case of architectures that do not contain a global track-and-hold circuit, bandwidth and clock timing mismatches are often critical [42].

Using digital signal processing techniques to address analog circuit mismatch in time-interleaved converter arrays is an active research area [43]. Basic algorithms that measure and remove gain and offsets in the digital domain have become mainstream (see e.g. [44]), while techniques that address timing and bandwidth mismatches are still evolving.

As with most digital enhancement techniques, the problem of dealing with timing and bandwidth mismatches consists of two parts: an algorithm that estimates the errors and a mechanism that corrects the errors. For the correction of timing and bandwidth errors, digital methods have been refined substantially over the years [45]. Nonetheless, the complexity and required energy of the proposed digital filters still seem to be beyond practical bounds, even for today's fine-line technology. As a consequence, timing errors in practical time-interleaved ADC are often adjusted through digitally adjustable delay lines [1, 2, 46], while the impact of bandwidth mismatches is typically minimized by design.

As far as estimation algorithms are concerned, there is a wide and growing variety of practical digital domain algorithms [43]. Particularly interesting are techniques that extract the timing error information "blindly," without applying any test or training signals [47, 48]. It is clear that algorithms of this kind will improve further and find their applications in practical systems.

4.5.3 Mismatch Correction

Assuming constant gate area ($W \times L$), transistor matching tends to improve in newer technologies. In matching-limited flash ADC architectures, this trend has been exploited in the past to improve the power efficiency by judiciously downsizing the constituent devices [14]. In order to scale such architectures more aggressively, and at the same time address new sources of mismatch in nanoscale technologies, it is desirable to aid the compensation of matching errors through digital means.

In flash ADCs, there are several trends in this direction. As illustrated in Fig. 4.9, the first and most transparent idea is to absorb offset voltages (V_{os}) in each comparator using dedicated "trim DACs" or similar circuits that allow for a digital threshold adjustment [49–51]. The input code for each DAC can be determined at

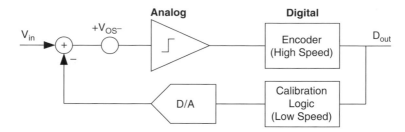

Fig. 4.9 One slice of a flash ADC showing a comparator with trimming-DAC-based offset calibration

start-up or whenever the converter input is in an idle condition. Alternatively, and for improved robustness, it is also possible to adjust the trim codes continuously in the background (during normal operation), e.g., using a chopper-based approach [52] or through a two-channel architecture [53].

An important aspect of the arrangement in Fig. 4.9 is that most of the digital circuitry required to control the trim DACs is either static during normal operation or can run at very low speeds. This means that there is often no concern about the digital energy overhead for calibration.

In modern technologies, and when near-minimum-size devices are being used, the comparator offsets may necessitate relatively complex trim-DACs to span the required range with suitable resolution. One idea to mitigate this issue is to include redundant comparators, and to enable only the circuits that fall into the (reduced) trim-DAC range after fabrication [54]. This scheme can yield good power efficiency as it attacks the mismatch problem along two degrees of freedom. Larger offsets can now be accommodated using smaller trim-DACs, which in turn imposes a smaller overhead in terms of area and parasitic capacitance introduced to signal-path nodes.

The idea of using redundant elements can be pushed further to eliminate the trim-DACs and the trim-range issue altogether. In [41], redundancy and comparator reassignment are employed to remove all offset constraints. A similar concept is proposed in [40] (see also [55]), but instead of a static comparator reassignment, a fault tolerant digital encoder is used to average the statistics along the faulty thermometer code. While this approach is conceptually very elegant, it requires a relatively large number of logic operations per sample in the high-speed data path. In light of the data from Fig. 4.8 (for low ENOB values), such a solution may be efficient only for very aggressively scaled technologies, i. e., 45 nm and below.

Thus far, the concept of using redundant elements has not yet found widespread use. It remains to be seen if these techniques will become a necessity in technologies for which the variability of minimum-size devices cannot be efficiently managed using trim-DACs. Extrapolating beyond flash ADCs, it may one day be reasonable and advantageous to provide redundancy at higher levels of abstraction, for instance through extra channels in a time-interleaved ADC array [56].

4.5.4   Digital Linearization of Amplifiers

As pointed out in Sect. 4.4, power-efficient and "minimalistic" design approaches
are typically unsuitable for high-resolution applications, unless appropriate digital
error correction schemes are used to enhance conversion linearity. A generic block
diagram of such a scheme is shown in Fig. 4.10. In [32], it was demonstrated that a
simple open-loop differential pair used in a pipeline ADC can be digitally
linearized using only a few thousand logic gates. In ADCs, the concept of digital
amplifier linearization has so far been applied to the pipelined architecture [32,
57–62]. However, it is conceivable to implement similar schemes in other
architectures, e.g., delta-sigma modulators [63].

One key issue in most digital linearization schemes is that the correction
parameters must track changes in operating conditions relatively quickly, preferably with time constants no larger than 1–10 ms. Unfortunately, most of the basic
statistics-based algorithms for coefficient adaptation require much longer time
constants at high target resolutions [59, 60]. Additional research is needed to extend
the recently proposed "split-ADC" [64, 65] and feed-forward noise cancelation
techniques [66] for use in nonlinear calibration schemes.

4.5.5   Digital Correction of Dynamic Errors

Most of the digital correction methods developed in recent years have targeted the
compensation of static circuit errors. However, it is conceivable that in the future
the correction of dynamic errors will become attractive as well.

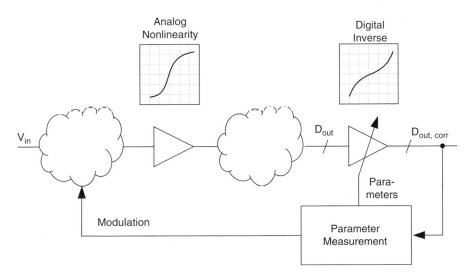

Fig. 4.10 Digital linearization of signal path amplifiers

One opportunity is the digital correction of errors due to finite slew rate, incomplete settling or incomplete reset of amplifiers [34, 58, 67, 68]. Another opportunity lies in correcting high-frequency nonlinearities introduced at the sampling front-end of high-speed ADCs [69–71]. Correcting high-frequency distortion digitally is particularly attractive in applications that already have a complex digital back-end that can easily provide additional resources (e.g., inside an FPGA (field-programmable gate array)). Such applications include, for instance, wireless base stations [69] and test and measurement equipment. Most digital high-frequency linearization schemes proposed to date are based on relatively simple, low-order non-linearity models. However, if digital capabilities in nano-scale technologies continue to improve, dynamic compensation schemes based on relatively complex Volterra series may become feasible [72].

4.5.6   System-Synergistic Error Correction Approaches

In the discussion so far, ADCs were viewed as "black boxes" that deliver a set performance without any system-level interaction. Given the complexity of today's applications, it is important to realize that there exist opportunities to improve ADC performance by leveraging specific system and signal properties.

For instance, large average power savings are possible in radio receivers when ADC resolution and speed are dynamically adjusted to satisfy the minimum instantaneous performance needs. The design described in [73] demonstrates the efficiency of such an approach.

In the context of digital correction, it is possible to leverage known properties of application-specific signals to "equalize" the ADC together with the communication channel [74–79]. For instance, the converter described in [77] uses the system's OFDM (orthogonal frequency-division multiplexing) pilot tones to extract component mismatch information (see Fig. 4.11). In such an approach, existing system

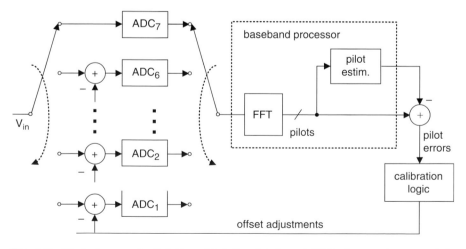

Fig. 4.11 Block diagram of a digitally assisted, time-interleaved ADC using a system-synergistic calibration approach

hardware, for instance the FFT (fast Fourier transform) block, can be re-used to facilitate ADC calibration.

4.6 Trends in Delta-Sigma A/D Converters

As considered in the previous sections, designers of analog circuits have to face various challenges associated with each new technology generation. New concepts at the system level as well as for sub-circuit blocks must be developed due to reduced supply voltage and reduced intrinsic gain per transistor while benefitting from increased transit frequency f_T. The following section illustrates recent approaches in delta-sigma ADCs in order to deal with these limitations. As will be seen, many of the concepts considered in Sects. 4.4 and 4.5 have already found their application in this kind of ADC, however, mostly as "proof of concepts" in prototypes implemented by research-orientated institutions, not in industrial mass-market products.

Delta-sigma ADCs have been with us for some decades. Today, they achieve medium resolution in signal bandwidths of several tens of megahertz, e.g., 12 bits in a signal bandwidth of 20 MHz [80], according to the well-known equation for the peak signal-to-quantization-noise ratio (SQNR) of an ideal modulator given by [81]

$$SQNR_{\mathrm{p,ideal}} = \frac{3\pi}{2}\left(2^B - 1\right)^2 (2N + 1)\left(\frac{OSR}{\pi}\right)^{2N+1}. \qquad (4.8)$$

In (4.8), N is the order of the loop filter and B the number of bits of the internal quantizer according to Fig. 4.12. The oversampling ratio OSR is defined as

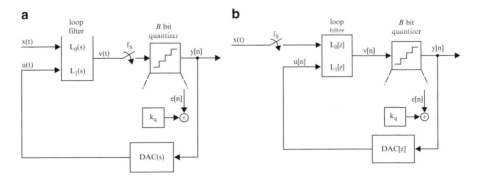

Fig. 4.12 System-theoretical model of a single-stage delta-sigma modulator, (**a**) Continuous-time modulator, (**b**) Discrete-time modulator

$$OSR = \frac{f_s}{2f_b} \qquad (4.9)$$

where f_s is the sampling frequency and f_b the signal bandwidth.

As illustrated in Fig. 4.12, the heart of any delta-sigma ADC, i.e., the delta-sigma modulator, may be implemented in either the discrete-time (DT) or the continuous-time (CT) domain. The loop filter is mostly implemented by means of active RC or gmC integrators in the case of CT implementations, or by means of switched-capacitor (SC) or switched–op-amp (SO) integrators in case of DT implementations. Thus, the specifications for the unity-gain bandwidth product of operational amplifiers for CT implementations are relaxed in comparison to SC implementations due to relaxed settling requirements, an important characteristic particularly in high-frequency applications. Consequently, CT implementations are considered to be more power efficient.

At first glance, this characteristic seems to be confirmed by Fig. 4.13a, where the FOM according to (4.1) is plotted versus the signal bandwidth. These data were collected from the survey provided in [9], extended by delta-sigma ADCs published in *IEEE Journal of Solid-State Circuits* from 2000 to 2010. Most outstanding is a factor of 10 between the FOM of the most efficient CT and DT implementation for a signal bandwidth of 20 MHz [80, 82]. However, in order to perform a fair comparison, i.e., following Sect. 4.2.1 by considering the FOM of implementations with similar bandwidth and resolution, Fig. 4.13b should be considered. Obviously, two pairs of designs achieve almost the same resolution in a bandwidth of 20 MHz, i.e., 50 dB and 63 dB [82–85]. Interestingly, however, and contrary to the above, the FOM of the DT implementation outperforms the FOM of the corresponding CT implementation by factors of 2 and 3, respectively.

How can this discrepancy be explained? First, the calculation of the FOM does usually not take into account that CT modulators offer inherent anti-aliasing filtering without additional power consumption. By contrast, this filtering must be explicitly implemented outside of a DT modulator, thus resulting in additional power consumption. Second, it is often not clearly stated in publications whether further sub-circuit blocks of a delta-sigma modulator are implemented on-chip or off-chip, e.g., clock generation, reference generation, or digital cancelation filters (DCFs) as needed for cascaded delta-sigma modulators (see Sect. 4.6.1). Even if so, the power consumption of each sub-circuit block is not always given separately. Indeed, the CT cascaded design achieving 63 dB included the implementation and thus the power consumption of the on-chip clock and reference generator as well as of the DCF [83]. By contrast, the calculation of the FOM of its DT counterpart was based on the power consumption of the modulator only [82]. Similar observations were made for the CT and DT designs achieving 50 dB [84, 85]. Thus, care must be taken even in comparing designs with similar characteristics. In order to clearly demonstrate or demystify the promoted power efficiency of CT implementations, it would be interesting to see a comparison between the implementation of a DT and a CT delta-sigma modulator

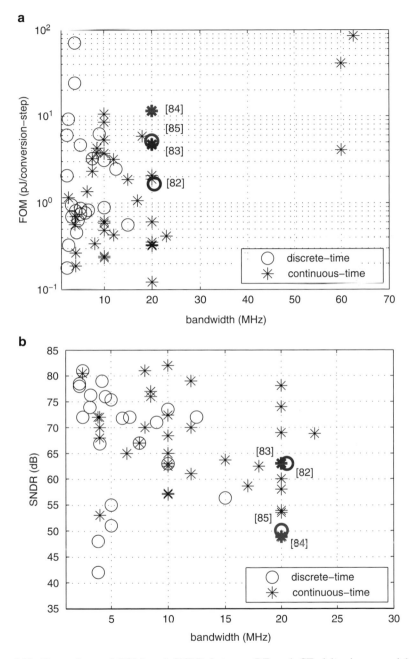

Fig. 4.13 Comparison of FOM and SNDR between DT and CT delta-sigma modulators exhibiting a bandwidth greater than 2 MHz. FOM (**a**) and SNDR (**b**) versus bandwidth

designed according to identical specifications. Furthermore, the same sub-circuit blocks should be implemented while being optimized according to the minimum requirements of a DT and a CT implementation.

For these reasons, it has yet to be seen in the future whether CT modulators will become the vehicle for low-power implementations or whether DT implementations keep pace with them. At present, CT architectures are the favored architectures for the implementation of broadband or high-speed delta-sigma ADCs as illustrated in Fig. 4.13. From this trend, it may be concluded that they will also become the architecture of the future. In the following, considerations are thus illustrated based on CT delta-sigma modulators.

From the linearized model shown in Fig. 4.12a, i.e., replacing the strongly nonlinear quantizer by a linear model consisting of a quantizer gain k_q and an additive white noise source e [22], the signal transfer function (STF) and noise transfer function (NTF) result in

$$
\begin{aligned}
y[n] &= y_x[n] + y_e[n] \\
&= STFx[n] + NTF[z]e[n] \\
&= \frac{k_q L_0(s)}{1 + k_q L_1[z]} x[n] + \frac{1}{1 + k_q L_1[z]} e[n]
\end{aligned}
\tag{4.10}
$$

where $L_1[z]$ is the discrete-time equivalent of the product of $DAC(s)$ and $L_1(s)$ due to the sampling process. Note that no argument for the STF is given since it is equal to a mixture between a continuous-time and a discrete-time transfer function. This aspect results in the anti-aliasing property, as will be discussed shortly.

Formerly, high resolution was achieved by applying a high OSR in a first-order modulator, i.e., larger than 100, while using a single-bit quantizer for the implementation of the internal quantizer. The latter was also of advantage since no additional dynamic element matching technique must be implemented in order to linearize the transfer characteristic of a multi-bit DAC. With steadily increasing demand for higher signal bandwidths, the OSR has become quite small. Today, an OSR between 10 and 20 represents a typical value in applications with signal bandwidths up to 20 MHz (e.g., [80, 86]). To compensate for the loss of resolution due to the reduction of the OSR, either the order of the loop filter or the resolution of the internal quantizer has been increased. However, both approaches have their limitations, because of either architectural or technology constraints, of which those for the loop filter will be considered first. At the same time, trends for overcoming these limitations will be outlined.

4.6.1 Loop Filter

The loop filter of a delta-sigma modulator is typically implemented either as a cascade of integrators in feedback (CIFB) or in feed-forward (CIFF) [22]. Both

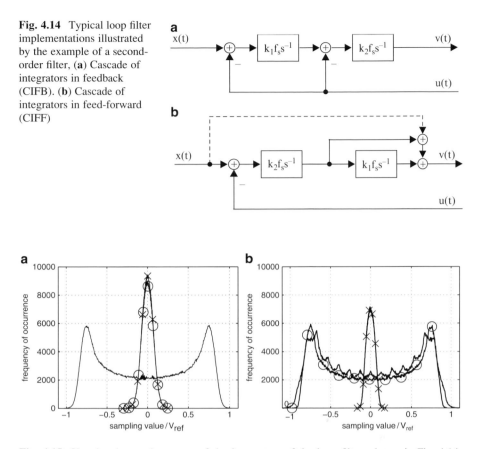

Fig. 4.14 Typical loop filter implementations illustrated by the example of a second-order filter, (**a**) Cascade of integrators in feedback (CIFB). (**b**) Cascade of integrators in feed-forward (CIFF)

Fig. 4.15 Signal swing at the output of the integrators of the loop filters shown in Fig. 4.14. (**a**) First integrator, (**b**) Second integrator (CIFB: *solid line*; CIFF: *solid line, circles*; CIFF with feed-in path from $x(t)$ to $v(t)$: *solid line, crosses*)

types are illustrated in Fig. 4.14 by means of a continuous-time second-order loop filter. An implementation of the same transfer function $L_1(s)$, and thus the same NTF according to (4.10), is possible by either architecture.

However, it is most likely that the CIFF architecture will become the dominant architecture in the near future. In comparison to a CIFB architecture, the output swing of each integrator is highly reduced, except for the one in front of the quantizer. If an additional feed-in path from $x(t)$ to $v(t)$ is inserted (Fig. 4.14b, dashed line), the loop-filter has to process almost only the quantization noise. Thus, the last integrator also exhibits a highly reduced signal swing at its output. These characteristics are illustrated in Fig. 4.15. For simulation purposes, a sine wave input signal was applied to the modulators ($f_{\text{sig}} = f_b$, $\hat{u} = -1.9$ dBFS (decibels relative to full scale), $OSR = 165$, $k_1 = 6/5$, $k_2 = 45/32$, 4-bit quantizer).

According to the simulation results, a reduction of the output swing by a factor approximately equal to five can be noticed. Thus, it can be concluded that this

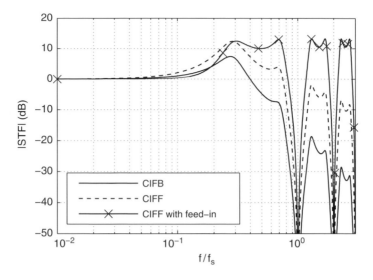

Fig. 4.16 Signal transfer functions of the different architectures based on the loop filters shown in Fig. 4.14

architecture offers not only a better robustness against distortion but also a high attractiveness for implementations in new technology nodes in which the supply voltage has been reduced. On the other hand, the suppression of the implicit anti-aliasing filter around multiples of f_s, as offered by the STF of any CT architecture, is less effective in a CIFF architecture than in a CIFB architecture. This characteristic is illustrated in Fig. 4.16.

Although the loop-filter architectures shown in Fig. 4.14 implement the same filter function $L_1(s)$, $L_0(s)$ results in

$$L_{0,\text{CIFB}}(s) = k_1 k_2 \left(\frac{f_s}{s}\right)^2, \tag{4.11}$$

$$L_{0,\text{CIFF}}(s) = k_2 \frac{f_s}{s} + k_1 k_2 \left(\frac{f_s}{s}\right)^2, \tag{4.12}$$

$$L_{0,\text{CIFF+feed-in}}(s) = 1 + k_2 \frac{f_s}{s} + k_1 k_2 \left(\frac{f_s}{s}\right)^2. \tag{4.13}$$

From (4.12) and (4.13), it can be seen that the signal path around the second integrator, or both integrators, acts as a short for frequencies larger than f_s. Thus, the suppression of high frequencies by $L_0(s)$ and thus the implicit anti-aliasing feature of a CIFF architecture is less effective than that of a CIFB architecture.

Taking the scaling of the loop filter into account, a more detailed analysis of the achievable SQNR of an ideal single-stage delta-sigma modulator results in [87]

$$SQNR_{p,scaling} = 10 \log (SNR_{p,ideal}) + 20 \log \left(k_q \prod_i k_i \right), \qquad (4.14)$$

i.e., the resolution also depends on the set of scaling coefficients and the quantizer gain. Typically, the product of the scaling coefficients is smaller than one, particularly in high-order single-stage modulators where less aggressive scaling must be applied for stability reasons [88]. The achievable $SQNR_{p,scaling}$ is thus reduced compared to $SQNR_{p,ideal}$ and single-stage modulators with loop filter order higher than four are not attractive.

It can be foreseen that, driven by the ever increasing demand for higher bandwidth and resolution in combination with reconfigurability demanded, for example, by new telecommunication standards, delta-sigma modulators must be pushed beyond their present performance limits in the near future. Towards these goals, cascaded delta-sigma modulators offer a promising alternative to single-stage architectures. In these architectures, as illustrated by the example of a CT 2-2 modulator in Fig. 4.17, the quantization errors of all stages except for the last are canceled via digital cancelation filters (DCFs) at the output $y[n]$. For the quantization error of the last stage, the cascade provides noise-shaping with the order of the overall modulator, i.e., four in the case of the 2-2 modulator. Taking scaling into account, the $SQNR_{p,scaling}$ of an ideal cascaded modulator consisting of M stages with order N_i in the ith stage results in

$$SQNR_{p,scaling} = 10 \log (SNR_{p,ideal}) +$$
$$20\log_{10} \left[\left(\prod_{i=1}^{M} \prod_{j=1}^{N_i} k_{ij} \right) k_{qM} \prod_{i=1}^{M-1} k_{iN_i(i+1)\,1} \right], \qquad (4.15)$$

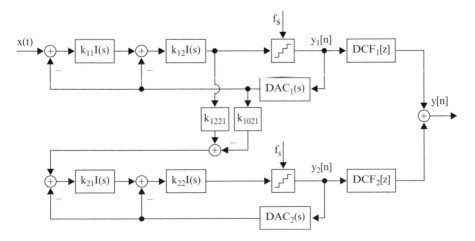

Fig. 4.17 Continuous-time 2-2 cascaded modulator

i.e., the performance loss is given by the product of the single-stage scaling coefficients, the quantizer gain of the last stage and the scaling coefficients of each signal path connecting a quantizer input to the input of the next stage.

Obviously, the major advantage of cascaded modulators consists in the cascading of low-order modulators. Thus, the performance loss due to scaling is less severe since each modulator in the cascade can be scaled more aggressively than a single-stage modulator whose order is equal to the overall order of the cascade. Furthermore, the values of the scaling coefficients between stages may become larger than one if multi-bit quantizers are used. Indeed, it has been reported that an ideal CT 3-1 cascaded modulator with four-bit quantizers outperforms a fourth-order single stage modulator in terms of resolution by three bits, and a CT 2-2 cascaded modulator by even more than four bits [89]. However, whether these values are achievable in an implementation, where several sources of non-idealities have to be considered as well, is still to be proven.

By switching later stages in or out of the cascade, cascaded architectures also offer a promising approach to reconfigurability in terms of bandwidth and resolution, as postulated in Sect. 4.5.6. The reasons why they have not yet replaced single-stage architectures despite all these advantages are manifold.

First, they rely on perfect matching between the analog and the digital parts of the modulator, i.e., the loop filters and the digital cancelation filters. Otherwise, low-order shaped quantization noise of earlier stages in the cascade leaks to the output, drastically reducing the resolution. However, since digitally assisted circuits may become available nearly "for free" in the future, these matching problems may easily be overcome. Different on-line and off-line approaches to their compensation have already been presented [90, 91].

Furthermore, a new architecture for a cascaded delta-sigma modulator has been presented recently [92]. As illustrated in Fig. 4.18, this kind of cascaded modulator performs a cancelation of the quantization error of the first stage without the need for digital cancelation filters. Thus, the non-ideality of a mismatch between the scaling coefficients of the analog and digital parts of a classical cascaded modulator is overcome. Additionally, it was proven by means of a first prototype that operational amplifiers with a DC gain as low as 35 dB may be used for a successful implementation, thus making it attractive for the implementation in a deep sub-

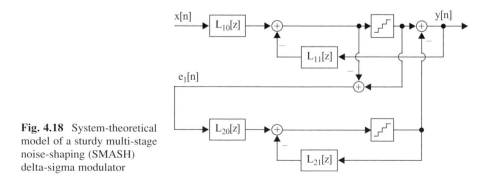

Fig. 4.18 System-theoretical model of a sturdy multi-stage noise-shaping (SMASH) delta-sigma modulator

micrometer CMOS technology with reduced intrinsic gain per transistor [93]. For these reasons, this kind of cascaded delta-sigma modulator is called a sturdy multistage noise-shaping (SMASH) delta-sigma modulator. However, at first glance, it seems that a SMASH modulator sacrifices the apparently most important advantage of a MASH modulator in comparison to a single-stage modulator: the guarantee of overall stability of the cascade by cascading low-order modulators. As is obvious from Fig. 4.18, the second stage is embedded inside the loop-filter of the first stage, its order increasing accordingly.

A second reason is given by the characteristic that cascaded architectures consume more power and need more area for their implementation since, besides the DCFs, at least one additional quantizer becomes mandatory.

These characteristics are even more apparent in a time-interleaved delta-sigma modulator, another alternative for the implementation of a broadband ADC applying small-to-moderate oversampling. Time-interleaved delta-sigma modulators consist of n identical delta-sigma modulators working in parallel, as illustrated in Fig. 4.19 (see also Sect. 4.5.2). Thus, the sampling rate f_s of each modulator can be reduced by a factor of n while the same resolution is provided as offered by a single-stage modulator clocked at f_s. This reduction becomes feasible since the outputs of all n modulators are combined in order to achieve the same resolution at an output rate f_s. However, only a few time-interleaved delta-sigma modulators have been presented so far, consisting of two paths only [84, 94–96].

Interestingly, the concepts for cascading and time-interleaving delta-sigma modulators have been known for more than 15 years [97]. However, they gained renewed interest only recently, particularly due to demand for broadband ADCs with the capability of reconfiguration in bandwidth and resolution. In the near future, it is to be expected that research on these architectures will be increased in order to overcome major non-idealities and to reduce their power consumption and area requirement, e.g., by op-amp sharing as presented in [84, 96].

Finally, first prototypes of minimalistic delta-sigma modulators as introduced in Sect. 4.4 were recently presented, i.e., the loop-filter was implemented based on simple inverters instead of traditional op-amps [27, 28].

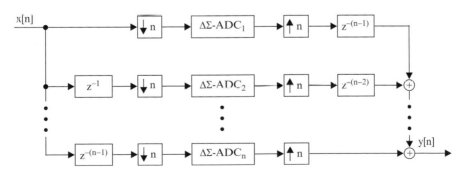

Fig. 4.19 System-theoretical model of a DT time-interleaved delta-sigma modulator

4.6.2 Quantizer

In the last subsection, trends for the implementation of the loop filter of a delta-sigma modulator were considered. Therein, special emphasis was placed on broadband applications, i.e., bandwidths in the range of several tens of megahertz, which allow only low oversampling to be applied. Furthermore, the aspects of highly scaled CMOS technologies were taken into account, i.e., reduced supply voltage and reduced gain per transistor. In the following, another key component of a delta-sigma modulator is considered: the quantizer.

Traditionally, the quantizer is implemented by means of a flash ADC. However, with upcoming supply voltages below 1 V, multi-bit flash quantizers can hardly be implemented. Considering the example of a 4-bit quantizer, the LSB (least significant bit) becomes as small as 62.5 mV. Since this value represents a typical value for the offset voltage of a CMOS comparator, either an offset compensation technique must be applied or a rather large input transistor must be implemented. Besides, multi-bit flash ADCs are very power consuming.

First attempts to overcome these hindrances covered the replacement of the multi-bit flash ADC by means of a successive approximation ADC or a tracking quantizer [98, 99]. However, the first approach requires an additional clock running at multiples of the original sampling frequency, which makes it less attractive for broadband applications. The second approach only reduces the number of comparators while still three comparators were needed for the replacement of a 4-bit flash ADC in order to guarantee good tracking capability.

Recently, two new concepts were presented that directly comply with the benefits of scaled CMOS technologies for digital applications. They thus may become the favorite architectures for the implementation of a multi-bit quantizer in the near future.

4.6.2.1 Voltage-Controlled Oscillator Based Quantizer

In the first approach, the flash ADC is replaced by a voltage-controlled oscillator (VCO), as illustrated in Fig. 4.20 [100]. The voltage V_{tune}, which is used to tune the frequency of the VCO, is equal to the output voltage $v(t)$ of the loop filter in Fig. 4.12. By means of registers, it is observed whether a VCO delay cell undergoes a transition or not within a given clock period. This is performed by comparing samples of the current and the previous states of the VCO by XOR gates. The resulting number of VCO delay cells that undergo a transition is a function of the delay of each stage as set by V_{tune}. Thus, a quantized value of V_{tune} is obtained.

The advantages of this approach with respect to future CMOS technology nodes are obvious. First, this implementation is a highly digital implementation; it relies on inverters, registers, XOR gates and one adder. An area- and power-efficient implementation thus becomes feasible. Second, any offset-related non-idealities are overcome since a CMOS inverter achieves full voltage swing at its output. For the same reason, the number of delay elements and thus the resolution can easily be increased.

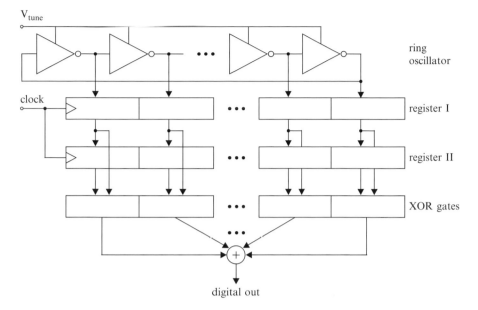

Fig. 4.20 Voltage-controlled oscillator based quantizer

Third, it is also to be expected that this architecture will take advantage of future technology nodes since the resolution versus sampling speed tradeoff directly improves with the reduced gate delays in modern CMOS processes. This characteristic makes it very attractive for high-speed operations with small latency. Furthermore, this architecture provides first-order noise-shaping for the quantization error, whereby the order of the loop filter can be reduced by one; thus, one op-amp can be saved. It also provides inherent data-weighted averaging for its digital output code, for which the linearity requirements imposed on the DAC in the feedback path become less stringent while making an explicit technique for dynamic element matching (DEM) obsolete.

A first prototype applying a VCO-based quantizer in a third-order delta-sigma modulator (0.13 µm CMOS, 1.5 V supply voltage) achieved 67 dB signal-to-noise-and-distortion ratio (SNDR) in a bandwidth of 20 MHz [100]. In a redesign, the SNDR was improved to 78 dB by quantizing the phase instead of the frequency of the VCO [101]. Thereby, harmonics generated by the nonlinear tuning gain of the VCO caused by its nonlinear voltage-to-frequency transfer characteristic were sufficiently suppressed. However, an explicit DEM technique became mandatory.

4.6.2.2   Time Encoding Quantizer

In the second approach, the multi-bit flash ADC is replaced by a time-encoding quantizer [102]. It consists of a 1 bit quantizer embedded in a feedback system followed by some filter blocks and a decimator, as illustrated in Fig. 4.21. The

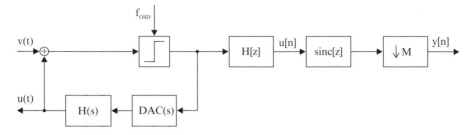

Fig. 4.21 Time encoding quantizer

feedback loop forces $u(t)$ to oscillate at frequency f_c with an amplitude A if no input signal is applied, i.e., $v(t) = 0$, and if the quantizer is not sampled with f_{OSD}. In order to establish this limit cycle, the filter $H(s)$ in the feedback loop must be designed according to

$$H(s) = A \frac{4f_c}{s} e^{-s/(4f_c)}. \tag{4.16}$$

However, if the quantizer is sampled with a frequency f_{OSD} sufficiently larger than f_c, then the feedback system can be considered as a quasi-continuous-time system. In this case, the limit cycle is still established and its frequency is almost not influenced if an input signal $v(t)$ is applied.

By means of this approach, most of the power of the quantization error of the 1-bit quantizer is located around f_c, which is designed to be much larger than f_b. The remaining power within the signal band resembles that of a multi-bit quantizer. Thereby, the 1-bit quantizer can be designed to function like an arbitrary multi-bit quantizer while overcoming any offset related non-idealities.

The output signal of the feedback system is filtered by an equalization filter $H[z]$. This filter is a digital replica of $H(s)$, where the transfer function from $v(t)$ to $u[n]$ is set equal to one within the signal band. Subsequently, the quantization noise power located around f_c is removed by a sinc decimation filter whose zeros are placed accordingly. Furthermore, the multi-bit encoding is performed by this filter block and the output rate is decimated to that of an equivalent multi-bit counterpart. Thus, the following further oversampling ratios have been defined for a delta-sigma modulator with an internal time-encoding quantizer:

$$COSR = \frac{f_c}{f_s} \tag{4.17}$$

$$ROSR = \frac{f_{OSD}}{f_c}. \tag{4.18}$$

First prototypes achieved an SNDR of 10 bits and 12 bits in signal bandwidths of 17 MHz and 6.4 MHz, respectively [103]. In both implementations, the time-encoding quantizer replaced a 4-bit flash quantizer. However, f_{OSD} was set equal to 1.632 GHz and thus became a factor of four larger compared to the sampling frequency f_s of an equivalent 4-bit implementation. The application of a time-encoding quantizer in a broadband modulator may thus be limited.

4.7   Conclusion

This chapter was intended to discuss trends in the context of low-power, high-performance A/D conversion. Using survey data from the past 13 years, it was observed that power efficiency in ADCs has improved at an astonishing rate of approximately $2\times$ every 2 years. In part, this progress rate is based on cleverly exploiting the strengths of today's technology. Smaller feature sizes improve the power dissipation in circuits that are not limited by thermal noise. In circuit elements that are limited by thermal noise, exploiting the high transit frequency f_T of modern transistors can be of help in mitigating a penalty from low supply voltages.

A promising paradigm is the trend toward minimalistic ADC architectures and digital means of correcting analog circuit errors. Digitally assisted ADCs aim to leverage the low computing energy of modern processes in order to improve the resolution and robustness of simplified circuits. Future work in this area promises to fuel further progress in optimizing the power efficiency of ADCs.

Overall, improvements in ADC power dissipation are likely to come from a combination of aspects that involve improved system embedding and reducing analog sub-circuit complexity and raw precision at the expense of "cheap" digital processing resources. These trends can already be observed in delta-sigma ADCs, which have been considered in detail.

References

1. Poulton, K., et al.: In: ISSCC Digest of Technical Papers, pp. 318–496. IEEE (2003)
2. Schvan, P., et al.: In: ISSCC Digest of Technical Papers, pp. 544–634. IEEE (2008)
3. Ahidi, A.: IEEE J. Solid State Circuits 42(5), 954 (2007)
4. Murmann, B.: In: Proceedings of Custom Integrated Circuit Conference, pp. 105–112. IEEE (2008)
5. Chiu, Y., Nikolic, B., Gray, P.: In: Proceedings of Custom Integrated Circuit Conference, pp. 375–382. IEEE (2005)
6. Kenington, P., Astier, L.: IEEE Trans. Veh. Technol. 49(2), 643 (2000)
7. Merkel, K, Wilson, A.: In: Proceedings of Aerospace Conference, Big Sky, pp. 2415–2427. IEEE (2003)
8. Walden, R.: IEEE J. Select. Areas Commun. 17(4), 539 (1999)
9. Murmann, B.: ADC performance survey 1997–2010. http://www.stanford.edu/_murmann/adcsurvey.html
10. van Elzakker, M., et al.: In: ISSCC Digest of Technical Papers, pp. 244–610. IEEE (2008)
11. Verbruggen, B., et al.: In: ISSCC Digest of Technical Papers, pp. 252–611. IEEE (2008)
12. Verbruggen, B., et al.: In: ISSCC Digest of Technical Papers, pp. 296–297. IEEE (2010)

13. Kawahito, S.: In: Proceedings of Custom Integrated Circuit Conference, pp. 505–512. IEEE (2006)
14. Verbruggen, B.: High-speed calibrated analog-to-digital converters in CMOS. Ph.D. thesis, Vrije Universiteit Brussel (2009)
15. Liu, C., et al.: In: Proceedings Symposium VLSI Circuits, pp. 236–237 (2009)
16. Hu, J., Dolev, N., Murmann, B.: IEEE J. Solid-State Circuits 44(4), 1057 (2009)
17. Devarajan, S., et al.: In: ISSCC Digest of Technical Papers, pp. 86–87. IEEE (2009)
18. Bult, K.: In: Proceedings of ESSCIRC, pp. 52–64 (2009)
19. Annema, A., et al.: IEEE J. Solid-State Circuits 40(1), 132 (2005)
20. Bult, K. The effect of technology scaling on power dissipation in analog circuits. In: Analog Circuit Design, pp. 251–290. Springer, Berlin/Heidelberg/New York (2006)
21. Annema, A., et al.: In: ISSCC Digest of Technical Papers, pp. 134–135. IEEE (2004)
22. Schreier, R., Temes, G.C.: Understanding Delta-Sigma Data Converters. IEEE Press, Piscataway (2005)
23. Vittoz, E.: In: Proceedings of International Symposium on Circuits Systems, pp. 1372–1375. IEEE (1990)
24. Nakamura, K., et al.: IEEE J. Solid-State Circuits 30(3), 173 (1995)
25. Draxelmayr, D.: In: ISSCC Digest of Technical Papers, pp. 264–527. IEEE (2004)
26. der Plas, G.V., Verbruggen, B.: IEEE J. Solid-State Circuits 43(12), 2631 (2008)
27. Chae, Y., Han, G.: IEEE J. Solid-State Circuits 44(2), 458 (2009)
28. van Veldhoven, R., Rutten, R., Breems, L.: In: ISSCC Digest of Technical Papers, pp. 492–630. IEEE (2008)
29. Ahmed, I., Mulder, J., Johns, D.: In: ISSCC Digest of Technical Papers, pp. 164–165. IEEE (2009)
30. Anthony, M., et al.: In: Proceedings of Symposium on VLSI Circuits, pp. 222–223 (2008)
31. Fiorenza, J., et al.: IEEE J. Solid-State Circuits 41(12), 2658 (2006)
32. Murmann, B., Boser, B.: IEEE J. Solid-State Circuits 38(12), 2040 (2003)
33. Nazemi, A., et al.: In: Proceedings of Symposium on VLSI Circuits, pp. 18–19 (2008)
34. Iroaga, E., Murmann, B.: IEEE J. Solid-State Circuits 42(4), 748 (2007)
35. Murmann, B.: Limits on ADC power dissipation. In: Analog Circuit Design, pp. 351–368. Springer, Berlin/Heidelberg/New York (2006)
36. Sangster, F., Teer, K.: IEEE J. Solid-State Circuits 4(3), 131 (1969)
37. Alon, E., et al.: IEEE J. Solid-State Circuits 41(2), 413 (2006)
38. Toifl, T., et al.: IEEE J. Solid-State Circuits 44(11), 2901 (2009)
39. Borkar, S.: IEEE Micro 19(4), 23 (1999)
40. Hesener, M., et al.: In: ISSCC Digest of Technical Papers, pp. 248–600. IEEE (2007)
41. Black, W., Hodges, D.: IEEE J. Solid-State Circuits 15(6), 1022 (1980)
42. Jamal, S., et al.: IEEE J. Solid-State Circuits 37(12), 1618 (2002)
43. Vogel, C., Johansson, H.: In: Proceedings of International Symposium on Circuits System, pp. 3386–3389. IEEE (2006)
44. Hsu, C., et al.: In: ISSCC Digest of Technical Papers, pp. 464–615. IEEE (2007)
45. Vogel, C., Mendel, S.: IEEE Trans. Circuits Syst. I, Reg. Papers 56(11), 2463 (2009)
46. Haftbaradaran, A., Martin, K.: In: Proceedings of Custom Integrated Circuit Conference, pp. 341–344. IEEE (2007)
47. Elbornsson, J., Gustafsson, F., Eklund, J.: IEEE Trans. Circuits Syst. I, Reg. Papers 51(1), 151 (2004)
48. Haftbaradaran, A., Martin, K.: IEEE Trans. Circuits Syst. II, Exp. Briefs 55(3), 234 (2008)
49. der Plas, G.V., Decoutere, S., Donnay, S.: In: ISSCC Digest of Technical Papers, pp. 2310 2311. IEEE (2006)
50. Tamba, Y., Yamakido, K.: In: ISSCC Digest of Technical Papers, pp. 324–325. IEEE (1999)
51. Verbruggen, B., et al.: In: Proceedings of Symposium on VLSI Circuits, pp. 14–15 (2008)
52. Huang, C., Wu, J.: IEEE Trans. Circuits Syst. I, Reg. Papers 52(9), 1732 (2005)
53. Nakajima, Y., et al.: In: Proceedings of Symposium on VLSI Circuits, pp. 266–267 (2009)

54. Park, S., Palaskas, Y., Flynn, M.: IEEE J. Solid-State Circuits **42**(9), 1865 (2007)
55. Frey, M., Loeliger, H.: IEEE Trans. Circuits Syst. I, Reg. Papers **54**(1), 229 (2007)
56. Ginsburg, B., Chandrakasan, A.: IEEE J. Solid-State Circuits **43**(12), 2641 (2008)
57. Daito, M., et al.: IEEE J. Solid-State Circuits **41**(11), 2417 (2006)
58. Grace, C., Hurst, P., Lewis, S.: IEEE J. Solid-State Circuits **40**(5), 1038 (2005)
59. Murmann, B., Boser, B.: IEEE Trans. Instrum. Meas. **56**(6), 2504 (2007)
60. Panigada, A., Galton, I.: IEEE J. Solid-State Circuits **44**(12), 3314 (2009)
61. Sahoo, B., Razavi, B.: IEEE J. Solid-State Circuits **44**(9), 2366 (2009)
62. Verma, A., Razavi, B.: IEEE J. Solid-State Circuits **44**(11), 3039 (2009)
63. O'Donoghue, K.,: Digital calibration of a switched-capacitor sigma-deltamodulator. Ph.D. thesis, University of California (2009)
64. Li, J., Moon, U.: IEEE Trans. Circuits Syst. II, Exp. Briefs **50**(9), 531 (2003)
65. McNeill, J., Coln, M., Larivee, B.: IEEE J. Solid-State Circuits **40**(12), 2437 (2005)
66. Hsueh, K., et al.: In: ISSCC Digest of Technical Papers, pp. 546–634. IEEE (2008)
67. Kawahito, S., et al.: In: Proceedings of Custom Integrated Circuit Conference, pp. 117–120. IEEE (2008)
68. Keane, J., Hurst, P., Lewis, S.: IEEE Trans. Circuits Syst. I, Reg. Papers **53**(3), 511 (2006)
69. Nikaeen, P., Murmann, B. In: Proceedings of Custom Integrated Circuit Conference, pp. 161–164. IEEE (2008)
70. Nikaeen, P., Murmann, B.: IEEE J. Sel. Topics Signal Proc. **3**(3), 499 (2009)
71. Satarzadeh, P., Levy, B., Hurst, P.: IEEE J. Sel. Topics Signal Proc. **3**(3), 454 (2009)
72. Chiu, Y., et al.: IEEE Trans. Circuits Syst. I, Reg. Papers **51**(1), 38 (2004)
73. Malla, P., et al.: In: ISSCC Digest of Technical Papers, pp. 496–631. IEEE (2008)
74. Agazzi, O., et al.: IEEE J. Solid-State Circuits **43**(12), 2939 (2008)
75. Namgoong, W.: IEEE Trans. Wireless Commun. **2**(3), 502 (2003)
76. Oh, Y., Murmann, B.: IEEE Trans. Circuits Syst. I, Reg. Papers **53**(8), 1693 (2006)
77. Oh, Y., Murmann, B.: In: Proceedings of Custom Integrated Circuit Conference, pp. 193–196. IEEE (2007)
78. Sandeep, P., et al.: In: Proceedings of GLOBECOM, pp. 1–5. IEEE (2008)
79. Tsai, T., Hurst, P., Lewis, S.: IEEE Trans. Circuits Syst. I, Reg. Papers **56**(2), 307 (2009)
80. Mitteregger, G., et al.: IEEE J. Solid-State Circuits **41**(12), 2641 (2006)
81. Ortmanns, M., Gerfers, F.: Continuous-Time Sigma-Delta A/D Conversion. Springer, Berlin/Heidelberg/New York (2006)
82. Paramesh, J., et al.: In: Proceedings of Symposium on VLSI Circuits, pp. 166–167 (2006)
83. Breems, L.J., Rutten, R., Wetzker, G.: IEEE J. Solid-State Circuits **39**(12), 2152 (2004)
84. Caldwell, T.C., Johns, D.A.: IEEE J. Solid-State Circuits **41**(7), 1578 (2006)
85. Tabatabaei, A., et al.: In: ISSCC Digest of Technical Papers, pp. 66–67. IEEE (2003)
86. Balmelli, P., Huang, Q.: IEEE J. Solid-State Circuits **39**(12), 2212 (2004)
87. Geerts, Y., Steyaert, M., Sansen, W.: Design of Multi-Bit Delta-Sigma A/D Converters. Kluwer Academic, Dordrecht (2002)
88. Norsworthy, S.R., Schreier, R., Temes, G.C.: Delta-Sigma Data Converters - Theory, Design, and Simulation. IEEE Press, Piscataway (1997)
89. Keller, M.,: Systematic approach to the synthesis of continuous-time multi-stage noise-shaping delta-sigma modulators. Ph.D. thesis, University of Freiburg (2010)
90. Kiss, P., et al.: IEEE Trans. Circuits Syst. II, Analog Digit. Signal Process. **47**(7), 629 (2000)
91. Schreier, G., Teme, G.C.: IEEE Trans. Circuits Syst. II, Analog Digit. Signal Process **47**(7), 621 (2000)
92. Maghari, N., et al.: IET Electron. Lett. **42**(22), 1269 (2006)
93. Maghari, N., Kwon, S., Moon, U.: IEEE J. Solid-State Circuits **44**(8), 2212 (2009)
94. Galdi, I., et al.: IEEE J. Solid-State Circuits **43**(7), 1648 (2008)
95. Lee, K., et al.: IEEE J. Solid-State Circuits **43**(12), 2601 (2008)
96. Lee, K.S., Kwon, S., Maloberti, F.: IEEE J. Solid-State Circuits **42**(6), 1206 (2007)
97. Aziz, P.M., Sorensen, H.V., van der Spiegel, J.: IEEE Signal Process. Mag. **13**(1), 61 (1996)

98. Doerrer, L., et al.: IEEE J. Solid-State Circuits **40**(12), 2416 (2005)
99. Samid, L., Manoli, Y.: In: Proceedings of International Symposium on Circuits Systems, pp. 2965–2968. IEEE (2006)
100. Straayer, M.Z., Perrott, M.H.: IEEE J. Solid-State Circuits **43**(4), 805 (2008)
101. Park, M., Perrott, M.H.: IEEE J. Solid-State Circuits **44**(12), 3344 (2009)
102. Hernandez, L., Prefasi, E.: IEEE Trans. Circuits Syst. I, Reg. Papers **55**(8), 2026 (2008)
103. Prefasi, E., et al.: IEEE J. Solid-State Circuits **44**(12), 3344 (2009)

Chapter 5
Interconnects, Transmitters, and Receivers

Bernd Hoefflinger

Abstract Interconnects on-chip between transistors and between functions like processors and memories, between chips on carriers or in stacks, and the communication with the outside world have become a highly complex performance, reliability, cost, and energy challenge.

Twelve layers of metal interconnects, produced by lithography, require, including the contact vias, 24 mask and process cycles on top of the process front-end. The resulting lines are associated with resistance, capacitance and inductance parasitics as well as with ageing due to high current densities.

Large savings in wiring lengths are achieved with 3D integration: transistor stacking, chip stacking and TSV's, a direction, which has exploded since 2005 because of many other benefits and, at the same time, with sensitive reliability and cost issues. On top of this or as an alternative, non-contact interconnects are possible with capacitive or inductive coupling. Inductive in particular has proven to be attractive because its transmission range is large enough for communication in chip stacks and yet not too large to cause interference.

Optical transmitters based on integrated III-V compound-semiconductor lasers and THz power amplifiers compete with ascending low-cost, parallel-wire transmitters based on BiCMOS technologies. Parallel mm-wave and THz transceiver arrays enable mm-wave radar for traffic safety and THz computed-tomography.

In spite of all these technology advances, the power efficiency of data communication will only improve 100× in a decade. New compression and architectural techniques are in high demand.

B. Hoefflinger (✉)
Leonberger Strasse 5, 71063 Sindelfingen, Germany
e-mail: bhoefflinger@t-online.de

B. Hoefflinger (ed.), *CHIPS 2020*, The Frontiers Collection,
DOI 10.1007/978-3-642-23096-7_5, © Springer-Verlag Berlin Heidelberg 2012

5.1 On-chip Interconnects

The interconnection by *wires* of millions and billions of devices, mostly transistors, on large-scale integrated circuits (LSICs) has evolved into the most serious obstacle to the performance increases expected of each new chip generation. The best assessment of the problem is the ITRS (International Technology Roadmap for Semiconductors) [1]. Regarding on-chip wiring requirements, we extract the data shown in Table 5.1.

The aggressive scaling of the first-level metal (M1) and the incredible wire lengths of the first six layers of a total of 12 layers, rising to 14 in 2020, present a massive resistive and capacitive load, which reduces speed and increases energy and cost significantly. Before we look at this load more closely, we point out the steps that promise to alleviate this burden:

- Increase the transistor/device efficiency, i.e., reduce the number of transistors/ devices per function (Sect. 3.6).
- Stack transistors in 3D on top of each other and use selective growth/deposition of Si/alloys/metal for local interconnects over distances of several nanometers as opposed to lateral wires/contacts, which are more than an order-of-magnitude longer and expensive (Sect. 3.7).
- Use TSVs (through-Si-vias) for 3D interconnects instead of 2D global wires to the chip perimeter.

The latter has finally received massive attention, but its effect on the kilometers projected in Table 5.1 has not yet been assessed more quantitatively. We notice that the global wiring pitch is not scaled at a level close to 1 μm, and there seem to be other accepted limits per chip:

Chip area <10 cm^2		
Pins on the chip perimeter	Total	$<$3,000
	Signals	$<$1,000
	Power	$<$2,000

The standard wire material is thin-film copper, whose characteristic parameters are listed in Table 5.2. The Cu resistivity in Table 5.2 includes the effects of grain boundaries and surface roughness. It clearly shows the limits of conventional metal interconnects and the motives for finding a replacement. We also show how large

Table 5.1 Roadmap for on-chip interconnects. First-level metal (M1), semi-global and global wires (From [1])

Year	2010	2015	2020
Node	32	18	10
M1 half pitch [nm]	45	21	12
M1 pitch [nm]	90	42	24
Length/cm^2 [km/cm^2] for M1 + 5 metal layers, signals only	2.2	4.7	8.4
Semi-global wire pitch [nm]	180	84	48
Global wire pitch [nm]	856	856	856

Table 5.2 Roadmap for Cu first-level metal interconnect and SiO_2 dielectric

	2010	2015	2020	Comments
M1 pitch [nm]	90	42	24	ITRS 2009
Aspect ratio	1.8	1.9	2.0	ITRS 2009
Cu resistivity [$\mu\Omega$ cm]	4.08	6.61	9.8	ITRS 2009
Resistance/μm [Ω μm^{-1}]	11	77	340	
Capacitance/μm [fF μm^{-1}]	0.19	0.18	0.16	ITRS 2009
Transistor intrinsic delay τ_S [ps]	1.6	1.2	0.8	Fig. 3.10
M1 length (μm) for combined delay $2\tau_S$ for $W/L = 1$	0.6	0.4	0.2	
$W/L = 10$	6.0	4.0	2.0	

the speed penalty is for Cu wires driven by transistors of the same generation and for two channel width-to-length ratios W/L. At the 2020 node, the minimum transistor with an intrinsic delay of 0.8 ps and with $W = L = 10$ nm, driving an M1 wire 200 nm long (a distance of two simple gates), would mean a total delay from transistor input to the end of the wire of 1.6 ps. Obviously, a transistor with $W = 100$ nm would provide ten times the current and it could drive a wire 2 μm long for a total delay of 1.6 ps. These delays are a part of the information in Fig. 5.1, which shows the expected delays for different interconnect lengths.

Figure 5.1 presents a summary of data on the delay times associated with various types of interconnects as a function of their length. We have studied Cu lines driven by NMOS transistors, and their characteristics are the superposition of two lines in our log–log diagram: The intrinsic delay of the 10 nm transistor from Sect. 3.4 (Fig. 3.10), and the intrinsic delay of a Cu wire with the parameters of Table 5.2, which is charged by this transistor with the maximum current of 190 μA μm^{-1} from Fig. 3.9 (center). The intersection of these two lines can be considered as the characteristic line length, where the total delay is twice the intrinsic delay of the transistor. Proper design of the CMOS driver would optimize (reduce) the total delay for a given wire length and load capacitance at its end: Choose a complementary pair as the driver with the input pair a minimum pair with width W_{ref} and output capacitance C_{ref}. Express the total capacitance of wire and load in multiples of C_{ref} as NC_{ref}. Choose the width of the driver pair as $N^{1/2} W$. We choose this design rule here in order to show that standard wiring with CMOS drivers can be tailored to provide a relative minimum of delay.

Figure 5.1 also shows reference characteristics for nanocarbon interconnects such as carbon nanotubes (CNTs) or graphene lines [2] in different transport modes. These fascinating structures have aroused tremendous research interest because of some outstanding fundamental parameters, particularly those for electron transport:

Mobility	20,000 cm^2 V^{-1} s^{-1}, 20 $\times$ Si
Peak velocity	8 $\times$ 10^7 cm s^{-1}
Diffusion constant	200 cm^2 s^{-1}
Ballistic transport length	3 μm

In his 2010 review paper [2], Jim Meindl, a leader in interconnect research, presented the nanocarbon delay times introduced in Fig. 5.1 with the parameters listed above. All nanocarbon delay lines are intrinsic delay times without any consideration of how to drive these lines.

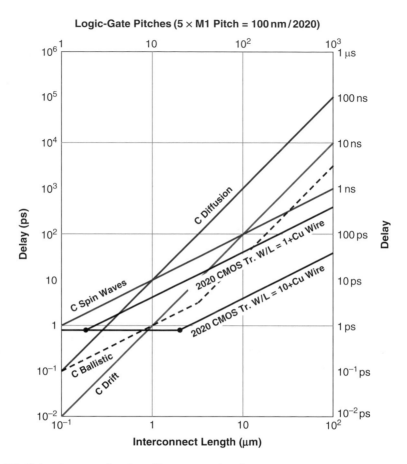

Fig. 5.1 Delay times as a function of interconnect length

The intrinsic delay time for the drift mode at 8×10^7 cm s^{-1} is obviously the shortest, rising with an increase in L^2 to cross the CMOS transistor-driven Cu line at 1 μm. The diffusion mode is an order-of-magnitude slower than the drift mode and also has an L^2 behavior. It is slower than the transistor driver beyond 300 nm. The ballistic mode starts out with a delay proportional to line length L and assumes an L^2 dependence at a few micrometers, when phonon scattering sets in. It is slower than a proper CMOS driver beyond 1 μm. For the spin wave, we have entered the characteristic of a high-speed wave at 10^7 cm s^{-1}. It has the longest delay in this comparative study and is slower than any Cu line driven by a CMOS transistor. The conclusion from this comparison in Fig. 5.1 is that delay times at the lower end are limited by intrinsic transistor speed at about 1 ps, and that for any interconnect length, the best and the working solution is the Cu line driven by CMOS transistors.

The line capacitances and the line delays in Table 5.2 are those for a SiO$_2$ dielectric with a relative dielectric constant $k = 4$. Research on low-k dielectrics for insulators between interconnects is an ongoing effort with some partial solutions.

It is interesting to note that the more radical solution, *air insulation*, is returning 40 years after its earlier appearance in beam-lead technology at Bell Labs [3]. Its renaissance is assisted by the large-scale use of air gaps in MEMS and NEMS (micro- and nano-electro-mechanical systems). Many examples appear in Chap. 14. As we move towards 2020, MEMS-type air bridges of 100 μm length are becoming long-distance interconnects, and gaining a factor of 4 in the reduction of wiring capacitance means a practical *quantum step* forward [4].

5.2   On-chip and Chip–Chip Communication

In this section, we consider various techniques to send and receive digital signals across a chip or between chips in an assembly by amplifier circuitry. In a way, we have already used this active-circuit technique in the previous section, when we drove the Cu lines by a CMOS inverter or inverter pair, and we saw the benefits or the necessity already for load capacitances on the order of femtofarads (fF). However, on ICs, we encounter loads 1,000 times larger, particularly between processor and memory elements, and another order-of-magnitude higher (10 pF) if we have to send or receive signals between chips assembled in a multichip package or on a chip carrier (CC) or printed-circuit board (PCB). For a data rate of 1 Gb/s and a voltage swing of 1 V between ONE and ZERO on a load of 10 pF, we need a power of 10 mW, and this power efficiency [mW/(Gb/s)] has become the figure-of-merit (FOM) for digital chip communication. It is, in other terms, the communication energy per bit: *energy/bit* $= CV^2$.

Remarkable progress has been made regarding this energy, predominantly by reducing the voltage to the typical level of 1 V in 2010 and by scaling the circuit dimensions, and here particularly the area of the contact islands, the pads on which electrical contacts are made to the outside of the chip. The state-of-the-art in 2010 is highlighted in Table 5.3.

Table 5.3 Speed and power efficiency of chip–chip transceivers

	Speed [Gb/s per pin]	Power efficiency [mW/(Gb/s) per pin]	Refs.
47-pin chip-to-chip, 45 nm node	10.0	1.4	8.1 in [5]
Clock-forwarded receiver, 65 nm	7.4	0.9	8.2 in [5]
Complete transceiver, 65 nm node	12.5	1.0	20.5 in [5]
Reconfigurable transceiver, 45 nm	25.0	3.8	20.7 in [5]
Inductive-coupling			
synchronous	1.0	0.4	[6]
asynchronous	11.0	1.4	[6]
4-channel optical transmitter	12.5		[7]
Optical Ge receiver	10.0	1.5	20.1 in [5]
24-channel 850 nm optical transceivers	15	6.0	[8]

We can summarize the achieved results as

Power efficiency in 2010 for chip–chip transceivers:	1 mW/(Gb/s) per pin
	= 1 pJ/bit per pin

For a typical voltage of 1 V, this means an effective capacitance of 1 pF, typical of fine-pitch μbumps with close alignment tolerances. The table also shows the 2010 status of contactless inductive transceivers [6]. Here, thin-film inductive spiral loops with presently 30 μm diameter serve as 1:1 transformer coils for vertical signal transmission with near-field ranges about equal to their diameter. This is a very promising technique for signal transmission in 3D chip stacks. It is one avenue towards further progress in the power efficiency of chip–chip communication. Fig. 5.2 contains a projection starting from the 2010 data, and the long-term history is plotted in Fig. 5.3.

From 2010 to 2020, another 100-fold improvement is feasible. The contribution to the power efficiency from reducing the voltage swing from 1 V to an outer limit of 300 mV would be a factor of 10, but this will require extensive use of LVDS (low-voltage differential signaling). The other tenfold improvement from reducing the effective capacitance implies further scaling and creative use of the third dimension, stacking devices and chips in order to avoid the penalties associated with lateral wiring distances.

A favorite in bridging lateral distances and reducing crosstalk has been optical communication across air or through fibers or SiO_2 waveguides. An example for the state-of-the-art in Ge detectors is listed in Table 5.2 with 10 Gb/s and a power efficiency of 1.5 mW/(Gb/s). Improvements are expected by operating the photodetector in the avalanche mode. On the transmitter side, 4-channel and 24-channel laser-diode transmitters have been announced with rates up to 15 Gb/s per diode channel [7,8]. The expensive optical interconnect will not readily replace the other alternatives, and its added cost will be carried only in specific application areas,

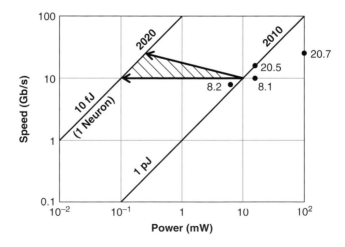

Fig. 5.2 The speed and power of digital chip–chip communication. The references (8.1 etc.) refer to papers at the ISSCC 2010 [5]

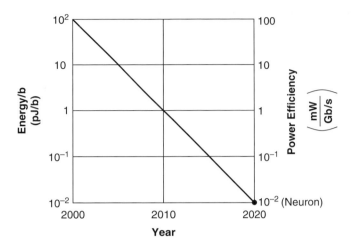

Fig. 5.3 The roadmap of power efficiency for chip–chip communication

such as large servers and supercomputers, where millions of such channels with lengths from centimeters to meters will be needed. A special example would be mask-less, direct-write nanolithography, where systems of more than 100,000 parallel beams would be driven at rates >100 terabit s^{-1} (Chap. 8).

The 2020 (short-range) communication target of 10 fJ/bit would take us to the typical *energy level of synapses of a neuron* (Chap. 18), where we find power levels of 1 pW in signals about 10 ms apart. If we revisit Table 5.1 to find a 2015 wire length of 5 km/cm^{-2} for six wiring levels with a total height of about 500 nm, we obtain a volume wiring density of *100,000 km cm^{-3}, a wire length that would stretchthree times around the earth*. We conclude that, in communications energy and wiring density, we are really approaching the characteristics of the human brain.

5.3 Wireless Transceivers

A highly important field for the connection of chips to the outside world, but also for the contact-less communication between chips, is the wireless transmission and reception of information. We give an overview, again with an emphasis on figures-of-merit and on how the trends in the world of digital mobile communication (Chaps.12 and 13) impact the required chip performance. We start with an overview of typical power requirements on wireless transmitters and receivers listed in Table 5.4. We have also listed the radio frequency (RF) power levels in dBm units, commonly quoted by RF professionals. This means power on a log scale relative to 1 mW:

$$Power(dBm) = 10 \cdot \log_{10}\left(\frac{Power(W)}{1mW}\right)$$

Table 5.4 RF power levels of wireless digital systems

33 dBm	2 W	Maximum output UMTS/3 G mobile phone (power class 1)
30 dBm	1 W	Maximum output GSM 800/1900 mobile phone
20 dBm	100 mW	Typical transmission power of a wireless router
15 dBm	32 mW	Typical WiFi transmission power of a laptop
4 dBm	2.5 mW	Bluetooth transmission power, class 2: 10 m
−10 dBm	100 μW	Typical maximum receiver signal of a WLAN
−70 dBm	100 pW	Typical wireless receiver in an 802.11 network
−120 dBm	1 fW	Receiver signal from GPS satellite

Table 5.5 Wireless transmitters and receivers. On-chip performance 2010 [5]

	Frequency [GHz]	Data rate	Microwave power [dBm]	PAE[a] [%]	Energy per bit [nJ/bit]	Ref.[b]
UWB impulse radio	3–5	1 Mb/s	−80...−92		0.92 transmit 5.3 receive	11.9
Receiver	2.4	0.5 Mb/s	−82		0.83	11.6
SiGeBiCMOS 16-ch. transmitter	60		13	4		11.3
Receiver 65 nm	60		−21			11.4
1.1 V Transceiver	56	11 Gb/s	1 mW Tx	3	0.06/14 mm	23.1
SiGeBiCMOS Transceiver	160		<5	1		23.2
1 V 65 nm Ampl.	60		17	11.7		23.6
1.2 V 90 nm Amp.	60		20	14.2		23.7
	60.5		18	4.9		23.8
1.2 V Amp.	85		6.6	3		23.3
3.3VSiGeBiCMOS	650		−54 Rx			23.9

[a]Power added efficiency
[b]References refer to papers at the ISSCC 2010 [5]

We choose some recent 60 GHz transceivers and the 2–5-GHz range to highlight the state-of-the-art performance in Table 5.5.

The reported results span a wide range of carrier frequencies with the emphasis on the 60 GHz band, which is considered for short-range radar and for high-data-rate (10 Gb/s) indoor wireless communication. The performance levels have been achieved with typically 1 V and 90–40 nm CMOS technologies with power-added efficiencies (PAEs), i.e., ratios of RF to DC power, of typically 10%. This is very promising for media-rich wireless, where these data rates are necessary.

Overall, we have to realize that bandwidth in broadband wireless is becoming limited (Fig. 5.4) and more precious so that bit-rate compression and coding require sustained R&D.

The data on the 650 GHz experiments also show that solutions for terahertz imaging are available with present 130 nm high-speed SiGe BiCMOS technologies, making up an impressive high-performance technology arsenal.

Nevertheless, the mobile-broadband volume has increased at such dramatic rates, Fig. 5.5, in particular for wireless video, that quantum-jump improvements in video data rates are needed to support the future mobile companions envisioned in Chap. 13,

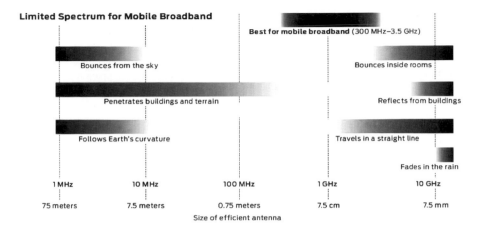

Fig. 5.4 Limited spectrum for mobile broadband [9] (© 2010 IEEE)

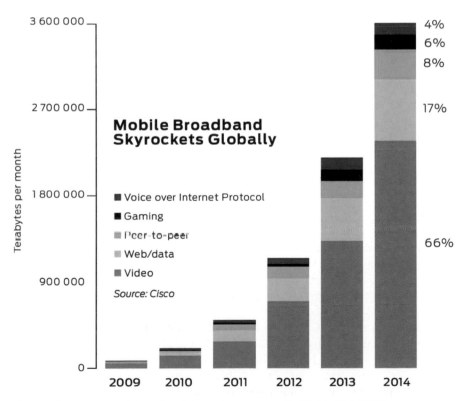

Fig. 5.5 Projected mobile-broadband data volume per month [9] (© 2010 IEEE)

and new, highly efficient video cameras with intelligent focal-plane frame processing (Chap.15) have to be developed.

References

1. www.ITRS.net/reports.html. 2009 Edition
2. Meindl, J., Naeemi, A., Bakir, M., Murali, R.: Nanoelectronics in retrospect, prospect and principle. In: IEEE ISSCC (International Solid-State Circuits Conference) 2010, Digest of Technical Papers, San Francisco, 7–11 Feb 2010, pp. 31–35. doi: 10.1109/ISSCC.2010. 5434062
3. Lepselter, M.P., Waggener, H.A., MacDonald, R.W., Davis, R.E.: Beam-lead devices and integrated circuits. Proc. IEEE **53**, 405 (1965). doi:10.1109/PROC.1965.3772
4. Johnson, R.C.: Air gaps move up from chip- to board-level. www.eetasia.com/ART_8800616596_480100_NT_3f74ff49.HTM. Accessed 19 Aug 2010. See also: Interconnect Focus Center at Georgia Tech (Prof. Paul Kohl) and Steve Hillenius, executive VP at the Semiconductor Research Corp. (SRC)
5. IEEE ISSCC (International Solid-State Circuits Conference). 2010, Digest of Technical Papers
6. Ishikuro, H.: Contactless interfaces in 3D integration. In: IEEE ISSCC (International Solid-State Circuits Conference) 2010, Forum 1: Silicon 3D Integration – Technology and Systems, San Francisco, 8–12 Feb 2010
7. Johnson, R.C.: Intel: silicon photonics to replace copper. Electronic Engineering Times, Asia, 29 July 2010
8. Schow, C., Doany, F., Kash, J.: Get on the optical bus. IEEE Spectr **47**(9), 32 (2010)
9. Lazarus, M.: The great spectrum famine. IEEE Spectr. **47**(10), 26 (2010)

Chapter 6
Requirements and Markets for Nanoelectronics

Bernd Hoefflinger

Abstract The semiconductor market grew 2010 by 70Bio.$ against 2009, more than in the previous 9 years taken together, and the semiconductor industry launched the biggest investment program in its history with 100Bio.$ over a 2-year period. This was the overture to a decade with great potential and great challenges. We look at the market segments and the required electronic functions, and we highlight four product and service areas:

- Approaching 6 Billion mobile-phone subscribers
- Access to education for any child
- One Carebot (personal robot) per family
- Efficient and safe personal mobility.

At the level of over four billion active mobile phones 2010, it is clear that mobile electronic companions have become the drivers of nanoelectronic innovations with growth only limited by the creation and support of new, attractive features and services. Energy, bandwidth, size and weight requirements of these consumer products provide the largest pressure for System-on-Chip (SoC) architectures.

Other exemplary new products are selected for their significance, some for their lengthy path into the market. Health care is such an example: The non-invasive glucose sensor and the portable ECG recorder" with automatic, neuroprocessor-driven event detection in the size of a quarter $ would serve hundreds of millions of people. Nanoelectronics for self-guided health is an area of public policy in view of the cost of "a posteriori" medical care.

Access to information and education for any child/student will be provided by 1$ tablets where service contracts and the spin-offs from surfing and cloud-computing will generate the revenue.

B. Hoefflinger (✉)
Leonberger Strasse 5, 71063 Sindelfingen, Germany
e-mail: bhoefflinger@t-online.de

B. Hoefflinger (ed.), *CHIPS 2020*, The Frontiers Collection,
DOI 10.1007/978-3-642-23096-7_6, © Springer-Verlag Berlin Heidelberg 2012

Personal robots, coined by the ageing Japanese nation as the key product after the PC and ridiculed by others, will arrive as *carebots* for education, entertainment, rehabilitation, and home-service, accepted as a large-scale need by 2020 in most developed countries including China.

Accident prevention systems on rail and road already would make millions of units per year, if required on all trucks and busses, and they would save many lives. For electric bikes, scooters and cars, there is no limit to more intelligent control and energy efficiency. Effective individual mobility, compatible with the environment, is another matter of global competition, of public well-being and of a related public policy.

6.1   Progression of the Semiconductor Market

The semiconductor market has seen a history of growth and cycles like no other industry. It reached the level of $1 billion ($10^9$) in 1970, and it grew by 15% per annum for the next 30 years, much faster than any other industry, to the level of $200 billion in 2000. This corresponded to about 100 billion chips per annum at an average selling price (ASP) of about $2. More than 80% of this market has been integrated circuits, and the average number of transistors on these circuits (chips) increased from about 100 in 1970 to about 100 million in 2000. This growth helped the information and communication industry (ICT) to grow from about $350 billion in 1970 to $3 trillion in 2000 at a rate of 7.5%/a (percent per annum), about half the rate of chips and yet about twice that of the world economy as a whole. This development is illustrated in Fig. 6.1 together with the first decade of the new millennium and an outlook towards 2020.

We see that between 1980 and 2000

– Chips quadrupled their share of the ICT industry to 7%
– The ICT industry doubled its share of the world economy to 10%.

The first decade of the new millennium has seen diminished growth on all fronts due to the burst of the internet bubble in 2000 and the financial crisis of 2008/2009, but also due to the slow start of third-generation mobile-phone products and services.

For the years 2010 and 2011, the semiconductor industry expects to grow by $100 billion, which is more than the growth in the past 9 years, to about $320 billion. As Fig. 6.1 shows, a conservative estimate will carry the semiconductor industry to $450 billion in 2020, and several sources predict an optimistic growth by 7%/a to $600 billion. At this level of chip revenue, the revenue of the ICT industry would reach $7 trillion in 2020, so that it would double its share of the world economy or, in other words, it would continue to support and to assure the growth of the world economy by at least 2.5%/a.

What are the market drivers among chips, and which are the key requirements on these chips? The fundamentals, in simple terms, have persisted over four decades, and they have settled recently at these parameters:

1. The cost per transistor drops by 50%/a.
2. The number of transistors per chip increases by 50%/a, which allows new functions and new applications.

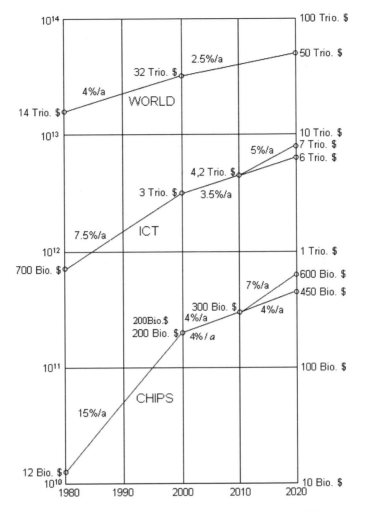

Fig. 6.1 The world economy (gross product of goods and services), the ICT revenue, and the semiconductor industry revenue (CHIPS) from 1980 to 2020 [1–5]

3. The number of chips increases by 15%/a.
4. Total available market (TAM) increases by 7%/a.

From a general functionality point of view, we can summarize the expectations/requirements for the decade 2010–2020 as:

- Memories: Increase the number of bits per chip-scale package 100–300-fold.
- Logic: Increase the operations per second per watt 300-fold.

These broad targets can be seen as general drivers for the advancement of the chip industry. In the following, we take a closer look at the market segments, the special functionalities on the chips, and on their relationships, as well as their significance for progress.

We notice in Fig. 6.2 that computers and communications make up about 65% of the chip market, with a trend towards communications. These two segments continue to drive the chip market. Consumer chips have changed by just 1% to 19%, with, however, the introduction of many new products within their segment, as we will see shortly (Fig. 6.3). Automotive chips have increased their share by

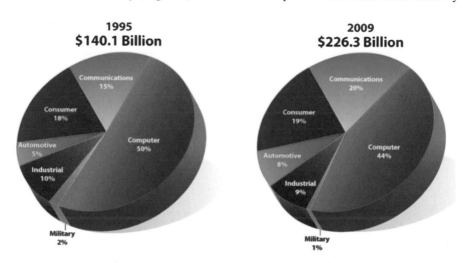

Fig. 6.2 Segments of the semiconductor market in 1995 and 2009 [5, 6] (©WSTS and EE Times)

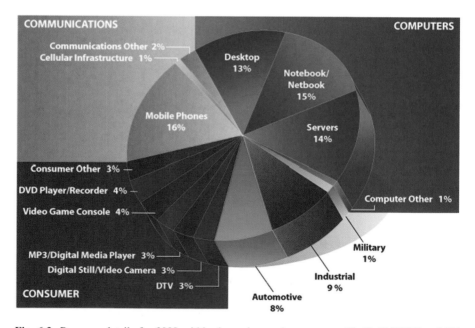

Fig. 6.3 Revenue details for 2009 within the major market segments [5, 6] (©WSTS and EE Times)

Table 6.1 Chip revenue ($billion) and 14-year growth by market segment

	1995	2009	Change (%)
Computing	70	99	+41
Communication	21	45	+114
Consumer	25	43	+72
Med./Industrial	14	20	+40
Automotive	7	18	+157
Military	3	2	−33
Total	140	227	+62

Table 6.2 Market segments (vertical) and special functions (horizontal) of nanochips

	Processor (Chaps. 10, 13, 16)	Memory (Chaps. 11, 18)	Analog/digital (Chap. 4)	Transmitter/ receiver (Chap. 5)	Sensor/actuator (Chaps. 14, 15, 17)
Computing	xxx	xxx		xx	
Communication	xx	xx	xxx	xxx	
Consumer	x	xxx	xxx	xxx	xx
Med./industrial	xx	xxx	xxx	xx	xxx
Automotive	xx	x	xxx	x	xxx
Military	xxx	xxx	xxx	xxx	xxx

60% and their revenue by 157% in that 14-year period. Table 6.1 shows that this period saw an overall growth in revenue of 62%. Besides automotive, consumer and communications sectors came in with growth beyond the chip average of +62%.

Within the major market segments, several product areas stand out, which have become the drivers, as we can see in Fig. 6.3.

Chips for notebooks/netbooks and for mobile phones have provided most of the revenue growth and will continue to do so. In the consumer segment, all areas have seen and continue to see remarkable growth. The actual progress will depend on the continuing, rapid improvement of the chip functionalities, which are listed in Table 6.2 in a matrix against the market segments. We have listed the chapters of this book in which the special functions are treated in detail, and we have highlighted their importance for the different markets. Here we continue by pointing out the potential and challenges for nanochips in the various markets.

6.2 Chips for Computers

The potential and challenges for nanochips in the computer segment are summarized in Table 6.3. Computers continue to mark the leading edge in processing power provided by the most complex chips, and Chap. 10 provides insight into the performance and progression of *mainframe* processors. Their ultimate assembly into supercomputers marks the high end of computing power,

Table 6.3 Challenges for nanochips in the computer segment

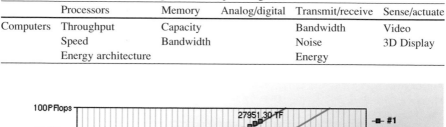

	Processors	Memory	Analog/digital	Transmit/receive	Sense/actuate
Computers	Throughput	Capacity		Bandwidth	Video
	Speed	Bandwidth		Noise	3D Display
	Energy architecture			Energy	

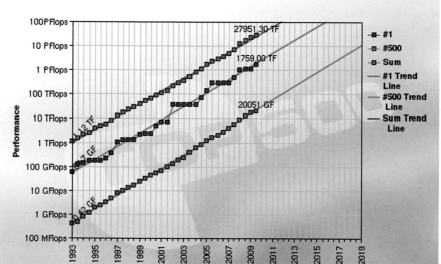

Fig. 6.4 Outlook in 2009 for the TOP 500 worldwide computing power measured in FLOPS (floating-point operations per second) [7] (©TOP 500)

continuously monitored by *TOP500* [7], the agency keeping track of the peak in worldwide computing power, as illustrated in Fig. 6.4.

We see that the world's leading supercomputer achieved 20 teraFLOPS in 2009 and that the industry believes in a trend line of $1,000\times$/decade, which predicts a performance of the number 1 supercomputer in 2020 of 20 peta(10^{15})FLOPS, equivalent to the total worldwide computing power in 2010. This performance will either be demonstrated in a *brute-force approach* of merging the processor power then available into such a powerful – and power-hungry – system, or by the DARPA-funded Program on Ubiquitous High-Performance Computing [8]. The required electrical power for the worldwide *processor farms* in 2020 will indeed be the number 1 problem of future high-performance computing, and it will be touched upon in Chap. 20.

Figure 6.3 tells us that the high-end computer chips, those going into servers and supercomputers, make up about one third of the total revenue from computer chips. Chips for desktops make up another, declining third, and chips for *portable* notebooks and netbooks represent the remaining, rising third. This sector is the promising, controversial, and embattled field of growth, because it has an army approaching from the mobile-phone camp from the communication segment with fourth-generation (4 G) smart phones and personal assistants/companions. This convergence is the most dynamic phenomenon of product development at the start of the decade 2010–2020. These mobile products have the dual requirement of *ultra-low-power and high-performance data and video processing*. We leave the video part to the sections on communication and consumer chips and just highlight the competition within the data-processing community:

– CISC versus RISC
– Cloud computing
– Processor-memory bandwidth

The CISC (complex-instruction-set computing) league, led by the Wintel alliance of Microsoft Windows and Intel, naturally stuck with high-power cores and broad communication between processing and memory elements, has faced a steadily growing, highly diversified league of RISC (reduced-instruction-set computing) companies, mostly licensing the architecture and the operating system from ARM (Advanced RISC Machines), today the world's most successful licensor of chip technology (see Chap. 21). In terms of shipments in 2010, there are *300 million PCs/a against 700 million mobile phones/a*, and the newcomers, *tablet PCs*. These are the new hardware/software/wireless-mobile medium for *e-books and e-newspapers/magazines*.

The progression of these wireless mobile media can be followed by watching ventures dedicated to provide computing and information access to children and students. This movement was pioneered by Professor Negroponte's, UN-endorsed project *OLPC – One-Laptop-per-Child*. It provided exemplary pressure on bringing the cost and power consumption of a PC down and solving the battery problem in remote locations (initially with a crankshaft dynamo). While the price for such an educational PC has been brought down to <$100, the wireless-mobility issue can now be served by third- and fourth-generation mobile-phone systems. Together with the user interface provided by the new generation of e-tablets/pads, the post-Gutenberg galaxy is becoming a reality, where OLPC takes on the new meaning of *Open Libraries for Practically all Children*.

We can see that educational e-books are spreading faster in newly developed countries than in the Old World, and that this global potential offers sufficient growth for computer chips, if their cost, performance, and power consumption make the big steps indicated above.

The trend towards mobile thin clients helps the promotion of *cloud computing*, the shift of complex-software-dependent and computationally intense tasks to remote servers in the internet. This system- and services-driven strategy with sizable investments in public and private infrastructure has evolved into government–industry

---

10×: Programs for cloud computing,

---

shifting the PC emphasis away from classical desktop/laptop PCs.

Advances in computing are more and more limited by cost and energy, mostly expended for the high-speed and highly parallel communication between processors and memories. This communication expense does not scale effectively with new technology generations, mostly because of the circuit topology and the interference between channels. Therefore, the pressure is rising on the classical monopoly of the von Neumann architecture of processor, data- and instruction-memory. This architecture will continue to be the best for mathematical tasks. However, for

– Databases
– Interconnection networks
– Search engines,
– Video and graphics processing

alternative architectures have advanced, such as *associative processing*, helped by scaling of transistors and memory cells. This is evident in a review of associative memories, also called content-addressable memories (CAMs) [9]. The more radical innovation strategy is a critical-size R&D effort on *silicon brains*:

10×: Programs Fast Analog Computing with Emergent Transient States, FACETS (Europe), and Systems of Neuromorphic Adaptive Plastic Scalable Electronics, SyNAPSE (DARPA) ([10, 11] and Chap. 18).

These programs mark a renaissance of cellular neural networks (CNNs), originally conceived in the 1960s and more broadly investigated in the 1980s and in some European centers continuously (Chap. 16). Now, it seems that there is a critical interest because of

– Roadblocks for von Neumann and
– Nanoscale devices and 3D integration necessary and now available for complex brain-like structures.

In any event, the results of these more radical programs will not only influence the computer segment of chips, but also any of the other segments. Of these, communications, with billions of subscribers, will certainly be the largest.

6.3 Communication

The potential and challenges for nanochips in the communication segment are summarized in Table 6.4. The communication segment has been driven by the digital mobile phone, first introduced in 1985 in Europe with the GSM standard and in Japan with the PCS standard. Global standardization evolved with code-division multiple access (CDMA) and then about 2000 with UMTS, the Universal Mobile Telecommunication Standard, also called 3G (Third generation). This has brought not only still images but also video to the mobile phone as well as e-mail and internet so that all mobile phones manufactured today have these features of *smart* phones. The demand on

Table 6.4 Challenges for nanochips in the communication segment

	Processors	Memory	Analog/ digital	Transmit/ receive	Sense/actuate
Communication	Energy	Capacity	Resolution	Bandwidth	Video
	Signal processing	Nonvolatile	Speed	Energy	3D display
	Architecture	Bandwidth	Energy		

- Transmission bit rate
- Video compression and coding
- Camera and display resolution
- Tactile interface

has put and continues to put tremendous pressure on the rate of progress regarding

- Processing power efficiency (GOPS mW^{-1})
- Transceiver bandwidth (GHz) and power efficiency (RF power/DC power)
- Overall power management (sleep vs active)

The processing-power efficiency, owing to the inclusion of video capability, had to be improved 300-fold between 2000 and 2008. This achievement led to the projection of another factor of 300 within the present decade, not only to bring high-definition TV (HDTV) resolution (2 k × 1 k) wireless anywhere, anytime to the user but real movie resolution of 4 k × 2 k pixels at rates of 25 frames per second. It is the live sports-event experience with personal zoom and possibly 3D that could eventually be demonstrated with this extreme technology specification.

The wireless, fiber, and satellite communication infrastructure will have to grow 50-fold to support this growth in data traffic, and also to meet the demand from the tablet computers mentioned in the previous section. Obviously, e-tablets will also be introduced by the leaders in the mobile-phone market as their up-scale product in this continuing convergence of PC and smart phone. The third party in this convergence is the TV, accounted for within the consumer segment, but really also a communication and computer product, considering its capabilities listed in the next section.

6.4   Consumer Segment

The potential and challenges for nanochips in the consumer segment are summarized in Table 6.5. The TV has advanced to a

- Server
- Broadband transceiver
- 3D movie theater

Table 6.5 Challenges for nanochips in the consumer segment

	Processors	Memory	Analog/digital	Transmit/receive	Sense/actuate
Consumer	Energy	Nonvolatile	Cost	Wireless	Video
	Cost	Cost			Audio
	Architecture	Capacity			3D display

These attributes make the TV a product as challenging as the computing and communication products discussed above. The requirements on the chips are as high as for PCs and smart phones, because, in any given technology generation, its video resolution will be 16–32 times higher, and it has to run virtually continuous-time, which tells us that we have an energy crisis unless we achieve the same levels of power efficiency and unless we can handle display drivers with the complexity of today's cinema projectors at the energy and cost level of a consumer commodity.

The silicon compact disk, a flash memory in a credit card, is about to replace CDs and DVDs, and solid-state drives (SSDs), having achieved 32GB capacity in 2009 and now heading towards terabytes (Ch. 11), will replace the disk drives in MP3 players.

HD graphics and video, heading towards 3D, represent a sizable and highly competitive market, justifying aggressive development, which will, however, fail the energy test unless quantum jumps in the efficiency of image processing are achieved. It is here that new architectures such as neuromorphic networks (Chaps. 16 and 18) have to be explored on a 10× level-of-effort.

These processing architectures will also help digital cameras to new heights of stand-alone video and image focal-plane processing and coding. In fact, with the global push towards 3D integration of device and chip layers, the retinomorphic, 3D integration of photoreceptor planes and processing planes (Fig. 3.33) will be introduced on a larger scale. This will be discussed in Chap. 15.

6.5 The Professional Segments of the Chip Market

The three segments medical/industrial, automotive, and military have a number of requirements in common that have forced the chip industry over decades to meet high standards. These mean significant levels of process R&D and rigorous quality-management systems (QMSs). Generally, there are higher operating requirements for chips in professional applications:

- Extended temperature range
- Electromagnetic protection and compatibility (EMC)
- Harsh environments
- Controlled degradation
- Guaranteed reliability levels (FIT).

A guarantee of 100 FIT (one failure in 10^7 h), a typical requirement for an application-specific chip (ASIC) with millions of transistors tells us that the assurance of the quality and reliability of chips is a major task.

A system of standards and certifications has evolved, which we summarize here to highlight the discipline in the chip industry that has been responsible for the unparalleled growth of this industry:

- *MIL-STD-883*: This has been the global procedure for the qualification of chips for 40 years.
- *QML* (Qualified Manufacturing Lines): Total plant qualification established and performed by RADC (Rome Air Development Center) since 1981 associated with the development of the chip factories manufacturing wafers and chips under contract with outside customers. The beginning of the *foundry* age.
- *ISO 90001* (1988) and *ISO 9000:2000*: International standards for total quality management of manufacturing enterprises.
- *CECC-EN 1001–14*: Certification of integrated-circuit manufacturing, specific for each technology generation.
- *QS 9000*: Qualification of chip suppliers by automotive customers (adopted by many professional industrial customers).

These qualifications have to be performed at least for each technology generation and mostly on a per-product basis, and the semiconductor company in turn has to certify all suppliers and service providers. Over the past 20 years, such quality chains have been established generally for most professional businesses worldwide. It is still a singular phenomenon that the chip industry has had this culture for 40 years, and, in the case of the rapid deployment of new *technology nodes,* has to do this every 18 months.

Naturally, those chip makers in the professional segments need a consistent, advanced R&D commitment. However, there is a large area of mutual interest between them and those in the volume segments.

We have seen that the three major segments of the chip market, making up more than 80% of the total market, have to make significant investments in video processing. On one hand, they can benefit from R&D conducted by the three remaining, professional segments of the market, where the system R&D level is naturally higher. On the other hand, the volume introduction of new chips in the big segments helps the professionals to bring the cost down for these innovations.

6.6 Medical/Industrial Segment

The potential and challenges for nanochips in the medical/industrial segment are summarized in Table 6.6. Industrial applications of microchips are to a large degree in electronic control. These often safety-critical tasks means that IC solutions had and have to go through elaborate approval procedures, for example, performed by UL (Underwriters Laboratories) or to obtain European certification (the CE mark).

Table 6.6 Challenges for nanochips in the medical/industrial segment

	Processors	Memory	Analog/digital	Transmit/receive	Sense/actuate
Medical	Performance	Non-volatile	Speed	Bandwidth	Multi-sense
Industrial	Architecture		Precision	Body-Area Net	Lab-on-Chip
					Power-drive
	Intelligence				
	Fail-safe				
	Fault-tolerant				

Historically, this meant that the certification laboratories went slowly and orderly from

– Mechanical to electromechanical (relays),
– To discrete-transistor boards,
– To small-scale ICs (operational amplifiers, then diode–transistor logic (DTL) and transistor–transistor logic (TTL)),
– To gate arrays,
– To microcontrollers,
– To application-specific chips,
  some of these levels taking more than 10 years. The message is that control hardware and software are slow to be certified and have long lifetimes once accepted. It means that *new computing paradigms* will take >30 years to be accepted in industrial control. Fuzzy logic as a parallel multi-input, rule-based control architecture, introduced in the early 1980s, has only made it into selective applications, and cellular neural networks (CNNs), after 20 years, are just beginning to make it into industrial control, although they have, for example, high levels of built-in fault tolerance. Essential properties of chips in industrial control have to provide

– High reliability under wide variation in operating conditions
– Fault-tolerant architecture
– Fail-safe performance (under degradation, go into a safe, defined state)
– FMEA (failure-mode and effect analysis)

These are just a few criteria intended to show that an aggressive scaling of technologies to levels where transistors perform marginally and with a large variance is not the strategy for progress in professional applications., Rather safe and effective new architectures should be explored that reduce the number of components and the power density in chips and packages, offering the largest benefits in cost, energy, and system lifetime/reliability.

Among the many exemplary industrial electronic products, *robots* stand out as the most complex systems and with the promise of being *the next big thing after the PC*, as predicted by Japanese experts in 2001 [12].

The personal robot, designated by the aging Japanese nation as the key product after the PC and ridiculed by others, has arrived in various shapes, from vacuum cleaners to humanoid marching bands and tree climbers. As of 2010, 50% of the world robotics market is already accounted for by home/service robots, and this market is predicted to grow threefold within the next 15 years [13], as detailed in Table 6.7.

Table 6.7 Worldwide robotics market growth 2010–2025 in $billion, after [13]

	2010	2025	Growth rate
Home	12	34	×2.8
Manufacturing	8	12	×1.5
Medical	2	9	×4.5
Public sector	2	8	×4.0
Bio-industrial	1	3	×3.0
Total	25	66	×2.6

Robot electronics today is composed of hundreds of outdated commodity sensors, controllers, and actuators, in industry owing to lengthy certification routines and otherwise often as a proof concept and devoid of an economy-of-scale. In order to really make robots adaptable, intelligent, and safe in real-life situations, a quantum jump towards adaptive, intelligent, fault-tolerant hardware with multisensor and multiactuator interfaces is needed, and neural architectures fit these tasks. They will also reduce significantly size, weight, power consumption, and cost. The new focus on *silicon brains* (Chaps. 16 and 18) will be key to this development.

Personal *carebots* for education, entertainment, rehabilitation, and home-services are accepted as a large-scale need by 2020 in most developed countries, including China, not only for the elderly in aging societies but also as companions and tutors for the young. Several exemplary projects illustrate the 10× level of these developments:

10× Programs for carebots:
– A Robot in Every Home by 2020 (Korea)
– Japan Robotics Association
– KITECH (Korea Institute of Industrial Technology)
– Taiwan [14]
– CARE (Coordination actions for robotics in Europe)

Different directions are being pursued. Many systems are special-function technical solutions while multifunction, companion-type carebots tackle the human-interaction and feel issue with humanoid, also called android, systems [15, 16], as shown in Fig. 6.5.

Robot learning and the robot–human interaction are continuing fundamental challenges [17, 18], but also interdisciplinary challenges, where nanochip strategists and designers have to get more involved given the formidable performance levels of functionalities that can be realized today and in the near future.

This call for interdisciplinary development work should also be heard more clearly regarding the vast opportunities in medical electronics. Pioneering

Fig. 6.5 The android robot Geminoid F by Hiroshi Ishiguro of Osaka University and ATR in 2010 [15] (© 2010 IEEE)

achievements such as the pacemaker of 1964 [19], based on a bipolar circuit just 4 years after the invention of the integrated circuit [20] highlights the importance of radical innovations for the formation of entirely new industries. In this case, it is the industry for electronic implants and prosthetics. Among the many systems being developed at present, we chose *retinal implants* for the restoration of vision as the most complex system. Its implantation in the retina of the eye and its interaction with the visual cortex was proposed in 1996. Experts in ophthalmology, physiology, neurology, microsurgery, biomedicine, physical chemistry, and microelectronics worked together, and after about 10 years of R&D [21], tests could be performed with patients worldwide. An account and perspective is given in Chap. 17.

We can only pinpoint a few areas where medical electronics has significant impact on public health and on growth within the chip market. Minimally invasive diagnosis, therapy, and surgery benefit most from the decreasing size and the increased performance of scaled-down chips. A key example for the state-of-the-art is the camera pill from Given Imaging [22], which can be swallowed, and, while traveling through the intestines, transmits images to a receiver on the body's surface (Fig. 6.6).

Personal, portable, non-invasive diagnostic and monitoring devices offer great benefits to patients and a potential mass market for medical electronics. The percentage reduction in size and weight of these devices would appear directly in the growth rate of the available market. As an example, the *Holter* tape today is a magnetic-tape recorder the size of a Walkman that monitors a patient's cardiogram over 24 h. It then has to be reviewed by the physician for critical events. This could be transformed into a device barely larger than the needed skin contact, containing a neural-network event-recorder chip, trained for the individual patient, and an event flash memory. This would deliver directly the critical events and the time when they

Fig. 6.6 The Given Imaging
autonomous camera capsule,
which transmits wireless
images from inside the
intestines after being
swallowed, has been used in
hospitals since 2002 [22].
Source: Given Imaging

occurred, and it could directly alert the patient or his supervision of any problems. Such a device would have a significant preventive effect and save a posteriori suffering of patients and expenses of (public) insurers. Therefore, the development of such a device would be in the public interest, and resources should be provided to accelerate their introduction into practice.

The same is true for the non-invasive, continuous monitoring of blood sugar for diabetics. About 20 million people worldwide suffer lifelong from type-I (genetic) diabetes, and 500 million acquire it due to lifestyle and age. The consequences of inadequate monitoring – blindness, amputation, and heart disease – are burdensome and expensive. It is also presently a big business for the pharmaceutical industry selling test strips, often 5 per patient per day, for intrusive blood tests, which, of course, are tough on patients and not good enough to detect the critical events of hypoglycemia. An ear clip with a multispectral, near-infrared spectrometer, continuously trained for the patient's characteristics, would offer continuous care-free monitoring. Research on this technology started in the 1980s, with an estimated total level-of-effort of < $100 million worldwide over 30 years, without and probably against the pharmaceutical industry. In 2010 there is no such system on the market. There are many other examples where a public research policy guided by long-term health requirements is needed, especially because lengthy public certification procedures limit the investment of venture capital and often are not survived by start-ups.

The remarkable advances of chips and microsystems in miniaturization and performance, driven by the big-three market segments, have created a new market with significant benefits and leverage: *Minimally-invasive surgical robots*.

The availability of microchips for

– Millmeter-size video cameras and 3D vision
– Microsensors and microactuators
– High-performance video processing
  has allowed quantum jumps in medical surgery and related fields [23].

3D medical imaging plays a key role in surgical robots, and it is driven to its highest levels in resolution and speed in *computed tomography* (CT), a sector with remarkable unit growth. It has extreme requirements for integrating large numbers of Si tiles on the torus of the equipment to sense and process the radiation arriving on their surfaces. It is now entering a new generation with the large-scale availability of the 3D integration of sensing and processing chips connected by thousands to millions of TSVs (Sect. 3.7).

6.7 Automotive Segment

The potential and challenges for nanochips in the automotive segment are summarized in Table 6.8. The automotive segment with a revenue of about $20 billion in 2010 is the fastest growing segment in the chip industry. In spite of the enormous pressure on cost put on the chip industry by the automotive companies, the chip content/car has doubled each decade to $400 in 2010, and it will double again by 2020, driven by the fundamental interest in *efficient and safe personal mobility*, which has, in the meantime, become a mass issue in the newly developed countries as well, with over 1 billion potential customers. This mobility issue and its support by electronics and computing became the topic of the largest-ever joint research effort in Europe:

10×: PROMETHEUS: Program for European Traffic with Highest Efficiency and Unprecedented Safety.

Fifteen car companies and over 70 research institutes started a joint, long-term research project in 1986, the centennial of the automobile, on all aspects of information and communication technology (ICT) to innovate road traffic [24]. Soon, automotive suppliers and ICT companies joined so that, at the peak of the program, about 50 companies and over 100 institutes participated in PROMETHEUS until about 1996, when it was continued as many, more specific programs. It was probably the largest seeding program for new automotive and traffic technology. Programs with similar goals were set up in the US and Japan, for example,

10×: IVHS: Intelligent Vehicle–Highway Society (USA),

an industry–government cooperative still in existence today. A 1993 special issue of IEEE Micro [25] reflects on some topics of these programs.

– Navigation

Table 6.8 Challenges for nanochips in the automotive segment

	Processors	Memory	Analog/digital	Transmit/receive	Sense/actuate
Automotive	Fail-safe	Non-volatile	Speed	Bandwidth	Multisense
	Architecture	Cost	Precision	Security	Power-drive
	Intelligence		Cost	Cost	Display
	Cost				Cost

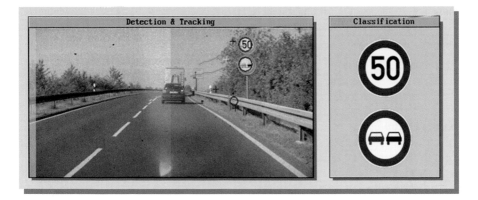

Fig. 6.7   Traffic-sign detection and classification [26] (©Springer)

– ESP (Electronic Safety Program) (Chap. 14)
– ACC (Autonomous Cruise and Anti-collision Control)
   are some of the prominent results in the market. A wealth of prototypes have
   seen tens of thousands of operating hours and hundreds of thousands of
   kilometers and are under consideration or waiting, indicative of the lengthy
   procedures necessary for certification or clarification of privacy or liability
   issues. Presently, accidents with considerable fatalities and damage, caused by
   trucks and buses due to driver fatigue or negligence, are rising steeply, while
   fatigue-warning, lane-keeping, and collision-avoidance systems are among the
   most tested functions [26] (see an example in Fig. 6.7).

However, governments are slow to make these systems mandatory, keeping
market demand low and cost high. Instead, their introduction should be mandatory
and temporarily subsidized in view of public-safety interests. Naturally, in the
introductory phase of these new systems, developed in the advanced countries,
high-level manufacturing jobs would be produced there, replacing the migration of
other manufacturing jobs to low-cost regions.

A recent area of high growth is the transition to hybrid and electric vehicles
because of the never-ending optimization of their electronic control. The intelligent
personalization of these cars to the driver and to his/her daily commute route will be
one application of the coming intelligent neuro-controllers (Chap. 16).

6.8   Military Segment

The potential and challenges for nanochips in the military segment are summarized
in Table 6.9. Military electronics, as we have seen, has been the leader in challenging specifications and the driver for fail-safe, high-quality, and reliable electronics.
This continues to benefit the other professional segments.

Table 6.9 Challenges for nanochips in the military segment

	Processors	Memory	Analog/digital	Transmit/receive	Sense/actuate
Military	Performance	Capacity	Precision	Bandwidth	Multisense
	Architecture	Bandwidth	Speed	Security	Power-drive
	Intelligence	Non-volatile			Body-area networks
		Secure			

References

1. www.EITO.com
2. www.Gartner.com
3. www.ICInsights.com
4. www.ISuppli.com
5. www.WSTS.org
6. Rhines, W.: Comment: is IC industry poised for consolidation? EE Times Asia, 7 April 2010. www.eetasia.com
7. www.TOP500.com
8. http://www.darpa.mil/Our_Work/I2O/Programs/Ubiquitous_High_Performance_Computing_%28UHPC%29.aspx. Accessed March 2011
9. Pagiamtzis, K., Sheikholeslami, A.: Content-addressable memory (CAM) circuits and architectures: a tutorial and survey. IEEE J. Solid-State Circuits **41**, 712 (2006)
10. facets.kip.uni-heidelberg.de/. Accessed March 2011
11. www.darpa.mil/Our_Work/DSO/Programs/Systems_of_Neuromorphic_Adaptive_Plastic_Scalable_Electronics_(SYNAPSE).aspx. Accessed March 2011
12. Japan Robot Association: Summary Report on Technology Strategy for Creating a "Robot Society" in the 21st Century (2001). www.jara.jp/e/dl/report0105.pdf. Accessed March 2011
13. Japan Robotics Association: Worldwide Robotics Market Growth, as quoted by many journals and companies such as The Robotics Report and Microsoft (2010). www.jara.jp
14. Taiwan aims to capture share of the robotics market. NY Times, 30 Oct 2008
15. See: Guizzo, E., Geminoid, F.: more video and photos of the female android. http://spectrum.ieee.org/automaton/robotics/humanoids/042010-geminoid-f-more-video-and-photos-of-the-female-android. Accessed March 2011
16. Division for Applied Robot Technology, KITECH, Ansan, Korea: http://news.nationalgeographic.com/news/2006/05/android-korea-1.html (2006)
17. Matarić, M.J.: Where to next? In: The Robotics Primer. MIT, Cambridge, MA (2007), Chap. 22
18. Marques, L., de Almeida, A., Tokhi, M.O., Virk, G.S. (eds.): In: Advances in Mobile Robots, Proceedings of 11th International Conference on Climbing and Walking Robots and the Support Technologies for Mobile Machines, Coimbra, Portugal. World Scientific, Singapore (2008)
19. Lillehei, C.W., Gott, V.L., Hodges Jr., P.C., Long, D.M., Bakken, E.E.: Transistor pacemaker for treatment of complete atrioventricular dissociation. JAMA **172**, 2006 (1960)
20. Noyce, R.N.: Semiconductor device-and-lead structure. US Patent No. 2981877, filed 30 July 1959. Issued 25 Apr 1961
21. Graf, H.-G., Dollberg, A., Schulze Spüntrup, J.-D., Warkentin, K.: "HDR sub-retinal implant for the vision-impaired", in *High-Dynamic-Range (HDR) Vision*. In: Hoefflinger, B. (ed.) Springer Series in Advanced Microelectronics, vol. 26, pp. 141–146. Springer, Berlin/Heidelberg (2007)
22. www.givenimaging.com
23. MTB Europe: surgical robot market share, market strategies, and market forecasts, 2008–2014. www.mtbeurope.info/bi/reports/

24. Hoefflinger, B.: Computer-assisted motoring. In: Calder, N. (ed.) Scientific Europe, pp. 28–33. Natuur &Techniek, Maastricht (1990)
25. Hoefflinger, B.: Special issue on automotive electronics, traffic management, intelligent control, roadside information, vision enhancement, safety and reliability. IEEE Micro. **13**(1), 11–66 (1993)
26. Knoll, P.M.: "HDR vision for driver assistance", in *High-Dynamic-Range (HDR) Vision*. In: Hoefflinger, B. (ed.) Springer Series in Advanced Microelectronics, vol. 26, pp. 123–136. Springer, Berlin/Heidelberg (2007)

Chapter 7
ITRS: The International Technology Roadmap for Semiconductors

Bernd Hoefflinger

Abstract In a move singular for the world's industry, the semiconductor industry established a quantitative strategy for its progress with the establishment of the ITRS. In its 17th year, it has been extended in 2009 to the year 2024. We present some important and critical milestones with a focus on 2020. Transistor gate lengths of 5.6 nm with a 3 sigma tolerance of 1 nm clearly show the aggressive nature of this strategy, and we reflect on this goal on the basis of our 10 nm reference nanotransistor discussed in Sect .3.3. The roadmap treats in detail the total process hierarchy from the transistor level up through 14 levels of metallic interconnect layers, which must handle the signal transport between transistors and with the outside world. This hierarchy starts with a first-level metal interconnect characterized by a half-pitch (roughly the line width) of 14 nm, which is required to be applicable through intermediate layers with wiring lengths orders of magnitude longer than at the first local level. At the uppermost global level, the metal pattern has to be compatible with high-density through-silicon vias (TSV), in order to handle the 3D stacking of chips at the wafer level to achieve the functionality of the final chip-size product. At the individual wafer level, the full manufacturing process is characterized by up to 40 masks, thousands of processing steps and a cumulative defect density of hopefully $<1/cm^2$.

7.1 History and 2010 Status of the Roadmap

The organization of the US semiconductor industry, the SIA (Semiconductor Industry Association), launched its first roadmap in 1992. From the experience with Moore's law of 1965 [1] and 1975 [2] (Fig. 2.10) and of the "scaling" law of 1974 [3] (Fig. 2.11), a forecasting scheme was derived with a 15-year horizon. That

B. Hoefflinger (✉)
Leonberger Strasse 5, 71063 Sindelfingen, Germany
e-mail: bhoefflinger@t-online.de

B. Hoefflinger (ed.), *CHIPS 2020*, The Frontiers Collection,
DOI 10.1007/978-3-642-23096-7_7, © Springer-Verlag Berlin Heidelberg 2012

effort was joined by the semiconductor industry in Europe, Japan, Korea, and Taiwan in 1998, and the first International Technology Roadmap for Semiconductors (ITRS) was issued in 1999. A fully revised roadmap has been issued since in every odd-numbered year. For our discussion here, we rely on the 2009 roadmap reaching out to the year 2024 [4]. Its aim is to provide the *best present estimate* (of the future) *with a 15-year horizon*.

The roadmap started out in the 1990s with shrink factors of $0.7\times/2$ years for minimum dimensions and other dimensional parameters such as film thicknesses, lateral tolerances within layers and between layers, etc., which had to be scaled concurrently. This target orientation had a tangible influence on progress, evident in the 1999 speed-up on scaling to $0.7\times/18$ months (Fig. 2.17). In the meantime, the roadmap process and structure has evolved into a highly complex, multidimensional effort with major enhancements:

2001: system drivers
2005: emerging research devices (ERD)
2007: emerging research materials (ERM).

As a result, numerous international working groups have been established. Table 7.1 is a listing, together with references to the chapters in the present book in which related issues are addressed by specialists.

Broad global assumptions and stiff models of exponential progress are characteristic of the roadmap, supported so far by 40 years of history. We start with the S-curve for the introduction of new products as shown in Fig. 7.1.

In Fig. 7.1, first conference paper means first product announcement (typically at the ISSCC). It is also characteristic that the specific equipment for a new generation becomes available in an early (alpha) version only 3 years before the mass production of chips. Furthermore, lead-time from samples to volume is less than 12 months. If the ITRS or the industry talks of a technology *node*, it means the start of volume production. For calibration: *2010 is the year of the 32 nm node*. The 32 nm here in the ITRS defines *the smallest half-pitch of contacted metal lines on any product*. Model assumptions are also made for the overall innovation process, as we see in Fig. 7.2.

Table 7.1 Structure of the 2009 ITRS and related chapters in the present book

System drivers	Chaps. 6, 10–13
Design	Chap. 9
Test and test equipment	Chap. 21
Process integration, devices and structures	Chap. 3
RF and analog/mixed-signal technologies	Chaps. 4, 5
Emerging research devices	Chaps. 3, 23
Front end processes	Sect. 3.2
Lithography	Chap. 8
Interconnects	Sect. 3.6, Chap. 5
Factory integration	Chap. 21
Assembly and packaging	Sect. 3.8, Chaps. 12, 21

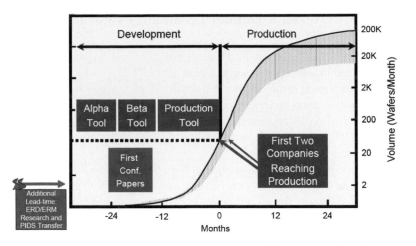

Fig. 7.1 Production ramp-up and technology-cycle timing [ITRS 2009]. *ERD/ERM* emerging research devices/materials, *PIDS* process integration and device structures (© SEMATECH)

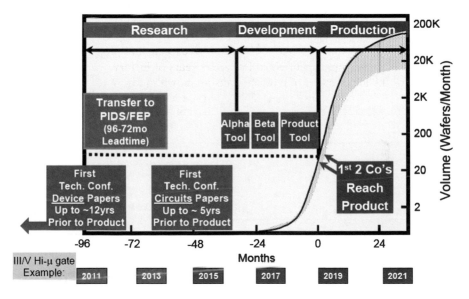

Fig. 7.2 Typical ramp from research to production. The example illustrates the assumption of introducing high-electron-mobility materials for the transistor channel by 2019 [4]. *FEP*: Front end processing (© SEMATECH)

The assumed R&D lead-time of 10 years in the example emphasizes that any new chip technology with an economic impact in 2020 should have been presented in a credible reduction to practice by 2010.

Based on these fundamentals, the core of the roadmap continues to be the overall roadmap technology characteristics (ORTC).

7.2 ORTC: Overall Roadmap Technology Characteristics

These characteristics are the smallest lengths for gates and linewidths of first-level metal introduced in a product in a given year (Table 7.2).

The trends expressed in this data are best illustrated by the lines in Figs. 7.3 and 7.4.

The basic observation on these trend lines is that progress is slowing: *The 0.7 × shrink, which classically happened every 2 years and even saw a speed-up in the decade 1995–2005 to 18 months, has slowed to 3 years and to almost 4 years for the physical gate length in logic chips.*

The 2009 roadmap draws a magic line at the *16 nm node in 2016*. The roadmap data is called near-term to this year (2016), and it happens to bear significance, because, after 2016 and the 16 nm node, the solutions to most of the technology issues related to a straight scaling strategy had to be declared as *unknown* in

Table 7.2 Key lithography-related characteristics by product and year [ITRS 2009]

Near-term years	2010	2012	2014	2016
Flash Poly Si ½ pitch [nm]	32	25	20	16
MPU/ASIC first metal ½ pitch [nm]	45	32	24	19
MPU physical gate length [nm]	27	22	18	15
Long-term years	2018	2020	2022	2024
Flash poly Si ½ pitch [nm]	12.6	10.0	8.0	6.3
MPU/ASIC first metal ½ pitch [nm]	15.0	11.9	9.5	7.5
MPU physical gate length [nm]	12.8	10.7	8.9	7.4

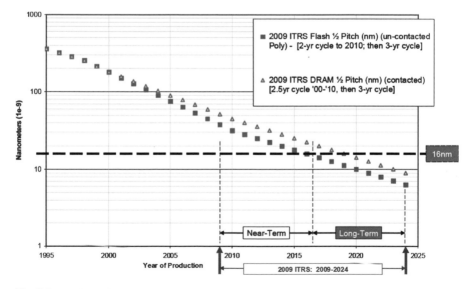

Fig. 7.3 ITRS trend lines for metal-1 half-pitch of DRAM and flash memory [4] (© SEMATECH)

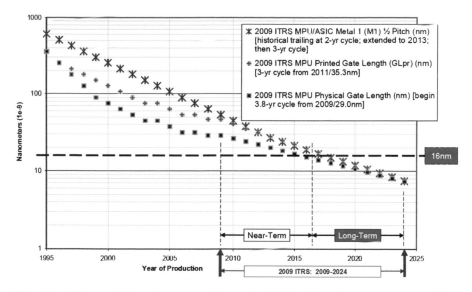

Fig. 7.4 ITRS trend lines for gates and metal-1 half-pitch in microprocessors and high-performance ASICs [4] (© SEMATECH)

2009, whereas with the lead-time model of Fig. 7.2, they should have been known since 2004–2006. As a case in point, lithography and transistor performance are highlighted here: As outlined in Chap. 8 on nanolithography, there is in 2010 *no known lithography technology for 16 nm and beyond other than low-throughput multiple-electron-beam (MEB) lithography*. As described in Sect. 3.2, MOS transistors with an equivalent oxide thickness (EOT) of 1 nm and $L < 15$ nm are a step back in current/μm, have *fundamentally large variances of threshold and current and mostly no intrinsic gain*.

In Sect. 7.4, we discuss to what extent emerging research devices can remove this roadblock by 2020. In any event, *the historic scaling strategy is coming to an end in a window of 16–10 nm*. Is this the end of the chip industry as the world leader in market growth? To answer this question, it is worthwhile forgetting the ORTC scaling trend lines as a self-fulfilling prophecy or driver for a moment and to look at the solely important product and service drivers.

7.3   System Drivers

The system-drivers working group was established in 2001 in order to focus on the requirements of the various chip markets, in a way a late recognition of a market-driven strategy (market pull) after the Y2K crash and after 40 years of an unparalleled technology push provided by a Moore's law implemented by scaling or, in other words, forceful 2-year successions of new technology generations.

The International Technology Working Groups (ITWGs) are a mix of market (horizontal) and specialty (vertical) groups in the sense of the matrix in Chap. 6 of this book, recalled here as Table 7.3. Our expert authors deal with the requirements and challenges in the specialty areas in the chapters indicated in the table.

The extensive listings of requirements and challenges of the 16 working groups of the ITRS show common denominators for the required rates of progress in terms of product performance:

- Double the computing/communication speed every 2 years
- Double the memory density every 2 years.

This cannot be achieved by scaling anymore, and, in fact, it has not been achieved by scaling since the mid-1990s, but rather by "engineering cleverness", as demanded by Moore in the 1980s. The multiprocessor cores in computing and the multilevel storage/transistor and stacked memory chips are striking examples of this cleverness.

Clearly, the energy per function, CV^2, demands continuous and forceful reduction of the capacitance C and the operating voltage V, and the decisive contributions have to come from

- The reduction of wire lengths per function
- The reduction of transistor sites per function
- The reduction of the bandwidth between processors and memories
- The reduction of voltage swing (e.g., low-voltage differential signaling, LVDS).

The progress on these issues is critical, and it will be addressed again in Sect. 7.4.

As to system requirements, we have to look some more at the $2\times/2$ years rate, as aggressive as it may be. It promises a $32\times$ progress per decade. Just looking at the personal mobile assistant of 2020 (Chap. 13), we see that advances *300× between 2010 and 2020* have been demanded and have been considered to be feasible. This strategizing clearly requires quantum jumps in performance beyond the $2\times/2$ years improvement. The multi-bit/cell flash RAM is the most prominent example of this phenomenon (Fig. 7.5).

Operations per second per milliwatt is the key metric for graphics and image processing, and it is here that rich media and 3D vision require a 300-fold improvement between 2010 and 2020. HIPERLOGIC in Sect. 3.6 and digital neuro-processing in Chap. 16 are indicative of such quantum jumps in performance

Table 7.3 Markets (vertical) and special functions (horizontal) of nanochips

	Processor Chaps. 10, 13, 16	Memory Chaps. 11, 18	Analog/ digital Chap. 4	Transmitter/ receiver Chap. 5	Sensor/actuator Chaps. 14, 15, 17
Computing	xxx	xxx		xx	
Communicating	xx	xx	xxx	xxx	
Consumer	x	xxx	xxx	xxx	xx
Med./Industrial	xx	xxx	xxx	xx	xxx
Automotive	xx	x	xxx	x	xxx
Military	xxx	xxx	xxx	xxx	xxx

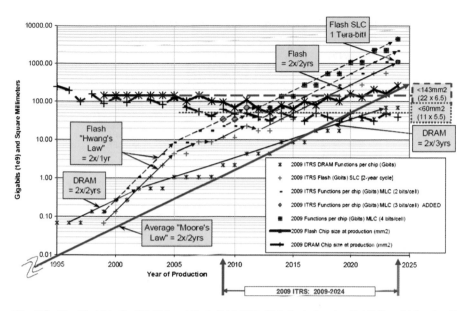

Fig. 7.5 Gigabits/chip for DRAM and flash RAM [4]. *SLC* single-level cell, *MLC* multi-level cell (© SEMATECH)

beyond just an extension of the mainstream by 2×/2 years. In view of the 16 nm barrier, the ITRS puts high hopes on emerging research devices.

7.4   ERD: Emerging Research Devices

Numerous keywords for future devices and new information paradigms are mentioned by this working group. With reference to the innovation curve in Fig. 7.2 as applied to product introduction by 2020, we focus on *enhanced MOS transistors with gates 10 nm long*. The roadmap shows a possible progression, reproduced in Fig. 7.6.

The most important, 10×, R&D tasks are clearly identified in this figure:

1. Transition to metal gate
2. Transition to high-*k* (high-dielectric-constant) gate dielectric
3. Transition to fully depleted silicon-on-insulator (FD SOI)
4. Transition to multiple-gate (MuG) transistors

While items 1 and 2 have been introduced into production, continuous improvement is necessary. Item 3, thin Si films on insulator, with thicknesses <5 nm, fully depleted, are a matter of complex trade-offs and closely related to the challenging introduction of multiple-gate transistors, either as

– Dual-gate transistors or as
– Surround-gate transistors.

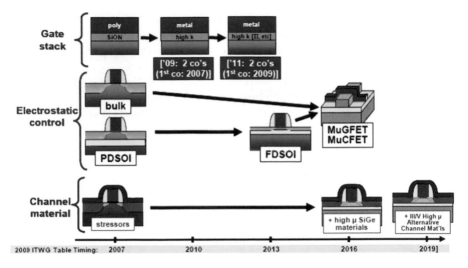

Fig. 7.6 A possible progression to a 10 nm transistor in 2020 [4] (© SEMATECH)

These transistor types upset the physical-design topographies so much engraved in the industry that their introduction by any company will really mean a quantum step in the industry. While dual-gate and tri-gate transistors have been introduced in 2011, and several interesting and powerful functionalities have been described in Sect. 3.5, the surround-gate transistor is a long-term topic closely associated with the nanotube-transistors (Sect. 3.8).

Figure 7.6 also points out new materials for the transistor channels. Strained Si and SiGe have already been introduced to achieve higher electron mobilities and hole mobilities coming closer to those of ideal electrons for speed and symmetrical operation. III–V materials for MOS transistor channels are considered with optimism to benefit from a fivefold or higher mobility. However, we should keep in mind that what counts in switching circuitry is maximum current, and this is determined by maximum velocity, (3.15), and this would be at best two times higher. While this might still be worth the effort, the processing and compatibility challenges and the loss of a technology with complementary transistors have to be weighed in.

The emphasis on speed is only justified for the gighertz and terahertz community of superprocessors and broadband communication. For all others, *energy/function* is the issue. Here we deal with *the world below 400 mV*. We studied this world extensively in Sects. 3.2 and 3.3, and the most important messages we learned are:

1. The maximum operating voltage for a 1 nm (EOT) gate transistor is 500 mV.
2. At <500 mV, MOS transistors operate in the barrier-control (diffusion, *bipolar*, or *leakage*) mode, where the potential barrier is controlled by both the gate and the drain voltages (3.17, 20).
3. The transconductance of a MOS transistor at room temperature is a decade of current per 120 mV at best, with a fundamental 4σ spread to <300 mV for $L = 10$ nm.

4. The intrinsic voltage gain for $t_1 = 1$ nm (EOT) and $L = 10$ nm is unity, i.e., no gain without channel engineering (3.22).
5. The maximum current is again determined by the maximum high-field velocity and not by low-field mobility (3.16, 21).
   Again, the mobility advantage of III–V materials and Ge is not relevant. Furthermore, the temperature dependence of currents in the barrier-control mode is rather strong. This means that on–off current ratios shrink rapidly at higher temperatures, a problem for Si and likely the *out* for Ge as a channel material.
6. The fundamental variance of nanotransistors due to Poisson distributions of charges (acceptors, donors, channel charges) requires that any signal regenerator/amplifier has to be made up of complementary transistors, and, in fact, differential signaling (LVDS), and cross-coupled complementary transistor pairs are required (Sect. 3.3) for safe logic operations.

These stringent criteria have to be observed in the evolution of scaled-down MOS transistors but also for any possible *replacements* in the real world of chips for the masses, which is at 20°C and mostly significantly higher on-chip.

Beyond the scenario of Fig. 7.6 and a 2020 horizon, *emerging research* pursues many directions in the absence of a winner. The ERD committee has assessed more than 15 different technologies in a multidimensional performance evaluation. We select the evaluation of two types of transistors, which we discussed in Sect. 3.8:

– The carbon-nanotube (CNT) MOSFET
– The single-electron transistor (SET)

Their evaluation, as shown in Fig. 7.7, shows eight well-chosen performance criteria:

– Performance
– Energy efficiency
– Gain
– Scalability
– Compatibility with CMOS architecture
– Compatibility with CMOS technology
– Operational temperature
– Reliability

Apart from the quantitative assessment of the criteria, the important observation on these diagrams is the evolution of the ratings over time: From green in 2005 through blue in 2007 to red in 2009. Although a span of four years is short on a research time scale, the apparent trend is that almost all values of 2005 looked optimistic compared with the assessment in 2009. This appearance of more *realism* is frequent in the 2009 issue of the roadmap. It is an indicator that more fundamental, radical innovations are seen as having an impact farther in the future than assumed in recent years. This does not mean that the progression of electronic chips has slowed, but that other disruptive innovations have taken center stage, which are less nano and quantum-driven. Interconnects are a superb example of this phenomenon.

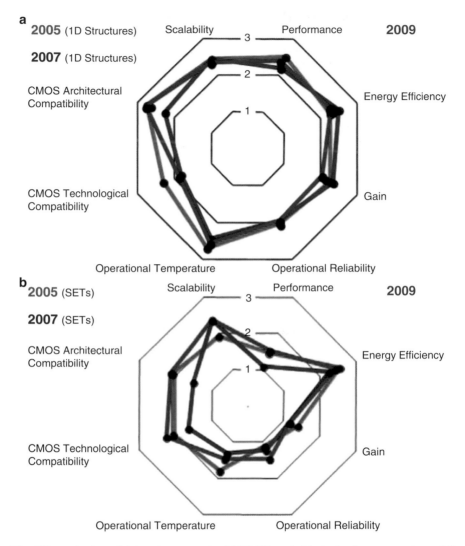

Fig. 7.7 Evaluation of **(a)** carbon nanotube MOSFETs and **(b)** single-electron transistors [4] (© SEMATECH)

7.5 Interconnects

For the integrated circuit functions on a chip, thousands to billions of transistors have to be interconnected by thin-film metal layers, each separated from those above and below by inter-level dielectric insulators. The schematic cross section in Fig. 7.8 of the surface of a chip illustrates the resulting structure.

At the bottom, we see a pair of complementary transistors, each in its well for isolation. Blue tungsten plugs reach up to the first metal interconnect layer. The manufacturing process up to this level is called the front-end of line (FEOL). What

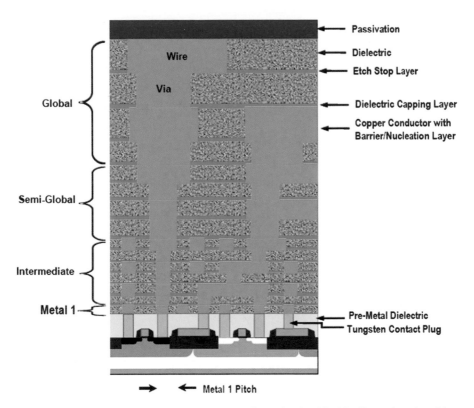

Fig. 7.8 Schematic cross section through the surface of a CMOS chip illustrating the wiring hierarchy. *Blue* and *orange*: metal, *gray*: insulators [4] (© SEMATECH)

follows in the example is three layers of intermediate-level wires, two layers of semi-global, and another five layers of global interconnects. This hierarchy visibly demonstrates the most serious roadblock for nanoelectronics:

The multi-level interconnect with its problems of

– Resistance, capacitance, inductance
– Loading the transistors
– Cross talk
– Area and volume
– Overlay tolerance
– Cost and yield
– Reliability

The scaling of this maze of inverted skyscrapers with skywalks at up to 14 levels is a highly complex challenge, and the roadmap demands a progression as shown in Fig. 7.9, for which manufacturable solutions beyond 2015 are not known or have severe drawbacks in resistance, crosstalk and reliability. In any event, the pressure to

– Reduce operating voltages
– Use low-voltage differential signaling (LVDS)

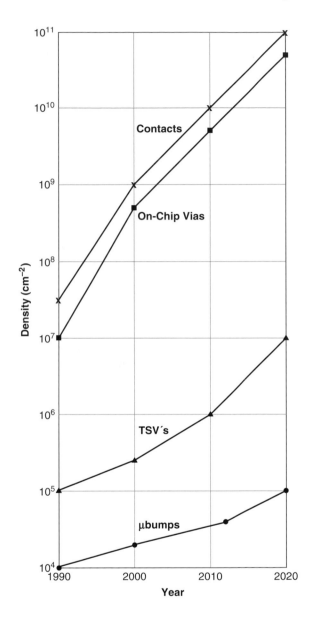

Fig. 7.9 Projected density of contacts, vias, TSV's, and bumps

mounts steadily.

At the same time, lateral wiring lengths are becoming excessive, because of the wider pitch, which has to be accepted at the higher, intermediate, and global levels, and also because of the slower pace of scaling transistors and, in fact, the end of the roadmap for lateral transistor dimensions.

This scenario has finally caused a widespread commitment since about 2005 to reduce lateral wiring lengths and chip sizes by stacking chips on top of each other,

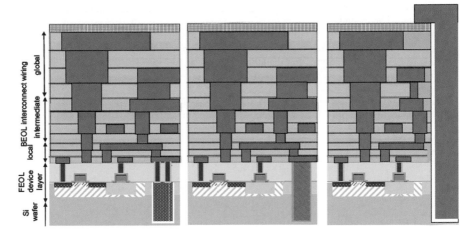

Fig. 7.10 Chip cross sections showing TSV first, at the end of FEOL, and last [4] (©ITRS)

as we discussed in Sect. 3.7. Many alternatives for stacking exist, depending on chip functionality and the manufacturers involved, leading to new networks and action chains. We highlight these possible scenarios with the illustration of three alternatives for producing vias, which have to go through the (thinned) chips to carry signals and power from one chip to the next (Fig. 7.10).

Assuming a SOI chip, TSV-first means that the via is formed literally before the formation of the transistor layer. In the center, the alternative TSV-middle is shown, where the TSV is formed at the end of the FEOL, together with the first metal. At the back-end-of-line (BEOL) on the right, a voluminous TSV-last is shown going through the chip, indicative of the aspect ratios that have to be realized by appropriate cutting and/or etching processes.

Bearing in mind the serious lithography limits on a volume-production level below 15 nm and the conductance limits of metal wires at this scale, the further scaling-down of interconnects is clearly the number one area for alternative wires/interconnects such as carbon nanotubes and graphene lines because of their fundamental advantages but also because they can be produced without lithography [5–7].

References

1. Moore, G.: Cramming more components onto integrated circuits. Electron. Mag **38**(8), 19 Apr 1965
2. Moore, G.: Progress in digital integrated electronics. IEEE IEDM (International Electron Devices Meeting) , Technical Digest, pp. 11–13 (1975)
3. Dennard, R.H., Gaensslen, F.H., Yu, H.N., Rideout, V.L., Bassous, E., LeBlanc, A.R.: Design of ion-implanted MOSFET's with very small physical dimensions. IEEE J Solid-St Circ **9**, 256 (1974)

4. www.itrs.net/reports.html. Semiconductor Industry Association. The International Technology Roadmap for Semiconductors, 2009 Edition. International SEMATECH: Austin, TX 2009
5. Appenzeller, J., Joselevich, E., Hoenlein, F.: Carbon nanotubes for data processing. In: Waser, R. (ed.) Nanoelectronics and Information Technology, 2nd edn. Wiley-VCH, Weinheim (2005)
6. Wolf, E.L: Chap. 5, Some newer building blocks for nanoelectronics. In: Quantum Nanoelectronics, Wiley-VCH, Weinhiem (2009)
7. Hoenlein, W., Kreupl, F., Duesberg, G.S., Grahem, A.P., Liebau, M., Seidel, R., Unger, E.: Carbon nanotube applications in microelectronics. In: Siffert, P., Krimmel, E.F. (eds.) Silicon: Evolution and Future of a Technology, pp. 477–488. Springer, Berlin/Heidelberg (2004)

Chapter 8
Nanolithography

Burn J. Lin

Abstract Nanolithography, the patterning of hundreds to thousands of trillions of nano structures on a silicon wafer, determines the direction of future nano-electronics. The technical feasibility and the cost-of-ownership for technologies beyond the mark of 22 nm lines and spaces (half pitch) is evaluated for Extreme-Ultra-Violet (EUV = 13 nm) lithography and Multiple-Electron-Beam (MEB) direct-write-on-wafer technology. The 32 nm lithography-of-choice, Double-Patterning, 193 nm Liquid-Immersion Optical Technology (DPT) can possibly be extended to 22 nm with restrictive design patterns and rules, and it serves as a reference for the future candidates.

EUV lithography operates with reflective mirror optics and 4× multi-layer reflective masks. Due to problems with a sufficiently powerful and economical 13 nm source, and with the quality and lifetime of the masks, the introduction of EUV into production has been shifted to 2012.

MEB direct-write lithography has demonstrated 16 nm half-pitch capability, and it is attractive, because it does not require product-specific masks, and electron optics as well as electron metrology inherently offer the highest resolution. However, an MEB system must operate with more than 10,000 beams in parallel to achieve a throughput of ten 300 mm wafers per hour. A data volume of Peta-Bytes per wafer is required at Gigabit rates per second per beam. MEB direct-write would shift the development bottleneck away from mask technology and infrastructure to data processing, which may be easier.

The R&D expenses for EUV have exceeded those for MEB by two orders of magnitude so far. As difficult as a cost model per wafer layer may be, it shows basically that, as compared with DPT immersion optical lithography, EUV would be more expensive by up to a factor of 3, and MEB could cost less than present

B.J. Lin (✉)
TSMC, 153 Kuang Fu Road, Sec. 1, Lane 89, 1st Fl., Hsinchu R.O.C. 300, Taiwan
e-mail: burnlin@tsmc.com

B. Hoefflinger (ed.), *CHIPS 2020*, The Frontiers Collection,
DOI 10.1007/978-3-642-23096-7_8, © Springer-Verlag Berlin Heidelberg 2012

optical nanolithography. From an operational point-of-view, power is a real concern. For a throughput of 100 wafers per hour, the optical system dissipates 200 kW, the MEB system is estimated at 170–370 kW, while the EUV system could dissipate between 2 and 16 MW.

This outlook shows that the issues of minimum feature size, throughput and cost of nanolithography require an optimisation at the system level from product definition through restrictive design and topology rules to wafer-layer rules in order to control the cost of manufacturing and to achieve the best product.

Optical patterning through masks onto photosensitive resists proceeded from minimum features of 20 μm in 1970 to 22 nm in 2010. The end of optical lithography was initially predicted for 1985 at feature sizes of 1 μm, and hundreds of millions of dollars were spent in the 1980s to prepare X-ray proximity printing as the technology of the future, which has not happened. The ultimate limit of optical lithography was moved in the 1990s to 100 nm, initiating next-generation-lithography (NGL) projects to develop lithography technology for sub-100-nm features and towards 10 nm. This chapter is focused on two of these NGL technologies: *EUV* (extreme-ultraviolet lithography), and *MEB* (multiple-electron-beam lithography). These are compared with the ultimate optical lithography, also capable of delineating features of similar size: *DPT* (double-patterning, liquid-immersion lithography).

8.1   The Progression of Optical Lithography

Optical lithography has supported Moore's Law and the ITRS (Chap. 7) for the first 50 years of microelectronics. The progression in line-width resolution followed the relationship

$$\text{Resolution} = k_1 \frac{\lambda}{NA}.$$

Shorter-wavelength optical and resist systems, and the continuing increase of the numerical aperture (NA) as well as reduction of the resolution-scaling factor k_1 provided a remarkable rate of progress, as shown by the data on the period 1999–2012 (Figs. 8.1 and 8.2).

The medium for optical patterning was air until 2006, when $k_1 = 0.36$ at the 193 nm line of ArF laser sources was reached. The single biggest step forward was made when the lens wafer space was immersed in water, and the NA effectively increased by 1.44× according to the refractive index of water at 193-nm wavelength, so that, with the inclusion of restricted design rules and more resolution-enhancement techniques (RETs), the 32-nm node could be reached in 2010.

Year	Node	Lens parameters	RET/OPC/Designs
1999	180 nm	$\lambda = 248$ nm, $k_1 = 0.58$	OAI; HLC; Single hole size
2000	150 nm	$\lambda = 248$ nm, $k_1 = 0.50$	RB OPC, aggressive HLC for line-end shortening
2002	130 nm	$\lambda = 248$ nm, $k_1 = 0.46$	AF, Model-derived OPC, MB OPC, reduced HLC
2004	90 nm	$\lambda = 248$ nm, $k_1 = 0.38$	MB OPC, HB OPC, HLC eliminated
2006	65 nm	$\lambda = 193$ nm, $k_1 = 0.36$	Hot spot check in DFM using lithography simulations
2008	45 nm	$\lambda = 193$ nm, $k_1 = 0.39$	Immersion lithography, forbidden pitch, S2E DFM
2010	32 nm	$\lambda = 193$ nm, $k_1 = 0.31$	Dipole illumination, MB AF, single orientation, on-grid, RDR
2012	22 nm	$\lambda = 193$ nm, $k_1 = 0.28$	Double dipole, double patterning, decomposable layout

Fig. 8.1 Lens and reticle parameters, OPC, and RETs employed for the technology nodes from 1999 to 2012 [1]. OPC: Optical proximity correction, OAI: Off-axis illumination, HLC: Handcrafted layout compensation, RB OPC: Rule-based OPC, MB OPC: Model-based OPC, HB OPC: Hybrid-based OPC, AF: Assist features, DFM: Design for manufacturability, S2E DFM: Shape-to-electric DFM, MB AF: Model-based assist features, RDR: Restricted design rules. © 2009 IEEE

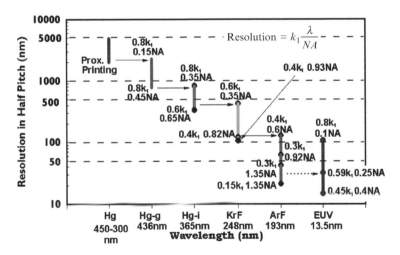

Fig. 8.2 Resolution for half pitch [(linewidth + space)/2] as wavelength reduction progresses [1]. © 2009 IEEE

With the introduction of DPT, the mask cost is doubled, the throughput is halved, and the processing cost more than doubled, making the total wafer-production cost unbearably high and causing serious economic effects.

A 2010 state-of-the-art optical-lithography system with ArF water-immersion DPT can be characterized by the following parameters:

Minimum half pitch	22 nm at $k_1 = 0.3$, $\lambda = 193$ nm, NA $= 1.35$
Single-machine overlay	<2 nm
Reticle magnification	$4\times$
Throughput	~175 wph (300 mm diameter wafer)
Price	~€40 M
Related equipment	~US$15 M
Footprint	~18 m^2 including access areas
Wall power	220 kW

This most advanced equipment is needed for the critical layers in a 32-nm process:

– Active
– Gate
– Contact
– Metal 1–n
– Via 1–n.

These layers form the critical path in the total process, which can involve more than 10 metal layers resulting in a total mask count >40. Liquid-immersion DPT can be pushed to 22 nm with restricted design rules, limited throughput, and a significant increase in cost. It is desirable to use less expensive alternative technologies at and beyond the 22 nm node. The following sections are focused on the two major contenders for production-level nanolithography at 32 nm and beyond: EUV and MEB. Nanoimprinting holds promise for high replicated resolution. However, fabrication difficulties of the 1× templates, limited template durability, expensive template inspection, and the inherent high defect count of the nanoimprint technology prevent it from being seriously considered here [7].

8.2   Extreme-Ultraviolet (EUV) Lithography

The International SEMATECH Lithography Expert Group decided in 2000 that EUV lithography should be the NGL of choice. The EUV wavelength of 13.5 nm means a $> 10\times$ advantage in resolution over optical, as far as wavelength is concerned, and sufficient progress overall, even though the k_1 and the NA with the necessary reflective-mirror optics would be back to those of the early days with optical lithography. The system principle is shown in Fig. 8.3.

The 13.5 nm radiation is obtained from the plasma generated by irradiating a Sn droplet with a powerful CO_2 laser. This is realized with an efficiency of 0.2–0.5% from CO_2 laser irradiation. The numbers in the figure illustrate the power at each component along the path of the beam, required to deliver 1 mJ/cm^2 of EUV light at the wafer at the throughput of 100 wph (wafers per hour). About 0.1% of the EUV power arrives on

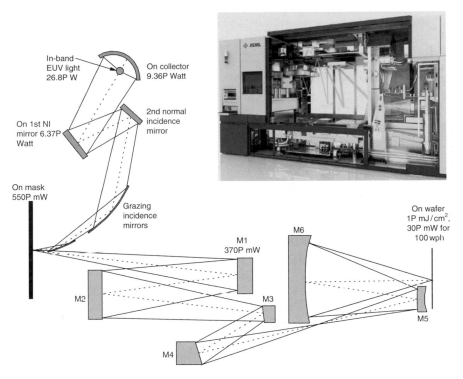

Fig. 8.3 EUV illuminator and imaging lens. It is an all-reflective system [1]. © 2009 IEEE

the wafer. Evidently, the overall power efficiency from the laser to the wafer is extremely small. This poses a serious problem: If we consider a resist sensitivity of 30 mJ/cm² at 13.5 nm, hundreds of kilowatts of primary laser power are needed.

Other extreme requirements [2] are:

- Roughness of the mirror surfaces at atomic levels of 0.03–0.05 nm, corresponding pictorially to a 1-mm roughness over the diameter of Japan.
- Reflectivity of mirrors established by 50 bilayers of MoSi/Si with 0.01 nm uniformity.
- Reticle cleanliness in the absence of pellicles and reticle life at high power densities.
- Reticle flatness has to be better than 46.5 nm, which is 10× better than the specification for 193-nm reticles.
- New technology, instrumentation, and infrastructure for reticle inspection and repair.

Nevertheless, EUV lithography has received R&D funding easily in the order of US$1 billion by 2010. Two alpha tools have been installed, producing test structures like those shown in Fig. 8.4.

In any event, EUV lithography puts such strong demands on all kinds of resources that, if successful, it can be used by only a few extremely large chip

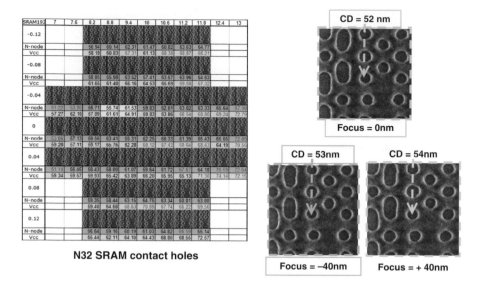

N32 SRAM contact holes

Fig. 8.4 Micrographs of SRAM test structures: 32-nm contact holes with a pitch of 52–54 nm [1]. © 2009 IEEE

manufacturers or consortia of similar size and technology status. Attaining lower imaging cost than DPT cannot be taken for granted. A 2012 system can be estimated to have the following characteristics:

Minimum half pitch	21 nm at $k_1 = 0.5$, $\lambda = 13.5$ nm, NA = 0.32
Single-machine overlay	<3.5 nm
Reticle magnification	4×
Throughput	125 wafers (300-mm dia. wafer)
Price	€65 M
Related equipment	US$20 M (excluding mask-making equipment)
Footprint	50 m^2 including access areas
Wall power	750–2,000 kW

If we consider that a Gigafactory requires 50 such systems for total throughput, we can appreciate that this technology is accessible to a few players only. Further data on these issues will be discussed in Sect. 8.4.

8.3 Multiple-Electron-Beam (MEB) Lithography

Electron-beam lithography has been at the forefront of resolution since the late 1960s, when focused electron beams became the tools of choice to write high-resolution masks for optical lithography or to write submicrometer features directly on semiconductor substrates. As a serial dot-by-dot exposure system, it has the reputation of being slow, although orders-of-magnitude improvements in through-put were achieved with

- Vector scan (about 1975)
- Variable shaped beam (1982)
- Cell projection through typically 25× Si stencil masks (1995)

A micrograph of a 1997 stencil mask with 200-nm slots in 3-μm Si for the gates in a 256 Mb DRAM is shown in Fig. 8.5.

Electron beams have many unique advantages:

- Inexpensive electron optics
- Large depth-of-field
- Sub-nanometer resolution
- Highest-speed and high-accuracy steering
- Highest-speed On-Off switching (blanking)
- Secondary electron detection for position, overlay, and stitching accuracy
- Gigabit/s digital addressing

The systems mentioned so far were single-beam systems, using sophisticated electron optics. A scheme to produce a large array of parallel nanometer-size electron beams is used in the MAPPER system [3], as shown in Fig. 8.6, where 13,000 beams are created by blocking the collimated beam from a single electron source with an aperture plate. The beam-blanker array acts upon the optical signal from the fiber bundle to blank off the unwanted beams. The On-beams are

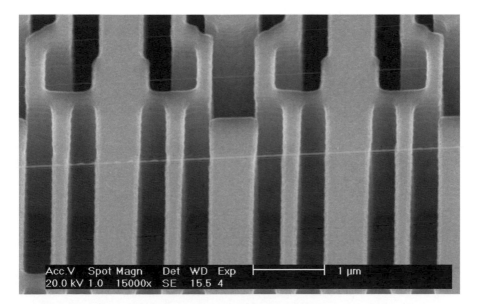

Fig. 8.5 SEM micrograph of a Si stencil mask with 200-nm slots in 3-μm Si for the gates in a 256 Mb DRAM (Courtesy INFINEON and IMS CHIPS)

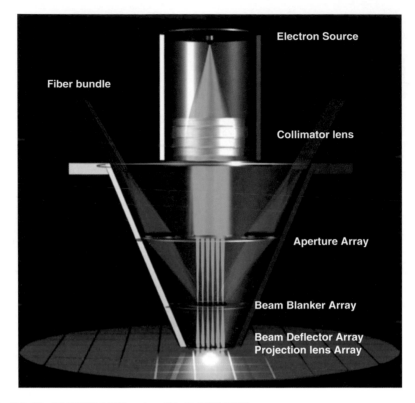

Fig. 8.6 The MAPPER MEB system [1]. © 2009 IEEE

subsequently deflected and imaged. All electron optics components are made with the MEMS (micro-electro-mechanical system) technology, making it economically feasible to achieve such a large parallelism [6].

The MAPPER write scheme is shown in Fig. 8.7. For a 45-nm half-pitch system, the 13,000 beams are distributed 150 μm apart in a 26×10 mm^2 area. The separation ensures negligible beam-to-beam interaction. Only the beam blur recorded in the resist due to forward scattering calls for proximity correction. The beams are staggered row-by-row. The staggering distance is 2 μm. The deflection range of each beam is slightly larger than 2 μm to allow a common stitching area between beams, so that the 150-μm spaces in each row are filled up with patterns after 75 rows of beam pass through. Each beam is blanked off for unexposed areas [4].

Massive parallelism is not allowed to be costly. Otherwise the cost target to sustain Moore's law still cannot be met. One should forsake the mentality of using the vacuum-tube equivalent of conventional e-beam optics and turn to adopting the integrated-circuit-equivalent highly-parallel electron optics, whose costs are governed by Moore's law. With cost of electron optics drastically reduced, the dominant costs are now the digital electronics, such as CPU (central processing

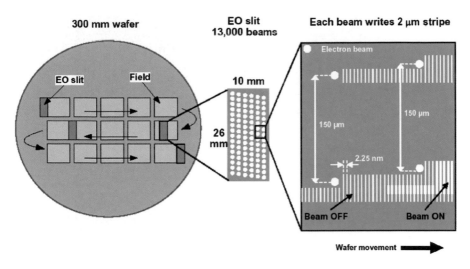

Fig. 8.7 MAPPER writing scheme (EO: Electron optics) [1]. © 2009 IEEE

unit), GPU (graphics processing unit), FPGA (field programmable gate array), and DRAM, required to run the massively parallel, independent channels. *For the first time, the cost of patterning tools is no longer open-ended but rather a self-reducing loop with each technology-node advance.*

Another multiple-beam system uses a conventional column but with a programmable reflective electronic mask called a digital pattern generator (DPG) as shown in Fig. 8.8.

The DPG is illuminated by the illumination optics through a beam bender. The DPG is a conventional 65-nm CMOS chip with its last metal layer turned towards the illuminating beam. The electrons are decelerated to almost zero potential with the Off-electrons absorbed by the CMOS chip while the On-electrons are reflected back to 50 keV and de-magnified by the main imaging column to expose the wafer. Even though there are between 1 M and 4 M pixels on the DPG, the area covered is too small compared to that of an optical scanner. The acceleration and deceleration of a rectilinear wafer stage to sustain high wafer throughput are impossible to reach. Instead, a rotating stage to expose six wafers together can support high throughput with an attainable speed. Again, all components are inexpensive. The data path is still an expensive part of the system. The cost of the system can be self-reducing.

The fascination with these systems, beyond the fundamental advantage that they provide *maskless nano-patterning,* lies in the fact that MEB lithography directly benefits in a closed loop from the

– Advances in MEMS and NEMS (micro- and nano-electro-mechanical systems) technology for the production of the aperture and blanking plates as well as arrays of field-emission tips, and
– Advances in high-speed interfaces to provide terabits/s to steer the parallel beams.

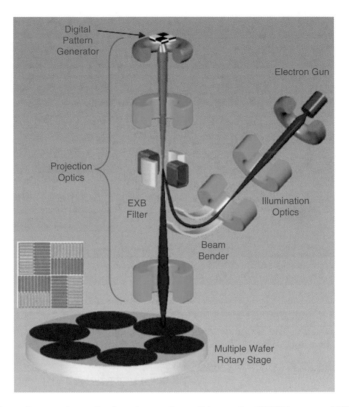

Fig. 8.8 Second generation reflective electron-beam lithography (REBL) system. EXB: charged particle separator $E \times B$

MEB direct-write lithography demonstrated 16-nm resolution (half-pitch) in 2009.

For high-volume manufacturing (HVM), the basic MEB chamber with 13,000 columns and 10 wph has to be clustered into 10 chambers to reach 100 wph, as illustrated in Fig. 8.9.

An MEB system with 10 chambers has about the same footprint as the 193-nm optical immersion scanner. A 2012 assessment would be as follows:

Minimum linewidth	22 nm
Minimum Overlay	7 nm
No reticle	Maskless
Number of e-beams	130,000
Throughput	100 (300 mm) wph
Data rate	520 Tbit/s
Estimated Price	US$50 M
Footprint	18 m^2

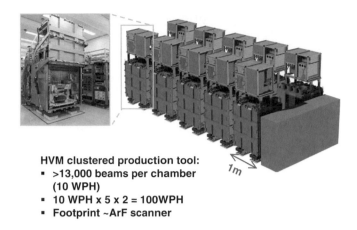

HVM clustered production tool:
- **>13,000 beams per chamber (10 WPH)**
- **10 WPH x 5 x 2 = 100WPH**
- **Footprint ~ArF scanner**

Fig. 8.9 MEB maskless-lithography system for high-volume manufacturing [1]. © 2009 IEEE

The following calculation of data rates illustrates the challenges and possible system strategies for these maskless systems [5]:

– Total pixels in a field (0 and 1 bitmap)

$$= (33 \, \text{mm} \times 26 \, \text{mm}/2.25 \, \text{nm} \times 2.25 \, \text{nm}) \times 1.1 (10\% \, \text{over scan})$$
$$= 190 \, \text{Tbits}$$
$$= 21.2 \, \text{TBytes}$$

– A 300-mm wafer has ~90 fields
– 10 wph $\Rightarrow$ <0.9 s/field
– 13,000 beams $\Rightarrow$ data rate >3.75 Gbps/beam!
– In addition to data rate, also challenge in data storage and cost
– Use more parallelism to reduce data rate

MEB lithography by 2010 has seen R&D resources at only a few percent compared with those for EUV, and no industry consensus has been established to change that situation. However, this may change in view of the critical comparison in the following section.

8.4 Comparison of Three Nanolithographies

The status in 2010 and the challenges for the three most viable nanolithographies for volume production of nanochips with features <32 nm are discussed with respect to maturity, cost, and factory compatibility. Table 8.1 provides a compact overview.

Table 8.1 Maturity, cost, infrastructure, and required investment

	Wafer cost	Mask cost	Infra-structure	Maturity	Investment
MEB ML2	Potentially low	0	Ready	Massive parallism needs most work	Badly needed
Immersion DPT	More than 2× SPT	Ever increasing	Ready	Mature	No question
EUV	From slightly below to much higher than DPT	Can be higher than DPT	Expensive, to be developed	Not yet	Plenty but never enough

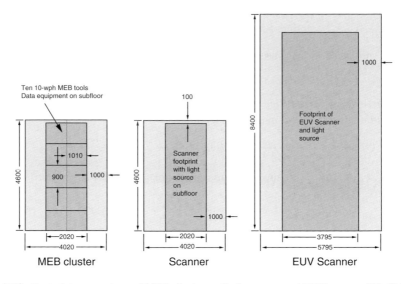

Fig. 8.10 Footprint comparison of MEB cluster, optical scanner, and EUV scanner [1]. © 2009 IEEE

Single-patterning technology optical immersion lithography (SPT) is taken as a reference. Given the 2010 status, neither EUV nor MEB is mature for production. Lithography has become the show-stopper for scaling beyond 22 nm so that serious investment and strategy decisions are asked for. Cleanroom footprint in Fig. 8.10, cost per layer in Table 8.3 and electrical power in Table 8.2 are used to provide guidance for the distribution of resources.

We see that the footprint for the throughput of 100 wph with optical and e-beam technology is about the same, while that for the EUV scanner is more than double.

Cost estimates, difficult as they may be, are important in order to identify fundamental components of cost and their improvement (Table 8.3). The numbers for optical DPT versus SPT are fairly well known. The exposure cost is a strong function of throughput, so it doubles for DPT. However, because of the extra etching step required by DPT, the combined exposure-and-processing cost more

Table 8.2 Electrical power (kW) for 32-nm lithography in a 150-MW Gigafactory for 130,000 12 in. wafers/month (HVM: High-volume manufacturing) [1]

kW	Immer scanner	EUV HVM			MEB HVM	
	Supplier estimate	Supplier estimate	30 mJ/cm² instead of 10 mJ/cm²	30 mJ/cm² resist + conservative collector and source efficiencies	Ten 10 wph columns	Share datapath
Source	89	580	1,740	16,313	120	120
Exposure unit	130	169	190	190		
Datapath					250	53
Total per tool	219	749	1,930	16,503	370	173
Total for 59 tools	12,921	44,191	113,870	973,648	21,830	10,222
Fraction of scanner power in fab	8.61%	29.46%	75.91%	649.10%	14.55%	6.81%
	130 k wafers per month 12″ fab, 150,000 kW					

Table 8.3 Cost per lithography layer relative to single-patterning (SP) water-immersion ArF lithography [1]

All costs normalized to SP exposure cost/layer	Imm SP	Imm SP	EUV goal	EUV possibility	MEB goal	MEB possibility
Normalized exposure tool cost	3.63E + 06	3.63E + 06	4.54E + 06	4.54E + 06	4.54E + 06	2.72E + 06
Normalized track cost	5.58E + 05	5.58E + 05	4.88E + 05	2.09E + 05	4.88E + 05	4.88E + 05
Raw thruput (WPH)	180	100	100	20	100	100
Normalized exposure cost/layer	1.00	1.80	2.17	10.37	2.17	1.37
Normalized consumable cost/layer	0.91	2.27	1.43	1.43	1.05	1.05
Normalized expo+consum cost/layer	1.91	4.07	3.61	11.80	3.22	2.42

than doubles. For EUV lithography, if the development goal of the EUV tool supplier is met, the combined exposure and processing cost is slightly less than that of DPT with ArF water-immersion lithography. If the goal is not met, either due to realistic resist sensitivity or source power, a fivefold smaller throughput is likely. It drives the total cost per layer to 3× that of optical DPT. With multiple electron beams meeting a similar throughput-to-tool-cost ratio as the EUV goal, a cost/layer of 80% against DPT is obtained. Further sharing of the datapath within the MEB cluster could bring the tool cost down by 40% so that overall cost/layer might drop to 60% of DPT.

The total electrical power required in a Gigafactory just for lithography will no longer be negligible with next-generation-lithography candidates. The optical immersion scanner as a reference requires 219 kW for a 100-wph DPT tool. A total of 59 scanners to support the 130,000 wafers/month factory, would consume 8.6% of the factory power target of 150 MW.

We saw in Sect. 8.2 that the energy efficiency of the EUV system is extremely small. Even the supplier estimate results in a power of 750 kW/tool or 44 MW for 59 tools, corresponding to 30% of the total fab power. If the resist-sensitivity target of 10 mJ/cm^2 is missed by a factor of 3 and the actual system efficiency from source to wafer by almost another factor of 10 from the most pessimistic estimate, then the EUV lithography power in the fab rises to unpractical levels of between 141 MW and 974 MW. The cost to power these tools adds to the already high cost estimate shown in Table 8.2. It is also disastrous in terms of carbon footprint.

The overall-power estimate for MEB maskless systems arrives at a total power of 370 kW for the 10-column cluster with 130,000 beams, of which 250 kW is attributed to the datapath. This would result in 22 MW for the fab, corresponding to 16% of the total fab power, also high compared to that of DPT optical tools. It is plausible to predict that the power for the datapaths could be reduced to 50 kW with shared data, resulting in a total of 173 kW for each cluster and a total lithography power in the plant of 6.8% or less.

References

1. Lin, B.J.: Limits of optical lithography and status of EUV. In: IEEE International Electron Devices Meeting (IEDM), 2009
2. Lin, B.J.: Sober view on extreme ultraviolet lithography. J. Microlith. Microfab. Microsyst. **5**, 033005 (2006)
3. www.mapperlithography.com
4. Lin, B.J.: Optical lithography: here is Why. SPIE, Bellingham (2010)
5. Lin, B.J.: NGL comparable to 193-nm lithography in cost, footprint, and power consumption. Microelectron. Eng. **86**, 442 (2009)
6. Klein, C., Platzgummer, E., Loeschner, H., Gross, G.: PML2: the mask-less multi-beam solution for the 32 nm node and beyond. SEMATECH 2008 Litho Forum. www.sematech.org/meetings/ archives/litho/8352/index.htm (2008). Accessed 14 May 2008
7. Lin, B.J.: Litho/mask strategies for 32-nm half-pitch and beyond: using established and adventurous tools/technologies to improve cost and imaging performance. Proc. SPIE **7379**, 1 (2009)

Chapter 9
Power-Efficient Design Challenges

Barry Pangrle

Abstract Design teams find themselves facing decreasing power budgets while simultaneously the products that they design continue to require the integration of increasingly complex levels of functionality. The market place (driven by consumer preferences) and new regulations and guidelines on energy efficiency and environmental impact are the key drivers. This in turn has generated new approaches in all IC and electronic system design domains from the architecture to the physical layout of ICs, to design-for-test, as well as for design verification to insure that the design implementation actually meets the intended requirements and specifications.

This chapter covers key aspects of these forces from a technological and market perspective that are driving designers to produce more energy-efficient products. Observations by significant industry leaders from AMD, ARM, IBM, Intel, nVidia and TSMC are cited, and the emerging techniques and technologies used to address these issues now and into the future are explored.

Topic areas include:

- System level: Architectural analysis and transaction-level modeling. How architectural decisions can dramatically reduce the design power and the importance of modeling hardware and software together.
- IC (Chip) level: The impact of creating on-chip power domains for selectively turning power off and/or multi-voltage operation on: (1) chip verification, (2) multi-corner multi-mode analysis during placement and routing of logic cells and (3) changes to design-for-test, all in order to accommodate for power-gating and multi-voltage control logic, retention registers, isolation cells and level shifters needed to implement these power saving techniques.

B. Pangrle (✉)
Mentor Graphics, 46871 Bayside Parkway Fremont, CA 94538, USA
e-mail: barry_pangrle@mentor.com

B. Hoefflinger (ed.), *CHIPS 2020*, The Frontiers Collection,
DOI 10.1007/978-3-642-23096-7_9, © Springer-Verlag Berlin Heidelberg 2012

– Process level: The disappearing impact of body-bias techniques on leakage control and why new approaches like High-K Metal Gate (HKMG) technology help but don't eliminate power issues.

Power-efficient design is impacting the way chip designers work today, and this chapter focuses on where the most significant gains can be realized and why power-efficiency requirements will continue to challenge designers into the future. Despite new process technologies, the future will continue to rely on innovative design approaches.

9.1 Introduction

Design teams find themselves facing decreasingly tighter power budgets while, simultaneously, the products that they are designing continue to require increasingly complex levels of integration and functionality. Consumer preferences, new regulations and guidelines on energy efficiency and environmental impact are key drivers in the market place. This in turn has generated new approaches in all IC and electronic-system-design domains from the architecture to the physical layout of ICs, to design-for-test, as well as for verification to insure that the design implementation actually meets the intended requirements and specifications.

This chapter covers key aspects of these forces from the technological and market perspectives that are driving designers to produce more energy-efficient products. The observations of key industry leaders from AMD, ARM, IBM, Intel, nVidia, and TSMC are cited, and the emerging techniques and technologies used to address these issues now and projected into the future are explored.

Product power efficiency significantly impacts the consumer-electronics industry. It is a major driver in terms of consumers' requirements and preferences for mobile and portable handheld devices. As far back as June 18, 2007, when Apple Inc. announced that the battery-life runtimes increased for its iPhone, Apple stock jumped 3.81% that day (As a reference comparison, the NASDAQ composite index closed down a fraction of a percent that same day). Consumers' preference for lower power devices has definitely commanded the attention of the chief officers that run large corporations producing consumer-electronic products.

The so-called "green" movement to create environmentally friendly products also has an impact on the industry and extends well beyond small handheld devices to computers, household appliances, and other electronics. The U.S. Environmental Protection Agency's (EPA) Energy Star program created in 1992 is one such example. The agency sets up requirements for Energy Star certification that allows companies to place the Energy Star logo onto their products. This program has gained consumer mind-share, and customers who want to save energy look for this label on products when comparison shopping. This program has proven so successful that it has also been adopted in Australia, Canada, Japan, New Zealand, Taiwan, and the European Union.

9.2 Unique Challenges

For the designers of electronic devices, the task of reducing power has introduced new challenges. Traditionally for CMOS designs, most of the logic could be designed under the assumption that the power and ground supply lines on the chip were constants. The only variations were mainly caused by coupling noise and voltage drop due to the inherent resistance of the supply and ground lines. In order to squeeze as much energy efficiency out of their designs as possible, designers now have to design logic blocks where the supply voltage may be set for multiple operating points and in some instances totally turned off altogether. This brings on a whole new set of challenges that did not exist under the old constant-supply-source design style.

When blocks of a design operate at different voltages, it is necessary to check that the signals that are used to communicate between these blocks are driven at appropriate voltage levels for correct operation. For example, if one block (Block A) has its supply voltage (V_{dda}) set to 0.8 V and it is sending a signal to another block (Block B) that has its supply voltage (V_{ddb}) set to 1.25 V, it is quite possible that a signal originating from A and arriving in B may not be able to send a logical '1' value that is recognizable by B. Since 0.8 V is only 64% of V_{ddb}, it is actually quite likely that B will see this as an "unknown" value, that is, not being a valid '1' or '0'. This can lead to major problems with the functionality of the design and will be discussed in more detail later in this chapter. The same applies to signals that cross between blocks that are powered up and powered down independently. Special precautions need to be taken so that input signals to blocks that are still powered ON are not just left floating. For blocks that are completely powered down, the designer also has to answer the question of what is the intended state of the block when it is powered back ON. Is it sufficient to just reset the registers in the block or should the original state be restored? Perhaps only part of the state needs to be restored. These and other questions need answers in order to produce a working design.

Joe Macri, CTO at Advanced Micro Devices, has said, "Platform power is as important as core silicon power" [1]. For IC engineers, it is important to understand that the chip power is only part of the story. At the end of the day, the chip has to go into a package, and that package has to go onto a board. The chip probably also has some software running on it. It is the power characteristics of the end product that the consumer is going to experience. Placing all types of hooks for power management into the IC is of little value if the software running on it doesn't make good use of those capabilities.

The diagram in Fig. 9.1 shows the interdependent relationship between system-level design, chip design, package design, and PCB design. The Open SystemC Initiative (OSCI) has released version 2.0 of its Transaction Level Modeling specification (referred to as TLM 2.0). Transaction level models are a way to let system designers create high-level models in order to explore the design space and perform power, performance, and area (PPA) analysis. The ability to separate the functionality from the power and timing allows system designers to focus on the models for

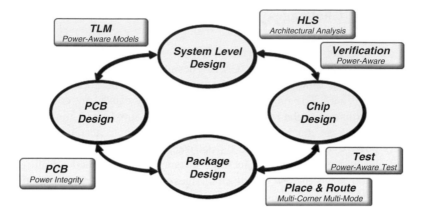

Fig. 9.1 Power impacts the whole system. *TLM* transaction level modeling, *PCB* printed circuit board, *HLS* high-level synthesis

the desired behavior and then perform successive refinements to tie the power and timing characteristics to specific implementations of that behavior. Using high-level synthesis (HLS) tools is one way to generate register transfer level (RTL) implementations from the behavioral models written in C or C++. These tools appear promising in terms of providing a way to quickly generate multiple implementations in order to search the design space for the best possible solutions.

Power-management techniques such as power-gating and dynamic voltage and frequency scaling (DVFS) that turn off the power to portions of the design and change the voltage level on the supply rails impose new requirements on the verification process. Power-aware verification takes into account the fact that logic cannot produce valid output unless it is powered ON. As previously mentioned, checks are also needed to insure that signals crossing power domains, i.e., portions of the chip that are running on different supply rails, have the appropriate voltage levels to insure that good logical '0' and '1' values are communicated. Likewise for test structures that create scan chains out of the registers for shifting in test vectors and then shifting out the results, it is important to understand the power domain structure on the chip to insure the test integrity.

Power management schemes that use multiple voltage levels increase the number of scenarios under which the design has to meet timing constraints. The number of modes of operation for power-efficient chips significantly increases the complexity of closing timing in a manner that the chip meets timing for all possible modes of operations and across all of the process, voltage, and temperature corners.

Power integrity is an important issue at the board level. Many ICs use off-chip voltage regulators placed on the printed circuit board (PCB). As the number of voltage levels increases per chip, the board complexity also increases. Another important factor from an integrity standpoint is the amount of current that needs to be delivered. Since dynamic power is proportional to the supply voltage squared, reducing supply voltages has been an important weapon in the designer's arsenal for fighting power. Back in the days of 0.5 μm CMOS when the supply voltage was

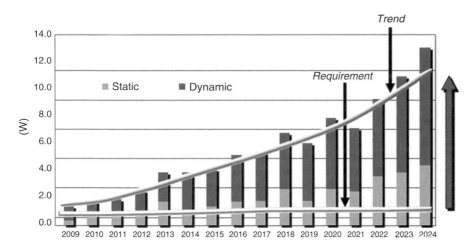

Fig. 9.2 ITRS 2009 SoC consumer portable power trend

typically set at 5.0 V, a 25 W chip would require 5 A of current. Today, a 25 W chip may be largely supplied by 1.0 V or less, thus requiring 25 A of current! For large-scale microprocessors and graphics processors, these current levels can exceed 100 A. Coupled with the speed of signals and other noise sources, maintaining PCB power integrity is challenging.

People often wonder if all of this additional effort for managing power is only a temporary phase that the industry is going through and whether some technological breakthrough will allow the return to a less complex design style. It seems inevitable that eventually some technology will come along and supplant CMOS. Unfortunately though, at the time of writing, it doesn't look like it will happen very soon. Figure 9.2 shows a chart from the 2009 International Technology Roadmap for Semiconductors (ITRS) [2] indicating the expected power trend for system-on-chip (SoC) consumer-portable ICs. The bars on the chart indicate static and dynamic power. Static power is the amount of power that is used just to have the device turned on without any useful work being performed. Typically in CMOS designs, this is due to current "leaking" in the transistors used as switches. In an ideal switch, the device is either "OFF" or "ON". When the ideal switch is ON, current can flow freely without any resistance and when it is OFF, no current flows at all. The real-world transistor implementation differs from the ideal, and one of the differences is that, when the transistor switch is OFF, some current can still get through the switch. As threshold voltages have decreased and as gate insulators have become thinner, the switches have become "*leakier*". Dynamic power refers to the power used to perform useful computation and work. For CMOS logic, this involves the transistor switches turning OFF and ON to perform the intended logic computations. CMOS found early success in low power applications, e.g., digital watches, because the vast majority of the power was used only for the dynamic switching portion of the logic computations. Ideally, as the transistor switches turn ON and OFF, current flows to charge up a given capacitive load or discharge it, but does not flow on a direct path from power to ground. This is accomplished by creating *complementary* circuits that only allow either the path to power or the path to ground to be activated for any given logic output, hence the name complementary metal oxide semiconductor (CMOS). The dynamic power for CMOS circuitry can be approximated by using a capacitive charge/discharge model, yielding the following common equation: $P_{dynamic} \approx \alpha f C V^2$, where α represents the switching activity factor. A major point of importance here is the quadratic impact of voltage on dynamic power.

Clearly, the trend in Fig. 9.2 is heading in the wrong direction, although you might detect brief periods where there are slight dips in the bars attributable to the projected use of technologies such as fully depleted silicon on insulator (FDSOI) or multi-gated structures such as Fin-FETs. While I won't hazard to make predictions 10+ years out in an industry as dynamic as the semiconductor industry, it is pretty clear that the expected trend going forward is for the power management problem to increasingly become even more challenging.

The Chief Technology Officer at ARM, Mike Muller, has presented a number of talks where he has mentioned "dark silicon" [3]. He points out that, by the year 2020, we could be using 11-nm process technology with 16 times the density of 45-nm technology and the capability of increasing clock frequencies by a factor of 2.4. Since dynamic power is proportional to clock frequency, roughly speaking we would need to reduce the power on a per device basis by approximately a factor of 38 (16 × 2.4). Unfortunately, Mr. Muller also points out that the expectation is that the 11-nm devices' energy usage will only drop to a third of the 45-nm level and is quoted as saying that this will leave engineers "with a power budget so pinched they may be able to activate only 9% of those transistors" [3]. Clearly, there is a need for more innovation in terms of getting the most capability out of the underlying capacity of the process technologies.

9.3 When Power Does Not Scale

The electronics industry has a history of pushing technologies until the limits from an energy-efficiency standpoint are reached, starting with vacuum tubes through bipolar-transistor emitter-coupled logic (ECL), n-channel metal oxide semiconductor (NMOS) to complementary metal oxide semiconductor (CMOS) (as a side note, I find it interesting that the industry stayed with the term MOS even after moving to polysilicon self-aligned gate processes and now, with the advantages of high-k metal gate [HKMG] processes, MOS is truly MOS again). In each case, the technologies were initially limited by the levels of integration. As the technologies advanced though, the ability to create higher-density designs exceeded the reduction in per-device energy usage. Eventually, power becomes a primary driver from a design perspective. If you want to go faster, you need to find ways to become more power efficient. All other things being equal, if one design uses less power than another, then the clock frequency on the lower power design can be

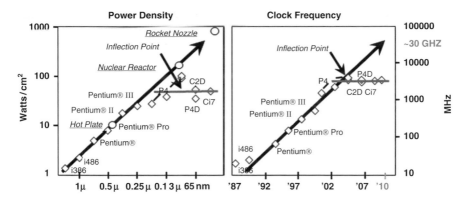

Fig. 9.3 Inflection point for power density and clock frequency

cranked up until the power matches the power of the higher-power design, thus reaping the benefits of the higher clock rate. Power budgets limit the amount of functionality that is integrated into a single chip. The more energy efficient a chip is, the more functionality that can be integrated into it. This is very relevant in light of Mike Muller's comments about activating only 9% of the transistors.

In many ways, the impact of power on CMOS design has been like the impact of driving along at 200 mph and hitting a brick wall, but that doesn't mean that this wasn't foreseen. The charts in Fig. 9.3 are an extension of two charts shown by Fred Pollack, an Intel Fellow, at Micro32 in 1999 [4]. His slides went up to the Pentium III processor and I have added more recent processors to the charts to illustrate what has happened since then. At the inflection point indicated in Fig. 9.3, the flattening of the power density curves corresponds to an equal flattening in the clock frequency curves. In 2002, then Intel CTO Pat Gelsinger predicted that by 2010 x86 processors would be running at 30 GHz clock frequencies [5]. In all fairness to the former Intel CTO, it is pretty clear to see how he arrived at that prediction. For roughly 15 years, processor designs were heading straight up the line on that log chart. Continuing that line to 2010 implies a clock frequency of ~30 GHz.

The brick-wall incident appears to take place around 2004 and not coincidentally at the 90 nm technology node. What happened? Going back to 0.5 μm again when the supply voltage was typically set at 5.0 V, the threshold voltage for NMOS transistors was in the neighborhood of 1.25 V. Leakage current for these devices was relatively negligible in the overall scheme of energy consumption. As the technology scaled down further to 0.35 μm, the supply voltage scaled down to 3.5 V, at 0.25 μm to 2.5 V, at 0.18 μm to 1.8 V. This scaling continued down to roughly 1 V at 100 nm, and the 100 nm node is a pretty reasonable marker for where the inflection point occurs. A general rule of thumb for performance reasons was to set the threshold voltage at approximately one fourth of the supply voltage so at a supply voltage of 1.0 V the threshold voltage would be somewhere in the vicinity of 0.25 V. This drop in threshold voltage from ~1.25 V at 0.5 μm to ~0.25 V at 100 nm has a huge impact on the leakage current of the transistors. In short, in order to limit subthreshold leakage current in the transistors, the threshold voltages have been held relatively constant and therefore so have the supply voltages at around roughly 1.0 V. Remembering that the dynamic power is proportional to the voltage squared, it becomes even more apparent why the inability to scale down the supply voltage at the same rate as the feature size is so important to the amount of energy consumed at successive technology nodes.

The graph on the right in Fig. 9.3 shows the sudden halt to increasing clock speeds. The industry had become accustomed to the exponential increase, indicated by the straight line on the log graph, in clock speeds over time. As was mentioned previously, dynamic power is proportional to the clock frequency at a given voltage. One trick that people who "overclock" processors use is to increase the supply voltage along with adding exotic cooling to remove all of the thermally dissipated energy. As was also mentioned earlier, dynamic power for CMOS circuits is proportional to the supply voltage squared. In effect, if it is necessary to increase the voltage in order to run at higher clock frequencies, then it can be said that the power increases proportional to the cube of the clock frequency. This approximation is at least often true for relevant operating regions, and you may see this relationship mentioned in other literature. It is no small wonder then that, after the release of the Pentium 4 570 J at 3.8 GHz in November 2004, there has been a long string of years without the introduction of $x86$ processors at a higher clock frequency. In fact, the clock rates for the subsequent generation of processors actually fell.

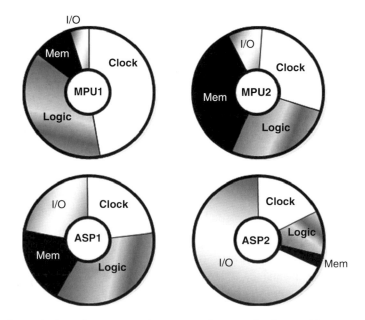

Fig. 9.4 Power budgets for two general-purpose and two application-specific processors (With kind permission from Springer Science+Business, © 2009 [6])

For processor designs, a significant portion of the energy consumed by the circuitry is actually used for clocking. Figure 9.4 shows power-budget pie charts for two general-purpose (microprocessor unit, MPU) processors and two application-specific (ASP) processors. This diagram appears in Slide 1.30 on p.20 in [6]. Across these four designs, the power budget for clocks is between approximately 20% to almost 50% with the values being higher for the MPUs than for the ASPs. For instance, a special-purpose processor designed for networking applications could easily have a higher percentage of its power budget allocated for I/O in order to deal with the transfer of large amounts of data. One thing that's interesting to note about all four of the processors shown is that the power budget for the clocks rivals the power budget for the logic.

A number of issues become more challenging as clock frequencies are increased. First, the constraints on skew and jitter are tighter, and this often leads to longer insertion delays and longer buffer chains that consume more power. Second, as the skew decreases, more registers are triggered in a narrower window of time causing more activity to occur virtually simultaneously. Higher activity levels create a need for more current to sustain the logic circuitry, and this can lead to noise and IR-drop issues. Third, higher current levels can exacerbate local hot spots and together with higher current densities can also adversely impact reliability. From a power stand-point, it's very important to pay attention to clocking schemes and to use efficient clock-tree synthesis tools and methods in order to generate energy-efficient designs.

9.4 Advanced Technology Adoption Drives Power-Aware Design

The halt in voltage scaling at about 100 nm creates significant challenges for designers designing with sub-100 nm technologies. Initially, this impacted only the design teams that were working on the very leading-edge technologies. As of the time of this writing, 32 nm processors are available off-the-shelf, and designers are working on 22-nm versions. The 22-nm node is four process generations beyond 90 nm. This implies that a lot of design teams are already designing at 90 nm and below. Often the perception is that the adoption of new technology nodes has slowed. This perception though depends upon the observer's point-of-view. If you take the viewpoint from silicon foundries, the demand for new technology nodes is as strong as ever. TSMC reported for Q2 2011 that 64% of their revenue was due to wafers at 90 nm and below and that 90 nm accounted for 9%, 65 nm for 29% and 40 nm for 26% [7]. From this data, revenue from the 40 nm node is on the verge of eclipsing 65 nm and 65 nm is already falling from a peak of 32% of revenue in Q1 2011. Since Q3 2009, the cumulative 90 nm and under percentage has gone from 53% to 64% of wafer revenue while total revenue has also increased significantly. Clearly from TSMC's viewpoint, there's been no slow down in the demand for new technology nodes. If the data is looked at from the perspective of new design starts, then a different picture is formed. Design starts in newer technology nodes appear to be decreasing. Certainly, the costs for designing in newer nodes are increasing, and this puts pressure on the number of projects that are economically viable at these nodes, so what we are really experiencing are fewer designs with higher volumes moving first to the new nodes. Another way that designers can make use of newer technology nodes, is to design for Field-Programmable Gate Arrays (FPGAs) that are implemented in these newer technology nodes. To the foundries though, an FPGA is one design. Intuitively, this would all seem to make sense, and it looks similar to other industries as they mature and start to consolidate.

The added complexity of elaborate power-management schemes is part of that additional cost of designing in more advanced technology nodes. One of the real changes that have occurred is how prevalent power issues are across the design spectrum once the 100 nm barrier is crossed. Originally, power was primarily a concern for designers that were designing chips for portable and battery operated devices. Before 130 nm, most other design teams were primarily concerned about meeting timing (performance) and area constraints. Power was pretty far down the list and was often left as an optimization exercise for subsequent versions of the design. In other words, the priorities were get the chip out meeting timing and area and then worry about the power later. This approach hit a number of teams very hard at the inflection point shown in Fig. 9.3. Chips have to go into packages and onto boards. As mentioned early, power is a platform issue and not just a chip issue. If a board has a 25 W power budget and a chip comes in at say 22 W, unless that chip is the only chip going onto the board, that's probably a problem. If a chip misses the maximum power limits for its intended package and the increased cost of the next available package that can accommodate the additional thermal power dissipation makes the combined part cost too expensive for its target market, that's a problem too. Chip area and yield fade a bit in importance when the cost of the packaging becomes significantly higher than the cost of the silicon. If the only way to get into a lower-cost package is to reduce the power consumption of the chip, suddenly power shoots up the list of design priorities. This became a concern for all types of designs and not just portable, battery operated applications.

The change in priorities has caused designers to re-think a number of strategies. Back in 1996 when Power Compiler ® was released, it had a feature for inserting clock-gating into designs. The prevailing thinking at that time was that adding any logic onto the clock tree was generally a "bad" idea. As might be guessed from the pie charts shown in Fig. 9.4, clock-gating is now a generally accepted design practice and can be a powerful technique for reducing dynamic power. Not using clock-gating often means that the design team is missing out on significant power savings. In response to increased leakage power that typically started to appear around 130 nm, foundries started to offer multiple cell libraries at the same technology node where each library corresponded to cells that were built using a threshold set for that library. As an example, one library might use transistors with a 0.35 V threshold and the other 0.25 V. The cells in the higher threshold voltage library leak less than the cells in the lower threshold library but at the expense of having slower timing characteristics. Basically, the idea is to use as many high-threshold cells as possible and only insert the faster, leakier low-threshold cells where necessary in order to meet timing. There could easily be a factor of 30 in the leakage between the two libraries implying that, if only 3% of the cells needed to be low-threshold cells, the overall leakage would still be roughly twice as high as a design only using high-threshold cells. Nonetheless, twice as high is still much better than 30 times as high. The technique could allow designers to meet timing and reduce leakage by an order of magnitude compared to using only low-threshold cells. High-threshold libraries are often referred to as "low-power" libraries. One needs to be somewhat careful about reading too much into that though. The high-threshold libraries are low power from a leakage standpoint but, from a performance and energy-efficiency standpoint, lower-threshold libraries will have better power characteristics for some designs. Remembering the "overclockers" trick of raising the voltage to get higher clock speeds, if in using a higher-threshold library it is necessary to raise the supply voltage to meet the same timing that could be achieved with a lower voltage using a low-threshold library, the lower-threshold library may win in terms of overall power consumption.

In response to the technological challenges of increased static and dynamic power in the smaller technology nodes, designers have turned to techniques for varying the supply voltages. One effective way to reduce static power is to turn off circuitry when it's not in use. This is often referred to as "*power gating*". At first, it may sound strange that there would be much circuitry not being used but for applications like mobile phones, most of the time, most of the designed functionality is not being used. For example, advanced mobile phones have features for making phone calls, playing music, watching videos plus numerous other personal computer-like applications. In reality though, the phone will likely spend most of its time in a mode that occasionally checks for incoming phone calls and perhaps nothing more. By cutting the power off to all of the circuitry that is not currently in use, designers have dramatically increased the standby lifetimes for these devices. At a finer level of granularity, even being able to turn off circuitry for time periods on the order of milliseconds can have a beneficial impact on the reduction of static power. For reducing dynamic power, the quadratic impact of voltage is the key. The ability to tune the supply voltage to match the necessary performance level enables significant potential dynamic-power savings. Portions of the design that use dynamic frequency scaling (DFS), i.e., are designed to run at different clock frequencies, are also good potential candidates for dynamic voltage scaling (DVS). By pairing up each of the operating frequencies with its own operating voltage, each frequency mode runs at a more efficient power level. The combination of the two is often referred to as dynamic voltage and frequency scaling (DVFS). Traditionally, power and ground had been assumed to be non-varying constants for most CMOS logic designs. The use of these power-gating and dynamic voltage-scaling schemes though makes those assumptions no longer valid and add additional complexities to the design and verification process.

Another technique for reducing static power is to dynamically change the threshold voltages of the transistors. Figure 9.5 includes a graph showing how the

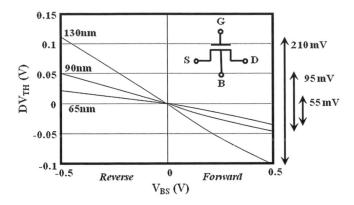

Fig. 9.5 Diminishing impact of body bias on the threshold voltage (With kind permission from Springer Science+Business, © 2009 [6])

threshold voltage varies with changes in the substrate bias along with a simple NMOS transistor diagram. This graph appears in Slide 2.13 on p.31 in [6]. MOS transistors are often thought about as three-terminal devices: source, gate, and drain, but the substrate is a fourth terminal that can also be controlled. Some designers have made use of this fourth terminal to adjust threshold and therefore correspondingly the leakage levels of transistors in their designs. It has the advantage that the savings can be had without modifying the power-supply lines for V_{dd} or ground lines in the design. Unfortunately, as the graph in Fig. 9.5 shows, this technique becomes less sensitive to changes in the substrate bias as the transistor sizes continue to shrink.

9.5 Multi-core Processors

Microprocessor clock frequencies have hit a wall but the performance capabilities have not. Designers continue to invent new ways to improve performance even in the face of hard limits on clock frequencies. The biggest factor in performance improvements since the halt in clock frequencies has been the incorporation of multiple cores onto the same piece of silicon. The thinking goes that if the performance can't be doubled by doubling the clock frequency then perhaps doubling the number of processor cores can do the trick. Figure 9.6 shows a progression in AMD Opteron designs since 2003, when the first single-core Opteron was released. I find it striking that we see the first dual-core processor at 90 nm just as the 100 nm threshold was crossed. Each successive node has led to more cores on a single piece of silicon, and it appears that this trend will continue for the foreseeable future with major processor manufacturers having already announced plans for designs with many more cores.

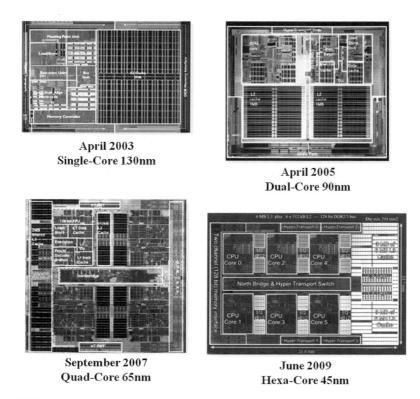

April 2003
Single-Core 130nm

April 2005
Dual-Core 90nm

September 2007
Quad-Core 65nm

June 2009
Hexa-Core 45nm

Fig. 9.6 Four generations of AMD Opteron processors (Photos courtesy of AMD)

Each core presents a unique opportunity from a power-savings standpoint. It is its own unit that can be individually controlled depending upon the processing needs at any given moment. From a logical viewpoint, each core can operate in a separate domain with its own supply voltage and clock frequency, which actually translates reasonably well into a physical implementation. In the server market, where some of these processors compete, energy efficiency is an important aspect of the design. Corporations with large server farms, such as Google or Amazon, have to provision these facilities to handle peak demand loads so that operations always run smoothly. These facilities may only experience peak loads for a relatively small portion of the time that they are in service and the power bills from running these facilities are a significant expense. Companies do not want to have to pay for peak-load power in off-peak hours. In order for server CPUs to be competitive, these CPUs need to scale back on energy usage, especially in nonpeak demand modes. Turning cores OFF as they are not needed is one way to accomplish this, as well as using DVFS schemes to reduce the power of selected cores that do not need to be running at top speed. It is interesting to note that both Intel and AMD have introduced capabilities to "overclock" individual processor cores, called Turbo Boost and Turbo Core respectively, when all of the cores on the chip are not simultaneously running. Since the part is designed for a specific thermal capability, if all of the cores are not running then there is some extra headroom for the cores that are. This extra margin can be utilized by bumping up the voltage and frequency of the running cores to improve their performance. These types of features can make benchmarking and comparisons of different parts quite challenging but definitely provide additional performance benefits for customers.

Multiple cores increase the level of design complexity. The number of different operating states or modes increases exponentially with the number of cores, if the cores are truly independent of each other in terms of power-up and -down conditions and their voltage and frequency settings. Even within a single core, it is possible to have sub-blocks that operate in different power and frequency domains. From an implementation standpoint, the challenge now becomes meeting all of the constraints across multiple modes simultaneously. Typically, each mode of operation has to be designed to function correctly across a range of process, voltage, and temperature "corners" to accommodate for variations in processing, voltage drop, and temperatures. Figure 9.7 [8] shows an example of a cell phone SoC that has more than 21 modes and corners that need to simultaneously meet design constraints.

Figure 9.8 depicts a diagram of a simple design with a core-power domain along with two other "island" domains that operate independently. The table in Fig. 9.8 illustrates how quickly the possible mode–corner combinations grow as more islands are added to the design. This has created a need in the marketplace for implementation tools that can efficiently design and evaluate across multiple modes and corners in order to close timing and other constraints.

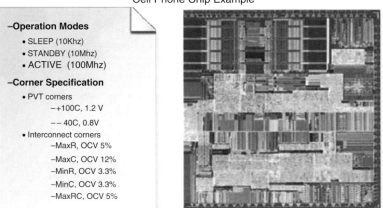

Cell Phone Chip Example

More than 21 mode/corner scenarios

Fig. 9.7 Multi-corner and multi-mode design example

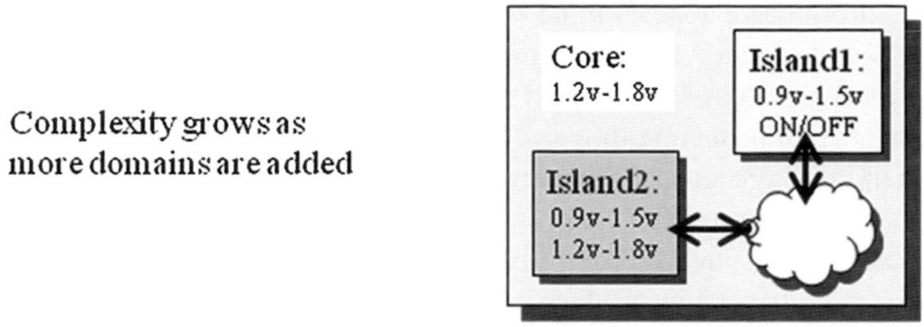

Complexity grows as
more domains are added

Core:
1.2v-1.8v

Island1:
0.9v-1.5v
ON/OFF

Island2:
0.9v-1.5v
1.2v-1.8v

	Single Core Design			Core + 1 Island				Core + 2 Islands				
	Lib	Core	RC	Lib	Core	Vdd1	RC	Lib	Core	Vdd1	Vdd2	RC
Setup1	Max	1.2	Max	Max	1.2	0.9	Max	Max	1.2	0.9	0.9	Max
Setup2	Max	1.2	Min	Max	1.2	0.9	Min	Max	1.2	0.9	0.9	Min
Hold1	Min	1.8	Min	Min	1.8	1.5	Min	Min	1.8	1.5	1.5	Min
Hold2	Min	1.8	Max	Min	1.8	1.5	Max	Min	1.8	1.5	1.5	Max
Setup1	—	—	—	Max	1.2	0	Max	Max	1.2	0	1.2	Max
Setup2	—	—	—	Max	1.2	0	Min	Max	1.2	0	1.2	Min
Hold1	—	—	—	Min	1.8	0	Min	Min	1.8	0	1.8	Min
Hold2	—	—	—	Min	1.8	0	Max	Min	1.8	0	1.8	Max
Setup1	—	—	—	—	—	—	—	Max	1.2	0.9	1.2	Max
Setup2	—	—	—	—	—	—	—	Max	1.2	0.9	1.2	Min
Hold1	—	—	—	—	—	—	—	Min	1.8	1.5	1.8	Min
Hold2	—	—	—	—	—	—	—	Min	1.8	1.5	1.8	Max
Setup1	—	—	—	—	—	—	—	Max	1.2	0	0.9	Max
Setup2	—	—	—	—	—	—	—	Max	1.2	0	0.9	Min
Hold1	—	—	—	—	—	—	—	Min	1.8	0	1.5	Min
Hold2	—	—	—	—	—	—	—	Min	1.8	0	1.5	Max

Fig. 9.8 Example illustrating the growth in modes and corners with design complexity

9.6 Electronic Design Automation Waves

At any given point in time, leading-edge CMOS technologies hold the promise of certain underlying capabilities that can only begin to be approached by the available tools and methodologies. When changes from one technology node to the next are relatively minor, the gap between available capability and serviceable capability tends to shrink, and the opposite happens when the changes are larger. The gaps that open when there are large changes, like those created for power management, present opportunities in the marketplace for new tools and methodologies to appear and once again close those gaps. For example, in the place-and route-market, gaps opened and were filled around moves towards SoC and timing-driven design with the latest now appearing around multi-corner, multi-mode (MCMM) designs.

The next big wave appears to be shaping up around design implementation. For many years, designers used schematic capture as a method for creating logic designs. Starting around the early 1990s, design teams started to transition from schematic capture to text-based design creation using hardware description languages (HDLs) such as VHDL and Verilog. This need for a more efficient design methodology to keep pace with the increasing capability of the underlying CMOS technology, often referred to as "Moore's law", created an opportunity for the successful entry of new products and methodologies into the marketplace. If, for arguments sake, we say that this transition to HDL-based design happened at 0.5 μm, then at 32 nm there have been seven generations of process-technology nodes as defined by the ITRS and a corresponding increase in capabilities of over two orders of magnitude. Using another rule of thumb, that the industry increases the level of integration roughly tenfold every 6–7 years, then you also reach the same conclusion. Pundits have been claiming for a number of years that the industry is on the verge of moving towards electronic system level (ESL) design and high-level synthesis (HLS) tools, and it appears that the pressure for higher designer efficiency to keep pace with the higher levels of available integration are making that happen.

The move to higher-level design tools and methodologies is also very relevant from a power-aware design and verification standpoint. As in typical design processes, the earliest decisions have the greatest impact on the design. At each successive stage of design refinement, options are eliminated as the design crystallizes towards its final implementation. This is also true for power optimization.

The table in Fig. 9.9 shows multiple implementations for an example inverse fast-Fourier-transform (IFFT) design and how the power, performance, and area change with each variation of the implementation [9]. This work was performed by a team at MIT using a high-level synthesis tool to quickly generate multiple designs for evaluation. There are two important concepts to note here: (1) the range of variation in power across the designs is almost an order of magnitude and (2) most

Transmitter Design (IFFT Block)	Minimum Frequency to Achieve Required Rate	Area (mm^2)	Average Power (mW)	
Comb (48 bfly4s)	1.0 MHz	4.91	3.99	3.99
Piped (48 bfly4s)	1.0 MHz	5.25	4.92	
Folded (16 bfly4s)	1.0 MHz	3.97	7.27	
Folded (8 bfly4s)	1.5 MHz	3.69	10.90	~8.6x
Folded (4 bfly4s)	3.0 MHz	2.45	14.40	
Folded (2 bfly4s)	6.0 MHz	1.84	21.10	
Folded (1 bfly4)	12.0 MHz	1.52	34.60	34.6

Fig. 9.9 IFFT example design showing power and area trade-off for constant throughput

design teams would not have the time and/or resources to evaluate this many implementations, if they had to code them all into a RTL HDL description by hand. The results back many industry expectations that high-level tradeoffs can lead to an 80% reduction in power. Further downstream optimizations account for a much smaller percentage.

The reproduction of a presentation slide shown in Fig. 9.10 [10] is from a keynote talk that Bill Dally, Sr. VP of Research and Chief Scientist at nVidia, gave at the 46th Design Automation Conference in July of 2009. During his talk, Dr. Dally stressed the need for tools for exploring power and performance tradeoffs at an architectural level. With the gap widening between the capabilities of the underlying silicon and designers' ability to fully make use of it, the need for better tools becomes increasingly important. Figure 9.11 is a chart from the 2009 ITRS Roadmap [2]. It predicts that in order for designers to deal with the graph shown back in Fig. 9.2 tools that enable that early architectural 80% power savings are going to be needed, and shows that by 2015 80% of the power optimization will need to happen at the ESL and HLS portion of the design. Even at the foundry level, the need for these high-level tools and flows is seen as critical. TSMC included ESL/HLS tools for the first time in 2010 as part of their reference flow in Reference Flow 11. Between the need for more productivity per designer to keep pace with the

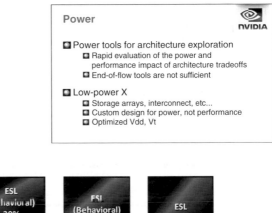

Fig. 9.10 Bill Dally's slide on the need for high-level exploration tools [10]

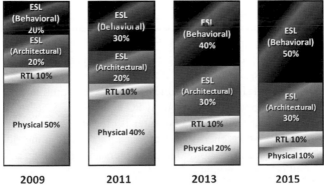

Fig. 9.11 ITRS 2009: growing importance of optimizing for power earlier in the design process

increased levels of integration and the need for better architectural analysis for power optimization, a near perfect-storm-like scenario is created for high-level tools to finally catch on in the marketplace.

9.7 Power-Aware Verification

Power-management schemes vary the V_{dd} supply-rail levels for portions of the chip, and this introduces new complexities for verification. At the register transfer level (RTL), many signals are represented by logical '1' and '0' values. The validity of those values depends on the underlying physical voltage levels. For example, let us assume that in order for the circuitry to recognize an input as a '1' or '0', the voltage level must be greater than or equal to 75% of V_{dd} or less than or equal to 25% of V_{dd} with the remaining 50% in the middle being an "unknown" area. So what happens if the core logic of the chip is running at 1.25 V and an island is put into a lower power mode and is running at 0.8 V? An output signal from the 0.8 V block at a logical '1' level will be at best at 0.8 V and probably slightly lower. If that signal is then fed as an input into the core running at 1.25 V there is a high probability of the signal not being recognized as a logical '1' value on the input. Even worse, if the signal is fed into a large buffer or inverter, it may cause both the PMOS and NMOS transistors to at least partially be "ON" simultaneously, creating a path for current to flow from V_{dd} to GND. Remember, earlier we mentioned that, ideally in CMOS designs, only the path from V_{dd} to the output or the output to GND is on at a given time. Because of the physical nature of the transistors implementing the logic, there may be brief periods of time where transistors on both sides are simultaneously conducting current across the channel and creating a path from V_{dd} to GND. This current is often called "short-circuit" or "crowbar" current and is typically lumped into the dynamic power equation since it is a function of the logic changing value. It is definitely not a desired condition. Back to our input signal, since it is at best at 0.8 V, it falls substantially below the threshold of 75% of 1.25 V or approximately 0.94 V, and won't properly be recognized as a '1'.

High-level simulators use logical representations of signals and not the actual physical voltage levels in order to increase their speed and design capacity. They work extremely well as long as the underlying assumption that logical '1' and '0' values have the proper physical voltage values associated with them. When two blocks communicating in a design run at different voltage levels, it is important that the simulator understands how these differences affect the communication between blocks. For example, suppose a hypothetical PC notebook designer puts together a system such that when the notebook computer is closed, it goes into a low power sleep mode in order to save energy and increase the run lifetime of the battery. As people will often take closed notebook computers and stick them into backpacks or briefcases with poor air circulation, the designer builds in a thermal detector to indicate if the sleeping notebook is getting too hot and should power itself down to avoid overheating. The thermal detection circuitry raises a '1' signal if it determines that the notebook is too warm. This warning signal then travels across the chip until it arrives at a part of the circuitry that is still awake to recognize it and power down the machine. What happens if the signal happens to pass through a block that has been placed into a lower-voltage mode to save power on the way to its final destination? Well, if the simulator only recognizes '1' and '0' and makes no distinction for signals crossing different voltage boundaries, everything in the simulation looks fine. Each block receives a '1' and passes it along to the next block until the '1' shows up at the final destination. What happens in the real physical world? The voltage on the signal drops below a recognizable '1' value and is never detected by the final destination, and so the notebook keeps running, possibly catastrophically.

Commercial simulators now incorporate power-aware capabilities to better reflect the physical realities. Designers use a format such as the IEEE Std 1801–2009 (also known as the Unified Power Format UPF 2.0) [11] to describe the power domains or voltage islands on the chip so that the simulator can determine where the block boundaries are and when signals are crossing them. The format has a wide range of capabilities and is architected to allow designers to overlay their power-management scheme on top of their functional descriptions. This allows the verification of the functionality of a block to occur independently of the power management, as is traditionally done, and then verify the correctness with the power management included. This enables better reuse of the design components since the power management does not have to be hard coded into the functional behavior of all of the components, and a block can then be reused in another design with a different power scheme by altering the UPF description.

Figure 9.12 shows a simulation waveform diagram for an example design going through a simulated power-down and then power-up sequence. It illustrates a

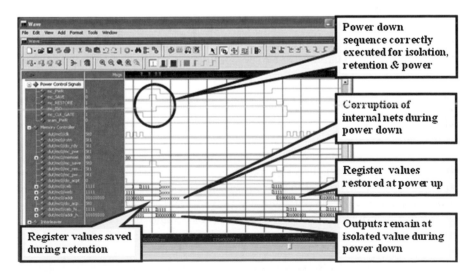

Fig. 9.12 Example waveform diagram for a simple power down/up sequence

number of conditions that a simulator has to handle in order to properly represent the design's behavior. In a typical power-down sequence, a number of actions need to occur before the power is shut off to the block. First, it is necessary to determine whether the state of the block needs to be saved. If it does, one popular way to save the state is to use so-called retention registers that have a separate low-leakage latch that stays powered ON when the rest of the block is powered OFF. There are numerous ways to implement retention registers. Some use separate "save" and "restore" signals, others share one signal for this capability. Some use the clock to save and restore values, others do not. For an RTL simulator to properly model this behavior, a description of the intended retention register's behavior must be supplied. In the example shown in Fig. 9.12, the first step in the process is to quiet the clocks by setting the clock enable signal low. The next step is to save the value of the registers and then isolate the output signals. Once these steps are completed, the voltage to the block can be turned off by setting the V_{dd} enable signal low. At this point, the V_{dd}-on signal goes low to indicate that the block is powered OFF. The block can now stay in this powered OFF state until it is ready to be powered ON again. Some items to note here are: the proper sequencing of the signals, the registered output values being saved and properly isolated (in this case to '0'), and the internal nets of the powered-down block going to corrupted "unknown" values. If any of the internal nets are somehow escaping and driving powered logic gates, the "unknown" values should propagate through the simulation and raise a red flag, indicating that something is wrong. Assertions could also be added to the simulation to provide additional checks for these conditions. To bring the block back ON, the process is reversed and V_{dd}-enable is set high to allow the voltage to again be supplied to the block. Once the voltage level has reached its proper operating point, V_{dd}-on can go high, indicating that the block is once again fully powered ON. The saved register values can now be restored and the isolation removed from the outputs of the block. The block is now back up, properly powered and in a good state so the clock-enable signal can go high, and useful activity for the block can begin again. The simulation output shows the restored values appearing in the registers, the clock starting and the isolation being removed from the outputs.

9.8 Testing

There are two main issues to address with regards to power for test: (1) creating proper test structures for the given power-management scheme (or test for low-power design) and (2) controlling the power while the chip is under test (or reducing power during test).

Automatic test-generation tools need to understand the power-architecture implementation much in the same way as the verification tools do, i.e., detect where the boundaries of different voltage regions are as well as how the voltages may vary and even be turned off for each region. Again, a standardized format, such as the IEEE Std 1801–2009, is useful for providing this kind of design information.

Given this description of the power architecture, the tools can make intelligent decisions about placing registers into scan chains used to shift in test vectors and shift out test results. In general, it is beneficial to keep scan chains contained in one domain. This eliminates concerns about level-shifter insertion on the chains or chains possibly not functioning correctly because of one domain being powered down during the test. In order for a scan chain to function correctly, it is necessary for power to be supplied to the registers in the chain.

The additional cells that are inserted for power management, such as power switches, retention registers, isolation cells, and level shifters require new tests in order to demonstrate that the devices are working properly. Retention registers not only need to function properly as a register when their power domain is powered ON but also must be able to recover a retained value correctly after going through a save, power OFF, power ON, and restore sequence. Isolation cells need to properly transmit input values when not in isolation mode and then, when in isolation mode, output the proper isolation value. Level shifters need to translate input signals to the appropriate output-voltage level so that the correct logic value is transmitted across the domain boundaries.

Managing power under test is another crucial step. For many designs, it is not uncommon for the switching-activity factor of the register to fall within the range of 15–20% per clock cycle. Designers use these activity levels to estimate the current necessary for the circuit to function properly and then size the power supply rails accordingly. Testers are expensive pieces of equipment, and the longer a chip has to sit on a tester for testing, the more expensive it is to test that chip. Tools that generate test patterns therefore try to reduce the number of patterns or vectors necessary to test the chip, and this implies increasing activity levels to try to test as much of the chip with each vector as possible. Increased activity levels imply higher current levels and higher power levels. The expectation for shifting in a random string of '1's and '0's into a scan chain is that about 50% of the registers will change value on each clock cycle. This is significantly higher than the typical 15–20% level the circuitry may see in normal operation. If the power supply rails cannot handle the extra current, the voltage may fall due to what is known as *IR*-drop. That is, the resistive nature of the supply lines along with the increased current will cause the voltage to fall along the power rail. If the voltage falls too far, then it is possible that the circuitry won't pass on valid '1' values. In situations like this, it is possible to have a chip that would operate perfectly in its intended environment but will fail on the tester. The end-effect is that "good" chips fail test and are thrown away. This is definitely not a desired outcome.

One way to reduce power under test is to not just randomly assign bit values to portions of the vector that do not impact the test. These bits are often referred to as "don't cares". Instead, if the filler values used for the don't care bits are set to produce long chains of the same value then when these bits are shifted into the registers the switching activity will be reduced. Registers will see long strings of the same value cycle after cycle and not have to switch as frequently. The tests and generated vectors can be optimized to reduce switching activity during load, capture, and unload phases of the testing.

Complex SoCs using power management operate in many different modes. Chips intended for smart mobile phones have functionality that is mutually exclusive, for instance playing an mp3 song and handling a phone call. The majority of the time, most of the functionality is OFF or in a low-power standby state waiting for an event to wake it up. When testing, it is important to test the chip in actual modes intended for operation and not to put it into modes that the device should never see in the final product, especially if those modes pull too much current. When the current levels are too high, as previously mentioned, a good chip could fail on the tester. That's bad. It is even worse though, if the excess power draw is so high that the elevated temperature actually causes the chip to fail in a destructive manner. In extreme cases, chips have been known to actually catch fire while on the tester. Power-management schemes have raised the bar for test-generation tools, and proper forethought is necessary in terms of partitioning the design and the tests to insure that all of the necessary functionality is fully covered in a manner that fits within the design constraints.

9.9 Packaging

Advances in packaging are promising to further raise the level of integration for electronic devices. Often thrown in the "more than Moore" category, the implication is that packaging allows additional levels of integration beyond the scaling that is occurring in silicon. A popular concept for packaging is the stacking of die commonly referred to as 3D packaging. It offers several potential benefits, amongst them being a reduction in the size of the delivered functionality. Two or more chips in one package uses less volume than two or more packages. Multi-chip solutions also enable better control of yields. Functionality across multiple chips is paired with known-good tested chips with less area than a corresponding one-chip solution containing all of the functionality. Semiconductor processes can also be better tailored for specific functionality, for example: logic, memory, and analog capabilities. Placing all of the functionality into one package reduces interconnection parasitics compared to a multi-package solution. Bernard Meyerson, VP/ Fellow at IBM, has said that, "about half of the dissipation in microprocessors comes from communication with external memory chips. If the chips are stacked together in 3D, communication energy cost might drop to a tenth." [12] That is a significant saving in power. As an example, if a microprocessor is using 100 W and half (or 50 W) is dissipated in memory communication then a 3D packaging scheme could potentially reduce the overall power to 55 W. That is an overall reduction of 45% and is huge in terms of the markets that these types of parts are targeted for.

Beyond the potential power savings are the thermal issues that now have to be modeled in order to insure that hotspots on one chip do not interfere with the proper operation of adjacent chips in the package. Designers are working with through-silicon vias (TSVs), edge-connect, and wire-bonding technologies to implement stacked package designs. Proper thermal modeling is important and parts that don't need large heat-sinks or fans are often more desirable from an overall packaging cost as well as consumer preference standpoint.

9.10 Boards

Power integrity for PCB designs, as well as thermal analysis, are increasingly more important. The use of multiple supply voltages on the same chip for power reduction also increases the complexity of the board design. Often the voltage regulators for the supply voltages are "off-chip" and placed onto the circuit board. The use of lower supply voltages also complicates matters. Back when 5 V for V_{dd} was common, if a chip operated at 25 W, then that implied 5 A of current. With chips now running at 1 V for V_{dd} or lower, the same 25 W power budget implies 25 A or more of current. Large processors can easily operate at over 100 W, implying the need to supply over 100 A of current. As the voltages are lowered so are the noise tolerances. More current means a larger impact due to IR-drop effects. Communications processors that have a large portion of their power budgets targeted for I/O, like the one diagrammed in Fig. 9.4 for example, are sensitive to changes in the activity levels of the I/O pins. A sudden increase in activity may require an additional 25 A of current in a matter of clock cycles. Large changes in current are limited by inductance in the current-supply paths and voltage levels drop based on $L(di/dt)$ effects.

Figure 9.13 shows a map for an example PCB design. High-frequency serial-communications links to reduce both power and pin counts on chips are more commonly used now, and they are also sensitive to the board layout and routing. Designers need tools to analyze signal integrity for the routes on the board and to make sure that the power-distribution networks (PDNs) are robust to the effects of changes in current demands and noise sources.

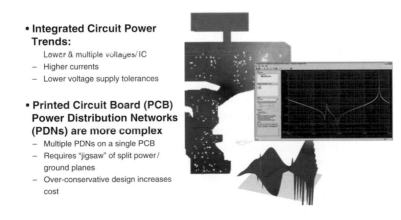

• **Integrated Circuit Power Trends:**
  Lower & multiple voltages/IC
  – Higher currents
  – Lower voltage supply tolerances

• **Printed Circuit Board (PCB) Power Distribution Networks (PDNs) are more complex**
  – Multiple PDNs on a single PCB
  – Requires "jigsaw" of split power/ground planes
  – Over-conservative design increases cost

Fig. 9.13 Example PCB distribution-network analysis results

Energy used in the circuitry is mostly dissipated as heat. Chips are designed to operate within specified temperature ranges based on their target environment. Requirements for consumer products are typically less stringent than for aircraft, military, or space applications but often are still quite challenging. For example, on the main PCB for a personal computer (often referred to as the mainboard or motherboard), there is a priority ranking for the chips going onto the board. Typically at the top of the list is the processor chip. This is in order to provide enough cooling for a high-power-consumption part. Often, other chips on the board then have to deal with the impact of the heat generated by the processor as well as other components that go into the case for the machine. With leakage power significantly increasing at 130 nm and below for current process technologies, keeping chips cooler is even more important. Leakage power increases exponentially with temperature, so as a chip uses more power and produces more heat, if that heat isn't carried away, it will cause the chip to use even more power and run even hotter. Left to itself, this leads to an unsatisfactory outcome. An important part of designing PCBs then is performing thermal analysis to determine the expected temperature ranges across the board so that each chip is guaranteed to operate within its specified thermal environment.

9.11 Summary

This chapter has presented an overview of power's impact on the design process starting from the initial system definition in terms of its architecture all the way through the implementation process to produce a complete design.

At the system level, architectural analysis and transaction-level modeling were discussed, and it was shown how architectural decisions can dramatically reduce the design power and the importance of modeling hardware and software together.

At the chip level, creating on-chip power domains for selectively turning power off and/or multi-voltage operation has an impact on the following areas: (1) chip verification, (2) multi-corner multi-mode analysis during placement and routing of logic cells, and (3) changes to design-for-test, all in order to accommodate power-gating and multi-voltage control logic, retention registers, isolation cells, and level shifters needed to implement these power-saving techniques.

At the process level: the disappearing impact of body-bias techniques on leakage control and why new approaches such as HKMG technology help but do not eliminate power issues that were covered in this chapter.

Power-efficient design is impacting the way chip designers work today, and this chapter focused on where the most significant gains can be realized and why power-efficiency requirements will continue to challenge designers into the future. Despite new process technologies, the future will continue to rely on innovative design approaches.

In the end, it will come down to economics more than technology. It won't be the technology that is the limiting factor but the cost of implementation.

Acknowledgments I thank Erich Marschner, Keith Gover, Chuck Seeley, TJ Boer, Don Kurelich, Arvind Narayanan, Jamie Metcalfe, Steve Patreras, Gwyn Sauceda, Shabtay Matalon, and Jon McDonald for their many discussions and help in compiling this information.

References

1. Magee, M.: AMD says fusion will knock Intel for six. Or five. TG Daily, 11 Nov 2009. http://www.tgdaily.com/hardware-features/44609-amd-says-fusion-will-knock-intel-for-six-or-five (2009)
2. International technology roadmap for semiconductors (ITRS). www.ITRS.net/reports.html (2009) Edition
3. Merrit, R.: Power surge could create 'dark silicon', EETimes, 22 Oct 2009. http://www.eetimes.com/news/semi/showArticle.jhtml?articleID=220900080 (2009)
4. Pollack, F.: Keynote address: new microarchitecture challenges in the coming generations of CMOS process technologies. MICRO-32. In: Proceedings of the 32nd Annual IEEE/ACM International Symposium on Microarchitecture, Haifa, 16 Nov 1999
5. Ross, A., Wood, J.: Spotlight processors: the next little thing. Comput. Power User **2**(6), 58 (2002)
6. Rabaey, J.: Low Power Design Essentials. Springer, New York (2009)
7. TSMC Q2: Quarterly Report (2011)
8. Rhines, W.C.: Keynote address: common wisdom versus reality in the electronics industry. In: DesignCon 2009, Santa Clara, 3 Feb 2009
9. Dave, N., Pellauer, M., Gerding, S., Arvind: 802.11a transmitter: a case study in microarchitectural exploration. In: Proceedings of the 4th ACM/IEEE International Conference on Formal Methods and Models for Co-design, MEMOCODE'06 (2006), Napa Valley, pp. 59–68. DOI: 10.1109/MEMCOD.2006.1695901
10. Dally, W.J.: Keynote address: the end of denial architecture and the rise of throughput computing. In: 46th IEEE/ACM Design Automation Conference, Anaheim, 29 July 2009, p. xv
11. IEEE standard for design and verification of low power integrated circuits, (a.k.a. unified power format 2.0), IEEE Standard 1801™-2009, 27 March 2009
12. Oishi, M.: 3D technology drives semiconductor evolution. Nikkei Electronics Asia, 4 June 2009. http://techon.nikkeibp.co.jp/article/HONSHI/20090527/170863/

Chapter 10
Superprocessors and Supercomputers

Peter Hans Roth, Christian Jacobi, and Kai Weber

Abstract In this article, we describe current state-of-the art processor designs, the design challenges faced by technology, and design scaling slow-down, problems with the new design paradigms and potential solutions as well as longer-term trends and requirements for future processors and systems.

With technology and design scaling slowing down, the processor industry rapidly moved from high-frequency designs to multi-core chips in order to keep delivering the traditionally expected performance improvements. However, this rapid paradigm change created a whole new set of problems for the efficient usage of these multi-core designs in large-scale systems. Systems need to satisfy an increasing demand in throughput computing while at the same time still growing single-thread performance significantly. The increase in processor cores poses severe challenges to operating system and application development in order to exploit the available parallelism. It also requires new programming models (e.g. OpenCL*). Furthermore, commercial server systems are more and more enriched with special-purpose processors because these specialty engines are able to deliver more performance within the same power envelope than general-purpose microprocessors for certain applications.

We are convinced that future processors and systems need to be designed with tight collaboration between the hardware and software community to ensure the best possible exploitation of physical resources. In the post-exponential growth era, hardware designs need to heavily invest in programmability features in addition to the traditional performance improvements.

P.H. Roth (✉) • C. Jacobi
IBM Systems & Technology Group, Technology Development, Schoenaicher Strasse 220, 71032 Boeblingen, Germany
e-mail: peharo@de.ibm.com; cjacobi@de.ibm.com

K. Weber
System z Core Verification Lead, IBM Systems & Technology Group, Technology Development, Schoenaicher Strasse 220, 71032 Boeblingen, Germany
e-mail: Kai.Weber@de.ibm.com

B. Hoefflinger (ed.), *CHIPS 2020*, The Frontiers Collection,
DOI 10.1007/978-3-642-23096-7_10, © Springer-Verlag Berlin Heidelberg 2012

10.1 Requirements in Computing, Business Intelligence, and Rich Media

Over recent decades, the world has seen computers penetrate almost every aspect of everyday life, and this trend is likely to continue. The rise of the internet, and more recently smart phones, resulted in an exponential increase of data. At the same time, companies and other organizations are striving to use the available data more intelligently in order to reduce operational risk and increase revenue and profit. The field of "business intelligence" or "business analytics" is becoming more and more important and is driving a change in the requirements on computer systems. Traditionally, data has been analyzed using database queries to generate reports. Business intelligence is using complex mathematical models to transform these descriptive reporting techniques into predictive and prescriptive models that allow forecasting of customer behavior or show optimization potential for business processes (Fig. 10.1).

In addition, the growth of bandwidth for internet connections has allowed users to create and share content types such as music and videos more easily. However, in order to help find interesting content, these music and video files need to be analyzed and tagged for search ability and converted to various formats for accessibility. This process is far more compute intensive than traditional text-based analysis and drives additional compute and bandwidth requirements into computer systems.

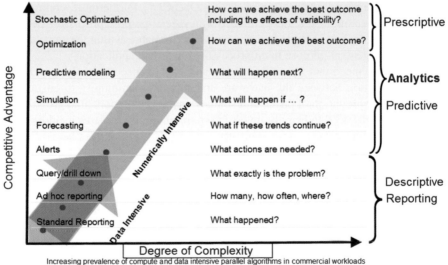

Fig. 10.1 Business analytics landscape [1]

10.1.1 State-of-the-Art of Multi-core Processors

Microprocessor design has come a long way in the last 20 years. The first POWER* processor POWER1*, for example, was released in 1990. It contained so much logic that it didn't fit onto a single chip, and had to be packaged as a multi-chip module (MCM). There were separate chips for the Instruction Cache, the Fixed-Point Unit, and the Floating-Point Unit, and the Data Cache was split across four chips. With today's technology, one could fit thousands of those processors on a single die.

The first high-end CMOS-based mainframe processor was released in 1997 in the IBM* S/390* G4 system. The entire processor fitted onto a single chip with roughly two million transistors. The processor was a simple in-order scalar pipeline, which was clocked at 300 MHz. 12 processors could be assembled on a MCM to build the symmetric multi-processing (SMP) system.

Numerous major steps happened between the G4 system in 1997, and today's high-end IBM zEnterprise* 196 processor. In 1998, branch prediction and binary floating point was added (traditionally, mainframes use hexadecimal floating point). In 2000, the IBM eServer* zSeries* 900 system introduced the 64-bit z/Architecture. The IBM eServer zSeries 990 system in 2003 was the first mainframe to break through the 1 GHz barrier. The processor implemented a super-scalar pipeline, with two Load/Store units, two Fixed-Point Units, and one Floating-Point Unit; execution was in-order, but memory accesses could be done out-of-order. It was also the first mainframe to have two full processors on a single chip ("dual core"). The IBM eServer zSeries 990 also introduced the "book concept", in which MCMs, each containing multiple CPU-chips, are packaged on a circuit board which can be plugged into a backplane. Up to four such "books" can plugged in like books in a bookshelf. This created a SMP with up to 32 processors; the CMOS technology advancements allowed this to grow to 54 processors in the IBM System z9* in 2005. Processor frequency grew continuously from 300 MHz in G4 to 1.7 G Hz in the IBM System z9.

From the G4 system to the IBM System z9, the basic processor pipeline was maintained, adding new features such as superscalar and limited out-of-order generation by generation. The IBM System z10* processor released in 2007 was a complete redesign. The old pipeline could no longer provide the required performance improvements, so a decision for an ultra-high-frequency pipeline was made; the system was released with the main pipeline running at 4.4 GHz, with four cores per chip ("quad core"), and with four books up to 80 CPUs in the system (of which 64 can be purchased by the customer, the remaining cores handle I/O or are used as spares). In order to achieve this high frequency, the pipeline was simplified to a two-scalar in-order pipeline.

The IBM zEnterprise 196 is the out-of-order successor of the z10 processor pipeline. The step to 32-nm CMOS technology allows this to run at 5.2 GHz, the fastest clocked commercial microprocessor ever. Figures 10.3 and 10.4 show the functional block diagram and the pipeline diagram, respectively. Figure 10.2 shows

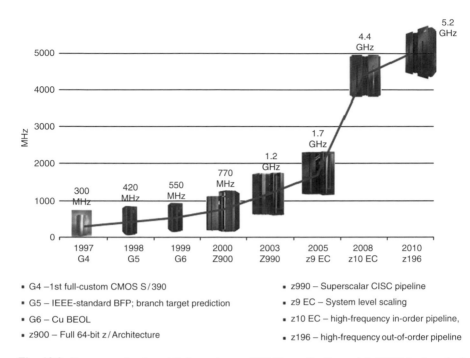

Fig. 10.2 Frequency development for system z. *BFP* binary floating point, *BEOL* back-end of line, *CISC* complex-instruction-set computer

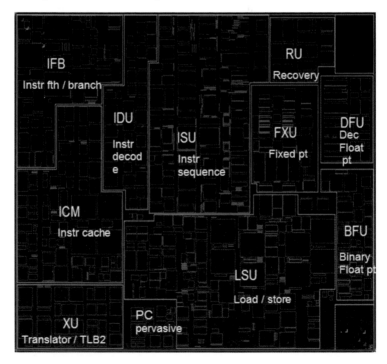

Fig. 10.3 Block diagram for the z196 processor

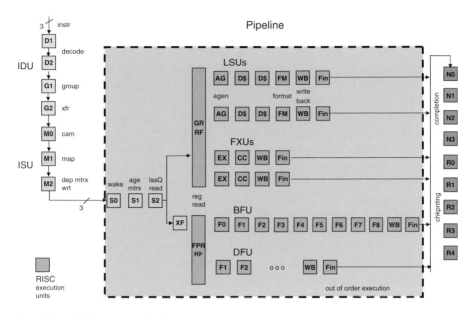

Fig. 10.4 z196 Processor pipeline diagram

the frequency development for all System z generations starting with the first CMOS machine G4. The Instruction-fetch/branch (IFB) unit contains an asynchronously running branch predictor, which feeds instruction fetch requests into the instruction cache (ICM). The instruction cache feeds through instruction buffers into the synchronous part of the pipeline, starting with the Instruction Decode Unit (IDU). Through three cycles, the instructions are decoded and grouped into three-wide dispatch groups. This is also the place where cracking for complicated CISC ops occurs; for example, a Load Multiple (LM) instruction is cracked into up to 16 micro-ops, each loading one register. A fourth IDU cycle is used to transfer the decoded and cracked micro-ops into the Instruction Sequencing Unit (ISU). There, the register renaming takes place before the micro-ops are written into the Dependency and Issue Matrices. The issue bandwidth is five micro-ops per cycle; the five RISC-like execution units are two Load/Store units (LSUs), two Fixed-Point units (FXUs), and one floating-point unit, which consists of two sub-units for binary and decimal floating point (BFU/DFU).

Instructions are completed in program order. At dispatch time, each micro-op is written into the ISU, but it is also written into the Group Completion Table (GCT), which keeps track of the execution status of all micro-ops. Once all micro-ops of an instruction have finished, the completion logic reads out the completion-table entry and completes the instruction. After completion, error correcting code (ECC) hardening of the state happens (see Sect. 10.3).

Since the z/Architecture is a CISC architecture, many complex instructions exist. There are various design features unique to the CISC processor to support this architecture:

- Many instructions use one operand from memory and a second operand from a register, with the result being a register again (so-called RX-instruction), for example R1 $\Leftarrow$ R1 + mem(B1 + X1 + D1). To prevent cracking of those instructions into two micro-ops, the concept of dual-issue was invented. A single micro-op in the ISU issues twice, first to the LSU to load the memory operand into the target register of the instruction; then a second time to the FXU to read that target register and the R1 register, perform the arithmetic operation, and write the final result back into the target register. This saves issue queue resources and thus allows for a bigger out-of-order window.
- Certain instructions operate directly on memory, for example Move Character (MVC) or Compare Logical Character (CLC) compare 1–256 bytes of two memory operands. These instructions are particularly common in commercial workloads (e.g., databases) and thus critical for the system performance. In order to provide best-possible performance, a sequencing engine is built into the LSU to execute those instructions at maximum cache bandwidth, without needing to crack them at dispatch time into many micro-ops.
- The concept to handle particularly complex instructions in a firmware layer called "millicode" is also employed in the z196 processor; this concept exists since G4. Instructions such as Compare String (CLST), which correspond to the cmpstr C-library routine, execute in this firmware; the hardware implements special "millicode assist instructions" to facilitate fast execution of the millicode.
- The z/Architecture provides special instructions for fixed-dictionary compression and for cryptography. These instructions are executed in millicode, but a hardware co-processor is invoked by millicode to accelerate the computations. This enables much faster compression and cryptography than would be possible by discrete instructions.

The instruction cache is 64 kB, the data cache is 128 kB. Both share a 1.5 MB second-level cache. Also, the translation lookaside buffers (TLBs) are large at 64×2 way and 256×2 way (I-TLB/D-TLB). The D-TLB also supports native 1 MB-pages in a separate 32×2 way 1 M TLB. The first-level TLBs are backed up by a shared second-level TLB, which also includes a hardware translation engine (XU). These big structures particularly help commercial workloads with their big instruction and data footprint. But running out-of-order at very high frequency also provides very good performance for compute-intensive workloads, as they become more and more important in today's business intelligence/business analytics workloads.

10.2 Errors, Recovery, and Reliability

The continuous decrease in transistor size has an increasing influence on the amount of transient errors that happen in a processor. The causes of these transient faults include radiation, aging, and voltage spikes. Researchers have shown that the soft-error rate increases around 8% per technology generation [2]. Recent research in the field of graphics processors used for the Folding@home* project showed memory error rates of about four fails per week [3]. These numbers show that the requirements for error-detection and correction are real today and will significantly increase over the next technology generations. Mainframes have a long history of built-in error-detection and recovery functionality, but in recent years even small and mid-sized servers as well as graphics cards used in high-performance computing saw the addition of ECC logic to their processors. These traditional approaches to error-detection and correction in hardware have been accompanied by some interesting developments in software-based reliability.

The 1997 system G4 already contained special circuitry for reliability, for example the Decode, Fixed- and Floating-Point Units were duplicated to detect hardware faults. Every result is checkpointed with ECC-hardening in the so called Recovery Unit. When an error is detected, checkpointing of incoming results is blocked, and the processor goes through a recovery sequence. During the recovery sequence, the R-Unit state is read, ECC-checked and corrected, and all architected registers [e.g., the general-purpose registers (GRs) and floating-point registers (FRs), the Program Status Word including the Instruction Address, ...] in the processor are refreshed from the R-Unit state. The caches, TLBs, etc. are purged to remove any possible parity fail in them. After the recovery sequence completes, the instruction processing resumes at the current instruction address that resulted from the last checkpointed instruction. Thus the recovery sequence cleans up hardware faults, completely transparent to software.

The concept of checkpointing and recovery still exists today in the z196 processor, and with shrinking technology it becomes more and more important. However, due to power limitations, the units are no longer duplicated. Instead, special checking circuitry is used to detect any hardware faults. Every databus is protected with parity; computational units either use parity prediction or residue-checking to detect errors in the computation. Control logic is checked by either local duplication, or by consistency-checking on the state machine and its transitions. For example, certain state machines are coded as one-hot states, and an error is detected should the state ever be not one-hot. (See also [4])

10.3 Scaling Trends

Figure 10.5 shows how each new technology generation is providing less relative performance improvement and requires alternative solutions to increase performance from one machine generation to the next. Some predictions actually show that technology may even have a negative impact on performance due to power density considerations.

Moore's law is still valid today, and it will probably stay valid for the next decade. However, Moore's law only addresses the density of the technology, not

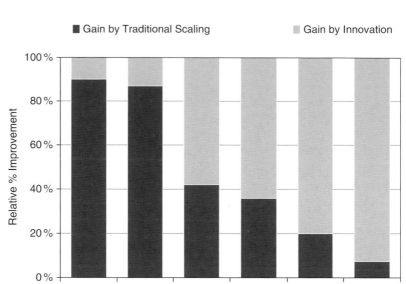

Fig. 10.5 Improvements by scaling and innovation

other attributes such as frequency or power consumption. However, threshold voltage is not shrinking at the historical rate, which overall leads to higher power and power density at constant frequency. This puts a limit on the frequency growth. The industry trend over the last 5 years clearly showed that the "frequency race" is over, at least for mainstream processors. For mainframes and other high-end servers, the situation is slightly different since more effort can be put into current delivery and cooling; but even with enormous efforts in power and cooling, the frequency growth will be limited by power consumption and power density.

Modern microprocessors such as POWER7* or the z196 CPU are very complex out-of-order processors that aim to extract as much instruction-level parallelism from the program as possible. However, this approaches the point of diminishing returns, since the inherent parallelism of single-threaded programs is limited. This leads to only small improvements in Instructions-per-cycle (IPC) compared to the past.

Given the limitations in frequency growth and IPC-growth, it seems obvious that the future lies in thread-level parallelism. Recent years brought multi-cores and symmetric multi-processing. The POWER7 processor chip for example contains eight cores, each running four threads in parallel. We expect this trend to continue: the number of cores will grow approximately with the CMOS shrinking factor, i.e., whereas in the past the CMOS density improvements went into more and more complex processors supporting higher and higher IPC and levels of SMT, we will now instead see an exponential growth in cores per chip.

The provided thread-level parallelism comes at a high cost: single-threaded programs cannot exploit the available compute power, and large investments into multi-threaded programming have to be made.

10.4 Accelerators

Another trend in the industry is the use of accelerators or specialty engines in order to address the performance needs of a subset of the workloads encountered on general-purpose processors. This trend started out in the field of 3D computer games, which require a large number of computations to correctly display a 3D game scene on the 2D computer screen. The increasing complexity of these game scenes drove the invention of additional processing capabilities in the Graphics Processing Units (GPUs) of PCs. It was also adapted in the Cell Broadband Engine* processor jointly developed by Sony*, Toshiba*, and IBM with its Synergistic Processing Elements. Today, the field of accelerators is still dominated by the use of graphics processor hardware to speed up computations on numerically intensive and parallel problems. These accelerators usually consist of several hundred or thousand rather simple processing elements and achieve the best performance if the computation task consists of a large number of independent calculations that can be done in parallel. Over the next couple of years, there will be an increase in all kinds of special- or general-purpose accelerator hardware, examples available today are IBM's Smart Analytics Optimizer*, Fixstar's GigaAccell 180* based on IBM's PowerXCell* 8i processor, IBM System z Crypto and Compression unit available on each System z processor, nVidia* Tesla* compute clusters, or Intel's "Intel Many Integrated Core Architecture"* (formerly codenamed "Larrabee"*). Furthermore, several companies and research projects are investigating the use of field-programmable gate arrays (FPGAs) for acceleration.

Accelerators suffer from two big problems today, first they need to be connected via a high-speed interconnect bus otherwise the potential performance improvement may be offset by the latency or bandwidth limitations of the interconnect. Secondly the integration of these accelerators requires special programming models in order to use their special purpose hardware. This means that a significant investment needs to be made in terms of software development in order to convert an existing application to use accelerator hardware directly. In order to shield the software developers from having to know the details of the accelerator hardware, accelerator vendors could provide libraries that abstract the accelerator function for ease-of-use in software. Especially with FPGAs, it becomes very hard to imagine that they will ever provide enough ease-of-use for software developers to become widely used.

For more generic accelerators such as GPUs but also general purpose processors, the industry is converging on a common programming model as well as an application programming interface (API) to use these devices and coordinate the parallel computations. This programming model is called OpenCL*. OpenCL solves many of the tasks required for parallel programming and different accelerator types. However, the big disadvantage OpenCL faces is that it is based on the smallest common set of capabilities of the different accelerator types, i.e., it limits the user to a programming model that works well on GPUs but leaves a lot of the capabilities of general-purpose processors unused. Examples of features commonly used in general purpose processors but not available in GPUs would be recursion, function pointers, and callback functions. The OpenCL standard does not support these constructs but vendors can define vendor-specific extensions to overcome these limitations. However, in that case the generic programming model becomes limited to using the hardware a vendor provides and therefore the application cannot be run on different accelerators easily.

10.5 Hardware–Software Collaboration

With the trends described above, significant effort needs to be put into collaboration between hardware and software groups to unlock the potential that modern microarchitectures provide. This collaboration happens at various levels.

Today's processor architects interlock regularly with the compiler groups. Modern compilers (be it static compilers such as C/C++ or dynamic compilers like a Java* just-in-time [JIT] compiler) internally model the processor pipeline, in order to influence instruction scheduling. The goal is to create a compiled program that fits almost perfectly to the internal CPU pipeline, to allow for best possible instruction-level parallelism. On the z196 processor for example, branch instructions end the three-wide dispatch groups. The compiler models this and tries to schedule branch instructions such that groups do not end prematurely.

New instructions are added to the architecture on almost every new generation of a processor. Often the suggestions for new instruction come from compiler or software groups. In order to exploit the new instructions, sometimes it is sufficient to simply change the compiler; in particular with dynamic compilers like the Java JIT this leads to rapid exploitation. Other enhancements to the architecture have a longer lead-time. For example, the Power6 and z10 system added hardware for Decimal Floating Point (DFP) arithmetic. It may take some years until software explicitly uses this feature, since current software's data types do not match the format. Again, Java is an early adopter, exploiting the DFP hardware for the BigDecimal class; this gives a vast variety of programs a performance boost, without rewriting those programs [5].

Multi-core and symmetric multi-threaded processors also pose exploitation challenges. If a computer runs enough independent programs to use the available hardware threads, exploitation is immediate (at the cost of higher n-way overhead in the operating system). Many important applications need to be rewritten to give the operating system enough tasks to exploit the many hardware threads.

Maybe the biggest exploitation challenges exist for accelerators. For example, it may be relatively easy for hardware to implement a Regular-Expression or XML (Extensible Markup Language)-parsing accelerator. However, there are many different software libraries for those tasks, and many software products even use ad hoc solutions instead of standard libraries. To exploit any specialized hardware, a rewrite of those libraries or ad hoc functions would be required. The example explains why on-chip accelerators are rare on today's mainstream processor chips. With slower improvements of single-thread performance, the investment in such features on the hardware and software side may be more viable to meet future performance targets. Middleware such as WebSphere Application Server could play an important role in such a development, since many applications call the Application Server for standard functions like parsing; this might enable early exploitation, similar to how Java can exploit DFP faster than other software.

Exploitation of GPUs for general-purpose computation as well as exploitation of FPGA accelerators may be boosted by the new OpenCL standard. OpenCL defines a language to describe algorithms device-independently, and an API for applications to communicate with the accelerator. The program is compiled to the available accelerator, which could be a rack of blade servers in a hybrid IBM zEnterprise 196 mainframe, or a rack of GPUs as in nVidia's Tesla compute clusters. One drawback of the OpenCL language is its restriction of commonly used general-purpose programming techniques such as recursion, function pointers, and callback functions. GPUs do not support those features; for general purpose accelerator programming, the restriction of those features could negatively impact OpenCL exploitation.

While there is no direct support for OpenCL from any of the major FPGA vendors, it is also conceivable to use the OpenCL API and the operating system services for managing OpenCL devices as an application access path to the FPGA accelerator.

10.6 Supercomputers

The world of high-performance computing and supercomputers differs from the general-purpose computing market in that applications on the supercomputer systems are highly optimized for the specific target architecture. The high degree of software optimization sometimes results in system architectures that would rarely be useful in typical commercial applications. One recent example of such a system-design point is the QPACE* project jointly developed by Deutsches Elektronen-Synchrotron (DESY), Research Center Juelich, University of Regensburg, and IBM Deutschland Research and Development GmbH [6]. This system is highly optimized for research in the area of quantum chromodynamics and uses a specialized high-throughput 3D torus interconnect for communication between the compute nodes. Even though this system may not be of interest for commercial applications, it is an excellent example of successful hardware/software co-design.

The majority of the systems listed in the TOP500 list for supercomputers [7] are built from industry standard components that are extended with specialized interconnects. In the 1990s, the majority of supercomputers were constructed based on vector-processor architectures [8], but, over time, the specialized vector processors almost completely vanished and were replaced by industry-standard processor architectures like *x*86 and Power. This trend, among other reasons, is owed to the fact that vector processors were not attractive for commercial applications, and the development of specialized processors for the field of supercomputing became too expensive. Another reason is that today's *x*86 and Power processors all contain Single-Instruction Multiple-Data Floating-Point Units (SIMD FPUs), which deliver the unprecedented floating-point performance required for supercomputers.

There is a clear trend in the TOP500 list towards hybrid systems consisting of standard processors coupled with specialized accelerators. The accelerators are either tightly coupled with the processor, e.g., the Cell Broadband Engine, or realized through PCI Express cards, e.g., nVidia Tesla. In comparison to commercial computing, the exploitation of accelerators is quicker due to the high degree of specialized and optimized software, and the inherent parallelism of many mathematical algorithms.

Regarding performance growth over the next years, it is safe to assume that the historic performance growth rates between 20% and 40% per 6 months will be maintained [9]. Owing to government funding, a lot of information on the future high-performance systems is publicly available, for instance the Lawrence Livermore National Laboratory plans to install a 20 petaflop/s system named "Sequoia" in 2012 [10,11]. The system will be based on the latest generation of IBM's System Blue Gene* called Blue Gene/Q*, and it will contain about 1.6 million processor cores.

10.7   Outlook

The field of processor development has evolved significantly over the last 40 years. While in the early days technology advancement defined processor performance, today power density is becoming the limiting factor and performance per watt the guiding metric. These changes are driving the requirement for innovation into other areas of processor and system design. Application performance can be influenced on the following five levels:

- Technology
- Micro-architecture/system design
- Architecture
- Compilers/operating systems
- Applications

With diminishing technology gains, the other four levels are becoming increasingly important and, in order to obtain optimal results, all of them need to be taken into account during the development cycle, e.g., it does not make sense to introduce new instructions to the architecture if they cannot be exploited efficiently by compilers and applications.

This new paradigm will also require hardware designers to get more and more involved in the software exploitation of the processors they design. Future performance gains can only be realized if a strong hardware–software collaboration is employed during the system design.

We are certain that successfully optimizing the whole stack of system design will result in another decade of exponential performance growth.

10.7.1 Trademarks

The following are trademarks of the International Business Machines Corporation in the United States and/or other countries.

AIX*	FICON*	Parallel Sysplex*	System z10
BladeCenter*	GDPS*	POWER*	WebSphere*
CICS*	IMS	PR/SM	z/OS*
Cognos*	IBM*	System z*	z/VM*
DataPower*	IBM (logo)*	System z9*	z/VSE
DB2*	Blue Gene	Blue Gene/Q	zEnterprise

*Registered trademarks of IBM Corporation

The following are trademarks or registered trademarks of other companies.

Adobe, the Adobe logo, PostScript, and the PostScript logo are either registered trademarks or trademarks of Adobe Systems Incorporated in the United States, and/or other countries.

Cell Broadband Engine is a trademark of Sony Computer Entertainment, Inc. in the United States, other countries, or both and is used under license there from.

Java and all Java-based trademarks are trademarks of Oracle, Inc. in the United States, other countries, or both.

Microsoft, Windows, Windows NT, and the Windows logo are trademarks of Microsoft Corporation in the United States, other countries, or both.

InfiniBand is a trademark and service mark of the InfiniBand Trade Association.

Intel, Intel logo, Intel Inside, Intel Inside logo, Intel Centrino, Intel Centrino logo, Celeron, Intel Xeon, Intel SpeedStep, Itanium, and Pentium are trademarks or registered trademarks of Intel Corporation or its subsidiaries in the United States and other countries.

UNIX is a registered trademark of The Open Group in the United States and other countries.

Linux is a registered trademark of Linus Torvalds in the United States, other countries, or both.

ITIL is a registered trademark, and a registered community trademark of the Office of Government Commerce, and is registered in the U.S. Patent and Trademark Office.

IT Infrastructure Library is a registered trademark of the Central Computer and Telecommunications Agency, which is now part of the Office of Government Commerce.

*All other products may be trademarks or registered trademarks of their respective companies.

References

1. Davenport, T.H., Harris, J.G.: Competing on Analytics: the New Science of Winning. Harvard Business School, Boston (2007)
2. Hazucha, P., Karnik, T., Maiz, J., Walstra, S., Bloechel, B., Tschanz, J., Dermer, G., Hareland, S., Armstrong, P., Borkar, S.: Neutron soft error rate measurements in a 90-nm CMOS process and scaling trends in SRAM from 0.25-μm to 90-nm generation. IEEE IEDM (International Electron Devices Meeting) 2003, Technical Digest, pp. 523–526 (2003). doi: 10.1109/IEDM.2003.1269336
3. Haque, I.S., Pande, V.S.: GPUs: TeraFLOPS or TeraFLAWED? www.cs.stanford.edu/people/ihaque/posters/sc-2009.pdf. Accessed 20 Sept 2010
4. Meaney, P., Swaney, S., Sanda, P., Spainhower, L.: IBM z990 soft error detection and recovery. IEEE Transactions on device and materials reliability, Vol. 5, No. 3, September 2005, pg. 419–427
5. Mitran, M., Sham, I., Stepanian, L.: Decimal floating-point in Java 6: best practices. www-304.ibm.com/partnerworld/wps/servlet/ContentHandler/whitepaper/power/java6_sdk/best_practice. Accessed Jan 2009
6. Pleiter D.: QPACE: QCD Parallel Computing on the Cell. www.itwm.fhg.de/hpc/workshop/mic/Qpace_%28Dirk_Pleiter_-_Desy%29.pdf. Accessed 28 Oct 2008
7. www.top500.org.Accessed 29 Sept2010
8. www.top500.org/overtime/list/35/procarch. Accessed 29 Sept 2010
9. www.top500.org/lists/2010/06/performance_development. Accessed 29 Sept 2010
10. https://asc.llnl.gov/publications/leadingHPC.pdf. Accessed 29 Sept 2010
11. https://asc.llnl.gov/computing_resources/sequoia/index.html. Accessed 29 Sept 2010

Chapter 11
Towards Terabit Memories

Bernd Hoefflinger

Abstract Memories have been the major yardstick for the continuing validity of Moore's law.

In single-transistor-per-Bit dynamic random-access memories (DRAM), the number of bits per chip pretty much gives us the number of transistors. For decades, DRAM's have offered the largest storage capacity per chip. However, DRAM does not scale any longer, both in density and voltage, severely limiting its power efficiency to 10 fJ/b. A differential DRAM would gain four-times in density and eight-times in energy. Static CMOS RAM (SRAM) with its six transistors/cell is gaining in reputation because it scales well in cell size and operating voltage so that its fundamental advantage of speed, non-destructive read-out and low-power standby could lead to just 2.5 electrons/bit in standby and to a dynamic power efficiency of 2aJ/b. With a projected 2020 density of 16 Gb/cm^2, the SRAM would be as dense as normal DRAM and vastly better in power efficiency, which would mean a major change in the architecture and market scenario for DRAM versus SRAM.

Non-volatile Flash memory have seen two quantum jumps in density well beyond the roadmap: Multi-Bit storage per transistor and high-density TSV (through-silicon via) technology. The number of electrons required per Bit on the storage gate has been reduced since their first realization in 1996 by more than an order of magnitude to 400 electrons/Bit in 2010 for a complexity of 32Gbit per chip at the 32 nm node. Chip stacking of eight chips with TSV has produced a 32GByte *solid-state drive* (SSD). A stack of 32 chips with 2 b/cell at the 16 nm node will reach a density of 2.5 Terabit/cm^2.

Non-volatile memory with a density of 10×10 nm^2/Bit is the target for widespread development. Phase-change memory (PCM) and resistive memory (RRAM) lead in cell density, and they will reach 20 Gb/cm^2 in 2D and higher with 3D chip stacking. This is still almost an order-of-magnitude less than Flash. However, their

B. Hoefflinger (✉)
Leonberger Strasse 5, 71063 Sindelfingen, Germany
e-mail: bhoefflinger@t-online.de

B. Hoefflinger (ed.), *CHIPS 2020*, The Frontiers Collection,
DOI 10.1007/978-3-642-23096-7_11, © Springer-Verlag Berlin Heidelberg 2012

read-out speed is ~10-times faster, with as yet little data on their energy/b. As a read-out memory with unparalleled retention and lifetime, the ROM with electron-beam direct-write-lithography (Chap. 8) should be considered for its projected 2D density of 250 Gb/cm^2, a very small read energy of 0.1 μW/Gb/s. The lithography write-speed 10 ms/Terabit makes this ROM a serious contentender for the optimum in non-volatile, tamper-proof storage.

11.1   High-Speed CMOS SRAM (Static Random-Access Memory)

The invention of CMOS technology in 1963 (Figs. 2.7 and 2.8) produced the first CMOS SRAMs by 1969. The classical 6 T memory cell (Fig. 3.18) is displayed here as Fig. 11.2, in order to compare it with the most recent 8 T cell, 40 years later [1], shown in Fig. 11.3.

The dominant issues for static RAM are

– Cell size
– Access time
– Standby power
– Power efficiency or energy/bit.

A representative recent layout is shown in Fig. 11.4 for a 90-nm design.

While the cell size in [2] is about 180 F^2 in a standard logic process, where F is the first-metal half pitch in the sense of a *node* in the ITRS roadmap, the cell size in [1] is 121 F^2 at the 45-nm node for this 8 T design in a more memory-specific process. In a 2D planar design, the area for a 6 T SRAM cell is >110 F^2. Any significant improvement of the cell size, i.e., the SRAM density, beyond the scaling to ~16 nm, will require the 3D integration of the transistors in the cell, as we will see shortly.

We consider next the speed of the SRAM, the primary reason for operating with this complex cell. We see in Fig. 11.2 that the current necessary for inverting the central nodes, carrying the cell information, into the opposite state in the write mode, is supplied through the V_{DD} and GND nodes per cell, providing a fast write. In the read mode, again, the current to charge or discharge the bit lines and the inputs of the sense amplifier at the end of a column is provided by these per-cell sources, allowing a fast read access time. Obviously, these times depend on the ON currents of the four transistors in the quad. In these operations, we face a difficult trade-off between the charging currents, which mean speed of access, rising with V_{DD}, and the switching energy CV_{DD}^2, which decreases with a lower V_{DD} (see. Sects. 3.2 and 3.3):

$$1/t_{access} \sim I_{Dmax}/V_{DD} \sim \exp(V_{DD} - V_T)/V_{DD}$$

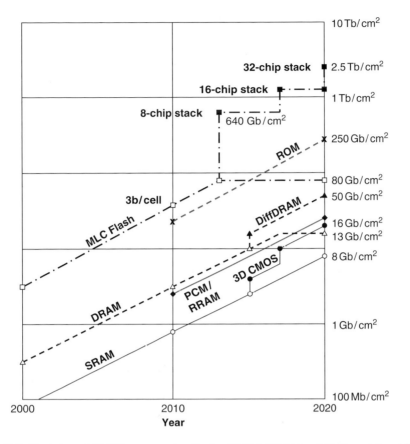

Fig. 11.1 Evolution of CMOS memory density for SRAM, DRAM and Flash memory. *MLC* multi-level per cell, *PCM* phase-change memory, *RRAM* resistive RAM

in low-voltage barrier-control mode or

$$\sim(V_{\mathrm{DD}} - V_{\mathrm{T}})^{1/2}.$$

in higher-voltage drift mode.

We have analyzed these relationships in Sect. 3.2, and in Sect. 3.3 on CMOS, we found the additional complication that, while we pursue an energy priority in lowering the supply voltage, the inherent DC standby current

$$I_{\mathrm{stdby}} = \sqrt{I_{\mathrm{max}}I_{\mathrm{min}}},$$

where I_{max} is the maximum transistor on-current and I_{min} the minimum off-current, and DC standby power

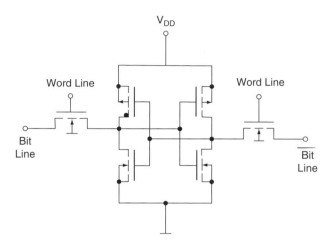

Fig. 11.2 The CMOS 6 T SRAM cell. Its core is the cross-coupled inverter pair, the *quad*. When the word line turns on the two access transistors, the content of the cell can be read as true and complement by a differential sense amplifier without modifying the content of the cell. In the write mode, the content of the cell can be confirmed or forced to the inverse

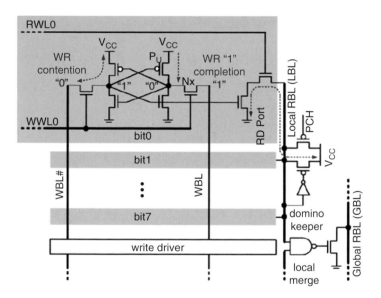

Fig. 11.3 A recent 8 T SRAM cell [1], in a 45-nm, CMOS technology. Two additional transistors form an asymmetrical path for a fast read at low voltages. RWL0 = read word line =, WWL0 − write word line 0, RD = read, RBL = read bit line, WBL = write bit line, LBL = local bit line, GBL = global bit line, PCH = pre-charge (© 2010 IEEE)

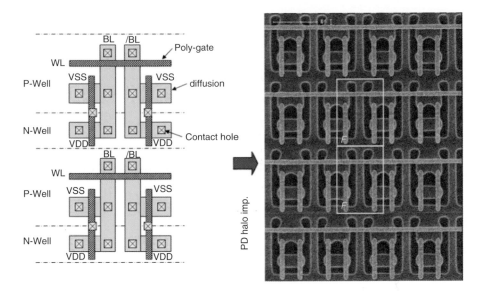

Fig. 11.4 Layout of a 6 T cell in 90-nm CMOS [2]. Memory cell sizes are usually given in multiples of half-pitch units F. The size of the cell shown here is about $12 \times 15 = 180\ F^2$. PD halo imp. = PMOS transistor halo implant. (© 2010 IEEE)

$$P_{DC} = V_{DD}\sqrt{I_{max}I_{min}}$$

means a constant loss, eventually becoming larger than the dynamic switching power

$$P_{dyn} = V_{DD}fI_{max},$$

where f is a duty factor $<1/4$. Within the total power

$$P_{tot} = P_{DC} + P_{dyn} = V_{DD}I_{max}\left(f + \sqrt{I_{min}/I_{max}}\right),$$

the on/off current ratio has to be >16 to keep the dynamic power the dominant component. We have seen in Sect. 3.2 that the low-voltage barrier-control mode offers larger on/off ratios than the higher-voltage drift mode and the lowest energies CV^2/bit anyway.

The price, however, is access time, and we take a closer look at the access time, selecting two publications cited before [1, 2]. We have plotted the log of the measured read access times against the supply voltage in Fig. 11.5.

In the semilog plot we see distinct slopes for the dependence of access time on supply voltage. At higher voltages, the memory cells work mostly in strong inversion, where current and access time have a power dependence on voltage, approximately $V^{1/2}$. At voltages <0.9 V, we see an exponential dependence

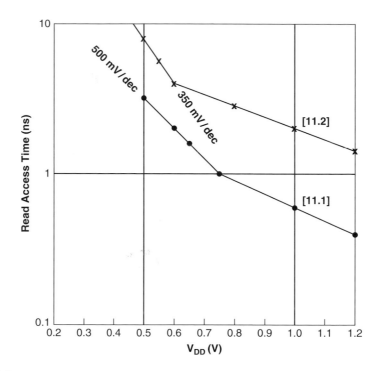

Fig. 11.5 Measured read access times versus supply voltage V_{DD}. [1] = 45 nm, [2] = 90 nm

characteristic of the sub-threshold or barrier-controlled current in the cell with slopes of 350 mV/decade and 500 mV/decade, respectively.

The energy/bit, also called the power efficiency/bit, has been reduced to <100 nW Gb^{-1} s^{-1}, or <0.1 fJ/b, for sub-1-ns access times, as shown in Fig. 11.6, and, by lowering the operating voltage to 0.57 V for the same 45-nm chip, can be as low as 0.03 fJ/b, while the access time increases to 3.4 ns.

The process design and the operating point of these SRAMs are the result of a sensitive balance including dynamic and static DC power as well as speed. The 45-nm design quoted here has a DC power of 0.6 µW/b compared with a dynamic power of 0.3 µW/b, indicating these compromises.

The robustness of state-of-the art SRAMs can be learned from [3], where data such as the static noise margin (NM) and the 10–90% corridor of minimum supply voltages were reported. We covered the fundamentals in Sect. 3.3, when we studied CMOS inverter transfer characteristics and the cross-coupled inverter pair (Fig. 3.17), the quad, and its noise margin (3.25). Figure 11.7 shows measured transfer characteristics on CMOS SRAM cells in a 32-nm technology [3].

It is impressive how these characteristics of 2009 still resemble the ideal model-curve of the 1960s (Fig. 3.15). The extracted noise margins correspond to average *voltage gains of 1.5 at 1.1 V and 2.5 at 07 V*, remarkable results for a 32-nm design and for the sub-threshold regime involved at 0.7 V. The paper also reports that the

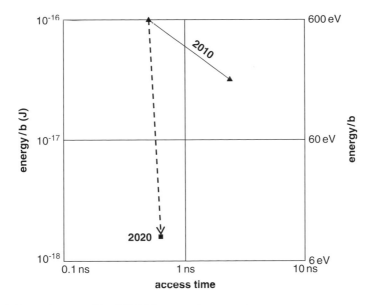

Fig. 11.6 Dynamic energy/bit of SRAMs versus the access time

Fig. 11.7 Transfer
characteristics of 32-nm
SRAM cells for three supply
voltages [3]. The static noise
margins NM are indicated as
the largest inscribed (*dotted*)
squares inside a characteristic
and its mirror image (© 2009
IEEE)

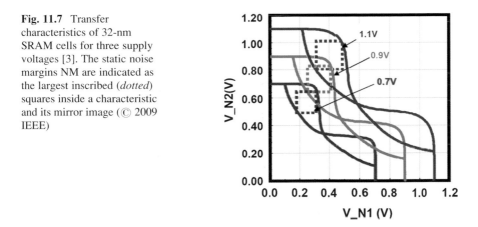

90% range of minimum supply voltages for dies on a single wafer reached from
0.79 V to 0.92 V or 130 mV, indicating the variance present at 32 nm and a sign
that this spread will widen quickly for smaller features in the future.

How much room is there for scaling SRAMs? We base our assessment on the
characteristics of the 10-nm nanotransistor treated in Sect. 3.2 and the CMOS
inverter and quad described in Sect. 3.3. Here is a listing of the characteristics of
the 6 T 10-nm cell:

10-nm CMOS 6 T SRAM 2020

$W = L = 10$ nm

$V_{DD} = 0.3$ V

$V_T = 0.3$ V
$I_{ON} = 25$ nA; $I_{OFF} = 0.25$ nA
$I_{DC} = 2.5$ nA
$P_{DC} = 0.75$ nW/b $= 750$ mW/Gb
Duty factor $f = 1/4$
$P_{dyn} = 1.8$ nW/b $= 1.8$ W/Gb
Access time $= 0.6$ ns
Energy/bit $= 1.1 \times 10^{-18}$ J/b $= 7$ eV/b
Power efficiency $= 1.1$ nW Gb^{-1} s^{-1}
Cell size: $120\ F^2 = 0.12 \times 10^{-9}$ cm^2/b $= 12$ mm^2/Gb
$P_{total} = 2.55$ W/Gb
Power density $= 21$ W/cm^2
$V_{DDstdby} = 0.15$ V
$P_{stdby} = 120$ mW/Gb

The characteristics listed here mark the 2020 performance level expected for a memory built with the average 10-nm transistors. The treatment of the fundamental variance of the CMOS nano-quads in Sect. 3.2.1 told us that $V_{DDmin} = 0.37$ V and $V_{DDstdby} = 0.21$ V would be the limits, and manufacturing tolerances would impose further allowances on this performance data.

We also recognize that these nano-SRAMs will operate at very high power densities (21 W/cm^2) and therefore in the very-low-voltage, deep-sub-threshold barrier-control mode. Nevertheless, ON/OFF current ratios are limited, and the power reduction in standby will be just a factor of 20. We have entered these 2020 data in Fig. 11.6. A sub-1-ns access time will be possible, and the energy/bit will be 100-fold smaller than in 2010, at just 7 eV/b. The active memory cell based on the CMOS quad maximizes speed and minimizes energy at the price of a larger cell size. Even so, a density of 8 Gb/cm^2 can be predicted for a 2020 2D SRAM.

We have seen in Sect. 3.4 that the SRAM cell is the top candidate for stacking complementary MOS transistors with a common gate and common drains on top of each other and that this can more than double the density with the same mask count and fewer metal contacts so that a *CMOS SRAM density of 16 Gb/cm^2 is possible in 2020.*

A bit more speculative is a future 6 T SRAM cell consisting of the complementary, vertical bipolar quad and NMOS select transistors proposed in Sect. 3.8. This quad would have the ideal transconductance of 65 mV/decade of collector current, offering the highest speed and the best ratios of dynamic power versus DC power and the relatively lowest standby power together with a very high density of $47\ F^2$, corresponding to *21 Gb/cm^2* at the 10-nm node.

The vertical-interconnect topography of the cells in Figs. 3.43 will also be ideal for:

– The flipped stacking of these SRAMs as local memories on top of arrays of processor elements (PEs).
– The 3D integration of these PEs with performance levels of 1 fJ per 16 $\times$ 16 multiply (Sect. 3.6), together with 3D SRAMs at 1 fJ/kb will advance information processing to truly disruptive innovations (Chaps. 13,16,18).

11.2   DRAMs (Dynamic Random-Access Memories)

DRAMs are the famous MOS memories with just *one transistor per bit*, invented in 1968 and in production since 1971. The basic elements of this memory are illustrated in Fig. 11.8.

In this electronic-charge-based memory cell, the key elements and the only elements in the cell are the storage capacitor C_S and the select transistor. A high number of electrons on the storage capacitor means a logical ONE, a small number a logical ZERO. If the Word signal on the transistor gate is low, the transistor, ideally, would be OFF, and the charge on the storage capacitor, i.e., the information, would stay in the cell indefinitely. In reality, it will leak through the transistor into the substrate and into the circuitry so that it has to be refreshed by *re-writing* it into the cell periodically by selecting the cell (raise the word-line voltage to High), reading the marginal cell content with the bit-line sense amplifier, and re-writing a perfect content into the cell. In the same way, with each *reading* of the cell, its content is *destroyed* and has to be *re-written* immediately. These requirements are a fundamental overhead on the operation of a DRAM so that write, read, and the full-cycle time of the DRAM make it slower than an SRAM of the same generation. What does *destructive reading* in a DRAM mean? The DRAM cell is a *passive* cell, in contrast to the SRAM, an *active* cell. Active means amplification and instant regeneration: The (differential) amplifier, the quad, inside the SRAM cell not only restores the charges in the cell immediately as a read access might destroy the touchy balance in the cell, it also provides charges to charge and discharge the bit lines as needed and provide healthy signal levels to the bit-line sense amplifier. By contrast, charges in the passive DRAM cell are lost by leakage or shared with other capacitors such as the bit-line capacitor C_{BL} once the select transistor is opened. In fact, if the bit-line sits at 0 V, the cell capacitor at a ONE $= V_{DD}$, and the cell is selected, charges would be shared, resulting in a ONE $= 0.5\ V_{DD}$ at the input of the sense amplifier, if $C_S = C_{BL}$. This voltage-divider effect on signal levels is the reason for making the storage capacitor large enough that leakage and the capacitance of long bit-lines cannot spoil signal levels too much.

In the early days, the area of the drain diffusion of the NMOS select-transistor was enlarged to produce a storage capacitance of ~30 fF. The rapid scaling-down of cell sizes has turned the realization of miniature capacitors of 30 fF and recently 20 fF into a formidable challenge. Because the footprint of this capacitor more-or-less determines the cell size, folded and deep-trench capacitors with small footprints have been developed, and the select-transistor has been manufactured

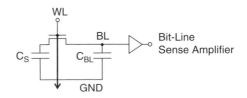

Fig. 11.8 Memory cell, bit-line, and bit-line sense amplifier of a MOS DRAM. *WL* word line, *BL* bit line

on top of the capacitor to save area, an example of 3D-integration at the device level. The making of dielectrics with a large relative permittivity (high-k dielectric) has become the show stopper for DRAMs, and they are trailing other chips on the roadmap by one generation for this reason and also because of

- Sub-threshold leakage of scaled-down select-transistors and
- Higher supply voltages to assure higher immunity against voltage-noise on bit-lines.

We now take a look at the 2010 state-of-the-art, analyzing two designs from opposite sides of DRAM applications. One is aimed at large capacity, by 3D-stacking of 4 2 Gb chips to produce an 8 Gb product [4]. The other is aimed at a high-speed embedded memory on a processor chip. Table 11.1 lists the technology, the density, the organization, the access-time, and the dynamic energy per bit in electron volts per bit, which illustrates *how many electrons are moved from 0 V to 1 V to store 1 bit of information*. We recall that $1 eV = 1.6 \times 10^{-19}\ V\ A\ s\ (J)$ or $1\ \mu W\ Gb^{-1}\ s^{-1} = 1\ fJ = 6.25 \times 10^3\ eV$.

The 2-Gb chip from [4] with its cell size of 20 F^2 has a density of 2 Gb/cm^2, and it requires a sizable dynamic energy of 120,000 eV/b. Its access time of 12 ns indicates that a larger string of cells, >512, are connected via one bit-line to a repeater/sense-amplifier.

The embedded DRAM [5] by contrast has a very short string of only 33 cells connected to a repeater-amplifier, thereby achieving a short access time of 1.35 ns, competitive with SRAM, while its density is a factor of eight higher at the same technology node. This advantage shrinks to 4 times compared with an SRAM built one technology node further. The energy/bit scales only mildly with voltage and is >10× that of an SRAM in the same technology.

The scaling-down of DRAMs is a problem not only of producing a high-density capacitor but also of making a very-low-leakage select-transistor. That is why a future DRAM, projected in [6], is based on the 22-nm node (Table 11.1), and the 20 fF capacitor would need an equivalent oxide-thickness (EOT) of 0.1 nm, which means a relative permittivity $k = 30$ and a physical thickness $t_1 = 3$ nm. SBT (strontium bismuth tantalite) would be a candidate. Even with the aggressive

Table 11.1 Performance of recent and future DRAMs

Chip	F[nm]	Area[F^2]	Bit/repeat	t_{access}[ns]	V_{DD}[V]	Energy/b [eV/b]	Density [Gb/cm^2]	Ref.
4 × 2 Gb	50	20 20 fF	512?	12	1.5	1.2×10^5 (20 $\mu W\ Gb^{-1}\ s^{-1}$)	2	[4]
4 × 292 kb	45	16 18 fF	33	1.35	1.0	5.6×10^4 (9 $\mu W\ Gb^{-1}\ s^{-1}$)	2.8	[5]
64 Gb proj.	22	4 20 fF	512		1.0	6.2×10^4 (10 $\mu W\ Gb^{-1}\ s^{-1}$)	13	[6]
256 Gb DiffDRAM proj.	16	8 2 × 5 fF	128	<1	0.5	8×10^3 (1.3 $\mu W\ Gb^{-1}\ s^{-1}$)	50	This book

cell size of 4 F^2, a target memory with 64 Gb would have a size of 5 cm^2 at an energy of 62,000 eV/b, 500 times higher than an SRAM at the same node.

The storage capacitor plus the bit-line capacitance, the associated voltage division, the bit-line noise, the transistor leakage and the required supply voltage seriously limit the advancement of DRAMs. While "a single transistor per bit" sounds attractive, in particular against the 6 T SRAM, it is timely to re-visit the architecture of these RAMs. The design of [5] points in a possible direction: It has one single-ended regenerator/33 bit and single bit-lines, while, at the other extreme, the SRAM has one differential regenerator/bit and two complementary bit-lines. Regarding the tremendous benefits of low-voltage differential signaling (LVDS), consider the differential DRAM in Fig. 11.9.

This differential DRAM stores each bit and its complement on two storage capacitors, it has two select-transistors and two complementary bit-lines per cell, and it has fully differential signaling and sense-amplifiers/repeaters. Because the information content is available as TRUE and COMPLEMENT, and because of the suppression of common-mode noise, and of differential amplification, the storage capacitors can be reduced to 5 fF and the voltage swing to 0.5 V, also considering a

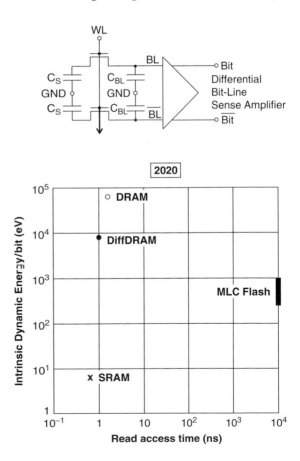

Fig. 11.9 A differential DRAM

Fig. 11.10 Energy/bit and access time for SRAM, DRAM, DiffDRAM, and NAND Flash

short string of just 128 cells per repeater, which amounts to a bit-line capacitance of 0.5 fF each at the 16-nm node, for which we find a line capacitance of 0.18 fF/μm in Table 5.2. Owing to the lower voltage and the differential-signal nature, we move one technology node further than the projected single-ended design [6]. The differential sense amplifier could look somewhat like the accelerator in Fig. 3.29, and it would provide a fast settling (high slew-rate) to full-swing signal levels. With the quad inside, it would also meet the transistor-variance concerns at 16 nm (see the preceding section on SRAM and Sect. 3.3, Fig. 3.17). The projected performance of this design with a cell size of 8 F^2 appears in the last line of Table 11.1: *A 16-nm fully differential DRAM (DiffDRAM) chip is projected to have a density of 50 Gb/cm^2, an access time of 1 ns, and a power efficiency of 1.3 μW Gb^{-1} s^{-1} = 8,000 eV/b.*

In a comparison with a normal single-ended DRAM design, the capacitor density (fF/μm^2) is reduced by 50% (EOT = 0.2 nm), the energy/bit eightfold, and the density/cm^2 is increased fivefold. Memory figures-of-merit are shown in Fig. 11.10.

The energy-efficiency of the future SRAM with just 7 eV/b shows the impressive benefit of the quad, differential amplification in each cell, at the price of cell sizes of 50 F^2 (3D) to 110 F^2 (2D), but with the potential of aggressive sizing-down to 10 nm and 3D-CMOS (Sect. 3.5). The DRAM is worst in energy per bit, presently optimal in the speed × density (16 F^2) product, but its down-sizing will stop soon at ~22 nm. Advancements such as DiffDRAM may offer a future optimum for speed × density. The ultimate in bit density is just a single transistor per bit and all transistors connected in series, that is, each transistor sharing its drain with the source of its neighbor. We treat this configuration in Sects. 11.3 and 11.4.

11.3 NAND Flash Memories

We can store information on an NMOS transistor in a string of NMOS transistors by setting its threshold voltage V_T to a high or low value. In Fig. 11.11, we show four such strings with vertical bit-lines per string and horizontal word-lines.

The memory transistors each show a floating gate, on which we assume that we have placed no charge or a negative charge, individually. If we want to analyze (read) the transistor (cell) under word-line WL$_1$ in the left string, we bias all non-selected word-lines to a high voltage, V_{GH} = 5 V, to make sure that all those transistors are turned on, and we set the gate voltage on WL$_1$ to the read voltage V_{GR} = 2 V. If the selected transistor had a negative charge on its floating gate, establishing a threshold voltage V_T = 2 V, the transistor would not conduct and there would be no current flow through the string. If the floating gate had no charge on it, V_T = 0 and the transistor would conduct, and current would flow through the string, because all the others have been turned on anyway. The topography of this type of memory is attractive, because each cell has just one minimum-size transistor and shares source and drain electrodes with its neighbors. No storage capacitor is needed, because we do not evaluate charge but rather measure a current. The minimum area would be 1 F^2 for the transistor gate, 2 × F^2/2 for source and drain,

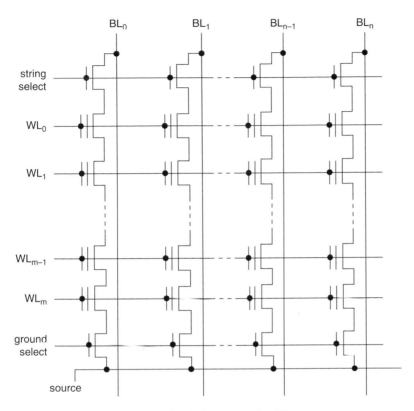

Fig. 11.11 NAND array architecture of a Flash memory after [7]

resulting in $2\,F^2$, and with the inclusion of the space F to the next string, the minimum total area would be $4\,F^2$.

The key is how to place the proper charge on the floating gate, and for re-programming, how to remove it. The dominant mechanism today for energizing channel electrons to jump over the potential barrier between Si and SiO_2 or SiON, to reach the floating gate, is the extremely high electric field of several 100,000 V/cm near the drain, producing hot-electron injection to the floating gate if a sufficiently high programming voltage is applied to the control gate of the selected transistor. For ON/OFF programming of the transistor, a difference $V_{T,OFF} - V_{T,ON} = 2$ V is produced by injecting ~30,000 electrons into the floating gate. In order to erase the negative charge on the floating gate, all NMOS memory transistors sit in p-type wells, typically 4 k cells in each well, and the well is biased positive (~+20 V) while the gates receive a negative bias. This causes a (non-destructive) breakdown of the oxide so that all electrons are driven into the well and into the n-type substrate in a *Flash*, giving the name to this type of non-volatile memory. A superb and detailed treatment of non-volatile memory technologies with emphasis on Flash has been given by Brewer and Gill [7].

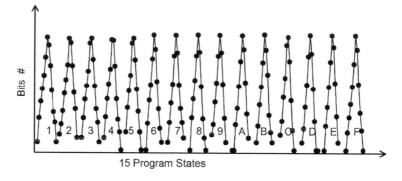

Fig. 11.12 The distribution of threshold voltages in the 4 b/cell 64 Gb Flash memory[9] (© 2009 IEEE)

We turn our attention to 1995 and paper [8], cited in Chap. 2 as a quantum jump for non-volatile MOS memories, those which do not lose their information when the power is turned off. Experiments had been made before to interpret different amounts of charge on a floating gate as 2 bits or more, but the Intel researchers in Folsom, CA, impressed the community in 1995 with a sophisticated, mature 32 Mb prototype, which could be programmed to four threshold voltages, representing 2 b/cell. Figure 2.19 shows the distribution of threshold voltages with their centers about 2 V apart. Gaining a factor of two or more in NV (non-volatile) memory density with multi-level-cell (MLC) Flash at the same technology node triggered an unparalleled worldwide development effort with a *2,000-fold increase in 2D NV memory capacity within 14 years*. A 64 Gb 4 b/cell NAND Flash memory in 43-nm CMOS was reported in 2009 [9], and we show the distribution of threshold voltages in Fig. 11.12.

The spacing of threshold voltages has shrunk to just 250 mV, equivalent to about 1,000 electrons. Only 2 years before, the multi-step programming algorithms, driving and sensing, had reached a resolution of 3,000 e^-/step [7], and in the meantime, the minimum programming steps must have come down to $<300e^-$ at speeds well below 1 ms. The state-of-the-art in 2010 is summarized in Table 11.2.

The table also contains estimates of the electrons/bit and of the intrinsic energy/bit, showing a spread of 64/1 between the lowest level (111) and the highest (000). It is worth noting that the intrinsic average energy/bit of $8000e^-$ is still a factor of seven better than that for DRAM and that the bits are non-volatile. However, the read-out time is 1,000-fold longer at several microseconds, which, for typical read-out currents of 1 μA, means a poor power efficiency of 10 mW $Gb^{-1} s^{-1}$.

The remarkable density gains of MLC NAND Flash memories are shown in Fig. 11.13 with a peak density of almost 30 Gb/cm^2 achieved in 2010.

We have added explicitly the prefix 2D to our discussion of NAND Flash so far, and we stay on the 2D scaling path a bit longer, before we address the 3D quantum jump. The end of the NAND Flash roadmap such as that for DRAM has been

Table 11.2 2010 MLC NAND Flash RAMs

Bit/chip	32–64 Gb
Technology	35–32 nm
Bit/cell	2–4
Cell area	$0.005\ \mu m^2 = 5\ F^2$
Bit/cm^2	26 Gb/cm^2
String	66 b
V_{DD}	1.8 V
V_{Supply}	2.7–3.6 V
Programming time	1.3–11.7 ms
Read time	10–100 μs
Electrons/b (3b/cell)	1000e$^-$–8000e$^-$
ΔV_T	250 mV
Intrinsic energy/bit	250–16,000 eV
Int. power efficiency	2.5 μW Gb^{-1} s^{-1}

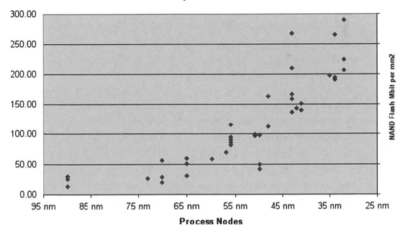

Fig. 11.13 The evolution of 2D MLC NAND flash density [Mb/mm^2] versus the process node [10] (©EE Times 2010)

predicted at 22 nm. The scaling of Flash is severely limited by the fact that the insulator thickness cannot be scaled because of charge losses from the floating gate due to tunneling. An insulator thickness of 6 nm continues to be demanded for long retention times, and the effective thickness to the control gate is even larger. Shortening the gate length massively enhances the influence of the drain on the transistor channel against the control intended to be dominated by the (control) gate. We treated this *drain-induced barrier lowering* in Sect. 3.2 and found that the threshold voltage becomes severely dependent on the drain voltage (3.18, 19). To show the effect, we simplify a bit by writing

$$V_T = V_{T0} - \frac{1}{A_V} V_{DS} \tag{11.1}$$

$$A_V = \frac{\varepsilon_I L}{\varepsilon_{Si} t_I} \sqrt{\frac{N_2}{N_1}} - 2 \tag{11.2}$$

A_V is the intrinsic voltage gain, obviously a strong function of the L/t_I ratio and the ratio of the permittivities, which is 1/3 for the Si/SiO_2 system. The doping concentration N_2 deeper in the channel should be higher than N_1 near the surface to suppress the drain effect. In any case, we see that $1/A_V$ becomes large quickly as L is reduced while t_I is not, giving the drain – source voltage a large influence on the threshold voltage. Depending on the position of a cell within a string, this can make thresholds overlap, causing an erroneous read. In addition, the variance of the charges on the gates will force a reduction of the number of distinguishable levels. The gate will lose control totally, as soon as field lines in the channel reach through directly from drain to source, causing a punch-through current to flow irrespective of the programming of the floating gate and the operation of the control gate. An optimistic projection towards 2020 is given in Table 11.3.

The concerns about scaling made NAND Flash a prime candidate for 3D integration by stacking chips with

– Through-silicon vias (TSVs) or
– Inductive coupling.

These technologies are very much in a state of flux. Remarkable results in 2010 were

– A 32 GB test chip with a stack of 8 chips, TSVs and 2 b/cell in 60-nm CMOS [11], which also indicates various architectural choices.
– A 2 Gb/s inductive-coupling through-chip bus for 128-die NAND Flash memory stacking [12].

Both publications are representative for the almost boundless advancements in the realization of ever larger memory densities and capacities on the chip scale. These interconnect schemes with power efficiencies of ~1 pJ/b = 1 mW Gb^{-1} s^{-1} (see Chap. 5) are basically compatible with the on-chip efficiencies of DRAMs and Flash. Beyond this prerequisite, the other issues for this progression strategy are

– Testing and known-good die
– Power density
– Reliability

Table 11.3 2020 projection for 2D MLC NAND Flash memory

Technology node	Bit/cell	Electrons/level	Density [Gb/cm^2]
22 nm	3	500	60
16 nm	2	250	80

– The manufacturing chain and product liability
– Cost

However, the basic notion is that 3D integration of memory by chip-stacking is a strategy with identifiable milestones and, therefore, that it is the most important path to reach the 2020 target of *DRAM and NAND-Flash capacity of several terabit on a footprint of a few square centimeters* (see Fig. 11.1).

11.4   Electron-Beam Permanent ROM (Read-Only Memory)

Silicon read-only memories with infinite retention times are of interest whenever the endurance of NV Flash memories is not sufficient or if the manipulation of the Flash data, modification or destruction, are an issue. So-called mask-programmable ROMs have filled this niche. On this type of memory chip, the memory transistors are set to a high or low threshold voltage by masking or not masking transistor gate areas. In the old days of metal gates, the thin-gate-oxide area of a transistor was either present in the gate mask or not. If not, the thick field-oxide would remain, the word-line gate metal would run across it as across all the others, but the field threshold voltage, being higher than the read voltage, would keep this transistor OFF forever. With the return of metal gates in nano-CMOS processes, this type of cross-bar memory with word-line gates orthogonal to source-drain bit-lines, becomes attractive again because of its *ultimate cell area of 4 F^2 and its ultimate scalability to <10 nm, offering a density of 250 Gb/cm^2* (Fig. 11.1).

The read-out of the transistor-string would have to detect a current ratio Hi/Lo, which can be optimized, contrary to the floating-gate transistor, even for 10 nm transistors by careful transistor and circuit engineering down to voltage levels of 0.3 V, where [13] is an exemplary paper. As a consequence, *energy/bit can be a factor of 20 smaller and read-out 100-fold faster than 16 nm Flash.* The penalty is read-only, but that would be acceptable or even desirable for important documents such as digital libraries or 3D high-quality movies as well as permanent embedded ROM.

Of course, mask-and manufacturing cost have been and could be prohibitive. However, with the general progress in technology, it pays to re-visit scenarios, and here it is maskless, direct electron-beam writing of the gate layer of the ROM. With a single beam, this has been a limited specialty in low-volume ASICs. But due to the emergence of massively parallel maskless e-beam MEB lithography (Chap. 8), this process for manufacturing giga-scale MOS nano-ROMs is feasible. We learn from Chap. 8 that e-beam systems with >100,000 microbeams at <10 nm resolution will be introduced by 2012. Such a system will be fed by data rates >100 Tbit/s. For the regular ROM pattern, the transistor grid is stored on the system, and only 1 bit will have to be transmitted per memory cell, namely an elementary shot or not. Therefore, *the e-beam direct-write programming speed can be 10 ms for 1 Tbit*, with no cost for masks and expenses just for machine time.

The capacity of the permanent ROM can be increased by 3D stacking of chips. The lifetime of this digital document can be enhanced by complete sealing, eliminating bond-related aging. Power could be supplied, and data could be read by inductive coupling, making this device a *digital Rosetta stone* [14].

11.5 Alternative Non-volatile MOS-Compatible RAMs

So far, we have dealt with electronic charge as the token for memory content and with MOS-memory types, which have been responsible for the relentless advancement of electronics. The non-volatile RAM with possibly unlimited retention time has continued to challenge researchers. Many solutions have appeared and will appear. We address those that have potential to make it onto the market by 2020.

Ferroelectric RAMs (FeRAMs): The dielectric in these 1 T per cell memories is lead zirconate titanate (PZT) or strontium bismuth tantalite (SBT), which change their electrical polarization as a function of temperature with a large hysteresis. Once they have been flipped into a state, they keep it over useful temperature ranges with long retention times.

Magnetic RAMs (MRAMs): Currents are the tokens in programming and reading the state of the cells in these memories via the electron-spin in a magnetic-tunneling junction (MTJ). The spin-torque-transfer (STT) variety is the most advanced.

Resistive RAM (RRAM): With the promise of the ultimate density $4\,F^2$ and possibly multiple-levels per cell (MLC), this crossbar-type memory continuously attracts inventors. The general direction is to place a material between orthogonal word- and bit-lines, whose resistance can be changed by programming-current pulses. In the simplest form, so-called half-voltage select is used for programming (writing) and reading. An alternative is to insert a select-diode or select-transistor together with a variable resistor. An example is the insertion of HfO_2 between TiN electrodes.

Phase-Change Memory (PCM): The material between the word-and bit-bars in this memory is chalcogenide glass, which changes reversibly from low-resistance crystalline to high-resistance amorphous when heated. An example is $Ge_2Sb_2Te_5$. The actual cell has a heater underneath and also a PNP bipolar transistor on top for cell-select.

A 2010 overview of the status of these RAMs is given in Table 11.4.

Table 11.4 2010 status of non-volatile RAMs

Type	Bit/chip	Technology node	Cell area	Density	Ref. (paper no. at ISSCC 2010 [15])
FeRAM	576 kb	130 nm	$41\,F^2$	150 Mb/cm^2	14.4
STT MRAM	64 Mb	130 nm	$41\,F^2$	150 Mb/cm^2	14.2
RRAM	64 Mb	130 nm	$10\,F^2$	0.6 Gb/cm^2	14.3
PCM	1 Gb	45 nm	$19\,F^2$	2.7 Gb/cm^2	14.8

The 2010 status of the PCM is write speeds comparable to NAND-Flash, however, with no data on write-energy, and random-access read times of 85 ns/b, again without data on energy/bit or the equivalent, power efficiency. We have entered the bit density in Fig. 11.10 and projected that with the indication that these crossbar-type resistive structures should be scalable to 10 nm, with some caution regarding the word-select devices.

Phase-change (PCM) and resistive (RRAM) memories are manufactured as back-end-of-line (BEOL) additions to CMOS so that they are envisioned as possible multi-layered additions generating a 3D-extension potential of their own, certainly with extreme challenges regarding power densities.

A chain of SETs (single-electron transistors), see Fig. 2.18, could form a serial memory. This chain with each transistor holding in its Coulomb box 0, 1, 2,... electrons under the condition of Coulomb blockade, would be a non-volatile memory. A serial read-out as in a multi-phase charge-coupled device (CCD), would yield its content but be destructive. Rewrite techniques as in the CCD days 40 years ago would have to be considered. At the 10×10 nm^2 node, the Coulomb energy would be about 20 meV, comparable with the electron thermal energy of 25 meV at room temperature. Therefore, cooling would still be required, but it shows the potential for this memory concept. Including the tunneling junctions, multi-phase clocking, and the re-write circuitry, this would offer a density of 10×20 nm^2 for a potentially multi-bit per cell, non-volatile serial memory at a hypothetical 5 nm node. However, contrary to the optimism until early in the first decade of the new century, no practical circuits have been realized by 2010 so that these innovations will become critical only after 2025.

Another revolutionary device is the memristor, which, as its name says, should make an impact on the realization of memories. The memristor M

$$M = \frac{d\Phi}{dq},$$
$$L = \frac{d\Phi}{dI},$$

is a passive two-terminal device, and it can be considered as the dual of the capacitor C:

$$1/C = \frac{dU}{dq},$$
$$R = \frac{dU}{dI}.$$

While a capacitor stores charge as a function of voltage, the memristor stores charge as a function of magnetic flux, and this flux results as the time-integral of voltage:

$$\Phi(t) = \int U(t) \cdot dt$$

This is a *monitor* of voltage over a certain time, reflecting its *history* as a time-continuous analog value, which can mean multiple levels, i.e., multiple bits per stored value. It is early in the exploitation of this device, and an obvious interest arises in its application as a synapse in electronic neural networks (Chaps. 16 and 18).

References

1. Raychowdhury, A., Geuskens, B., Kulkarni, J., Tschanz, J., Bowman, K., Karnik, T., Lu, S.-L., De, V., Khellah, M.M.: PVT-andageing adaptive wordline boosting for 8 T SRAM power reduction. In: IEEE ISSCC (International Solid-State Circuits Conference), Digest Technical Papers, pp. 352–353 (2010)
2. Nii, K., Yabuuchi, M., Tsukamoto, Y., Hirano, Y., Iwamatsu, T., Kihara, Y.: A 0.5 V 100MHz PD-SOI SRAM with enhanced read stability and write margin by asymmetric MOSFET and forward body bias. In: IEEE ISSCC (International Solid-State Circuits Conference) Digest Technical Papers, pp. 356–357 (2010)
3. Wang, Y., Bhattacharya, U., Hamzaoglu, F., Kolar, P., Ng, Y., Wei, L., Zhang, Y., Zhang, K., Bohr, M.: A 4 GHz 291 Mb voltage-scalable SRAM design in 32 nm high-k metal-gate CMOS with integrated power management. IEEE ISSCC (International Solid-State Circuits Conference) Digest Technical Papers, pp.456–457 (2009)
4. Kang, U., et al.: 8 Gb 3D DRAM using through-silicon-via technology. In: IEEE ISSCC (International Solid-State Circuits Conference), Digest Technical Papers, pp.130–131 (2009)
5. Barth, J., et al.: A 45 nm SOI embedded DRAM macro for POWER7™ 32 MB on-chip L3 cache. In: IEEE ISSCC (International Solid-State Circuits Conference), Digest Technical Papers, pp. 342–343 (2010)
6. Schroeder, H., Kingon, A.I.: High-permittivity materials for DRAMs. In: Waser, R. (ed.) Nanoelectronics and Information Technology, pp. 537–561. Wiley-VCH, Weinheim (2005). Chap 21
7. Brewer, J.E., Gill, M.: Nonvolatile Memory Technologies with Emphasis on Flash. Wiley, Hoboken (2008)
8. Bauer, M., et al.: A multilevel-cell 32 Mb flash memory. In: IEEE ISSCC (International Solid-State Circuits Conference), Digest Technical Papers, pp. 132–133 (1995)
9. Trinh, C., et al.: A 5.6 MB/s 64 GB 4b/cell NAND flash memory in 43 nm CMOS. In: IEEE ISSCC (International Solid-State Circuits Conference), Digest Technical Papers, pp. 246–247 (2009)
10. Young C.: 3 bit-per-cell NAND Flash ready for prime time.www.eetasia.com/ART_8800596524_499486.htm.27Jan2010.
11. Maeda, T. et al.: International Symposium on VLSI Circuits 2009, Digest, pp. 122–123 (2009)
12. Saito, M., Miura, N., Kuroda, T.: A 2 Gb/s 1.8pJ/b/chip inductive-coupling through-chip bus for 128-die NAND-Flash memory stacking. In: IEEE ISSCC (International Solid-State Circuits Conference) Digest Technical Papers, pp. 440–441 (2010)
13. Chang, M.-F., Yang, S.-M., Liang, C.-W., Chiang, C.-C., Chiu, P.-F., Lin, K.-F., Chu, Y.-H., Wu, W.-C., Yamauchi, H.: A 0.29 V embedded NAND-ROM in 90 nm CMOS for ultra-low-voltage applications. In: IEEE ISSCC (International Solid-State Circuits Conference) Digest Technical Papers, pp. 266–267 (2010)

14. Yuan, Y., Miura, N., Imai, S., Ochi, H., Kuroda, T.: Digital Rosetta tone: A sealed permanent memory with inductive-coupling power and data link. In: IEEE Symposium on VLSI Circuits, Digest Technical Papers, pp. 26–27 (2009)
15. IEEE ISSCC (Internatioanl Solid-State Circuits Conferenc.) Digest Technical Papers, Feb 2010

Chapter 12
3D Integration for Wireless Multimedia

Georg Kimmich

Abstract The convergence of mobile phone, internet, mapping, gaming and office automation tools with high quality video and still imaging capture capability is becoming a strong market trend for portable devices. High-density video encode and decode, 3D graphics for gaming, increased application-software complexity and ultra-high-bandwidth 4G modem technologies are driving the CPU performance and memory bandwidth requirements close to the PC segment. These portable multimedia devices are battery operated, which requires the deployment of new low-power-optimized silicon process technologies and ultra-low-power design techniques at system, architecture and device level. Mobile devices also need to comply with stringent silicon-area and package-volume constraints. As for all consumer devices, low production cost and fast time-to-volume production is key for success.

This chapter shows how 3D architectures can bring a possible breakthrough to meet the conflicting power, performance and area constraints. Multiple 3D die-stacking partitioning strategies are described and analyzed on their potential to improve the overall system power, performance and cost for specific application scenarios. Requirements and maturity of the basic process-technology bricks including through-silicon via (TSV) and die-to-die attachment techniques are reviewed.

Finally, we highlight new challenges which will arise with 3D stacking and an outlook on how they may be addressed: Higher power density will require thermal design considerations, new EDA tools will need to be developed to cope with the integration of heterogeneous technologies and to guarantee signal and power integrity across the die stack. The silicon/wafer test strategies have to be adapted to handle high-density IO arrays, ultra-thin wafers and provide built-in self-test of attached memories. New standards and business models have to be developed to allow cost-efficient assembly and testing of devices from different silicon and technology providers.

G. Kimmich (✉)
ST-Ericsson, 12, rue Jules Horovitz, B.P. 217, 38019 Cedex Grenoble, France
e-mail: georg.kimmich@stericsson.com

B. Hoefflinger (ed.), *CHIPS 2020*, The Frontiers Collection,
DOI 10.1007/978-3-642-23096-7_12, © Springer-Verlag Berlin Heidelberg 2012

12.1   Toward Convergence of Multimedia Devices

The advancements in silicon process technology over the last few decades enabled a significant increase of integration of about $2\times$ transistors every 2 years as stated in [1]. This increased density was enabled by smaller process geometries, which also resulted in smaller transistor gate lengths and gate oxide thickness, which increased the transistor and hence CPU (central processing unit) performances at similar rates as the density.

Higher performance levels combined with integration of multiple functions such as phone modem, CPU, video codecs (coder–decoders), graphic processors, image and audio processing hardware along with multiple interfaces for audio, display and connectivity into one piece of silicon are the basis for the convergence of multimedia devices.

The second important contribution for this convergence is the development of the software from the low level software drivers through the operating system used on the phone, up to the application software, which enables the user to run endless applications on the devices, including office automation, e-mail, web browsing, gaming, and mapping.

The sophisticated application software running on a fast processor and the hardware implementation for multimedia co-processing together with ultra-fast modem technology enable very rich combinations that are creating new forms of communication through social networking, online gaming, and information sharing on a local and global scale (Fig. 12.1).

Fig. 12.1 Multimedia convergence

12.1.1   Multimedia Performance Parameters

Similar to the personal computer evolution, the multimedia processor application software complexity leads to ever increasing requirements in CPU frequency, memory access bandwidth, and memory density. On top of the application software, the CPU can also be used to run some of the multimedia use cases such as mp3 decode or video encode/decode.

The CPU performance is commonly benchmarked in D-MIPS (million Dhrystone instructions per second). The base performance will be defined by the processor architecture such as instruction and data bus width as well as the CPU clock frequency. Clock frequencies passed the 1 GHz mark in 2010 and the D-MIPS performance only a couple of years behind the laptop CPU performances, a remarkable performance taking into account the stringent power and area constraints for handheld devices.

To increase the CPU performance, there are multiple solutions, which are optimized for the best trade-off between performance, software (SW) flexibility, power dissipation, and silicon area. A good trade-off was achieved with the introduction of the SMP (symmetric multi-processing) architecture which allows multi-tasking application software to be run and dynamic power reductions for lower MIPS use cases on a network of multiple CPUs.

To obtain the best multimedia performance, specific hardware, acceleration circuitry or co-processors have to be implemented on chip. Audio decode and encode is not very demanding in terms of MIPS and memory access bandwidth but needs to be optimized for power to enable long mp3 or similar digital format audio file playback times. Digital still camera functions require significant amounts of post-processing for image quality improvements. The image sensor resolution has quickly increased from 1 Mpixel for the first available camera phones to more than 10 Mpixels for the smartphones released in 2010.

New HD (high density) display standards for increased video quality increase the processing and memory bandwidth requirements. 1080p encode/decode with 30 frames per second are standard today on high-end smartphones, which require a total of about 3.2 GB/s bandwidth to the external LPDDR (low-power dual data rate RAM) memory. This evolution is not over yet; the frame rate will be doubled for the next-generation devices to allow slow-motion replay, the display sizes and hence resolutions will increase beyond the 1080p standard and, last but not least, new graphic applications require graphic processors that will require more than 10 GB/s of memory bandwidth alone to reach their full performance (Fig. 12.2).

12.1.2   Mobile Device Constraints

Portable devices need to meet, in addition to the performance standards, also constraints for volume and power. Small device package footprints enable smaller

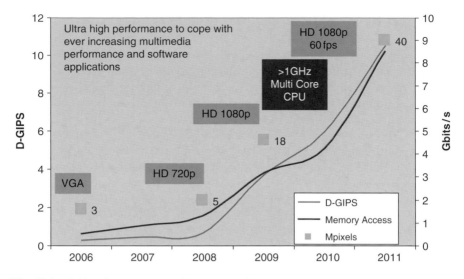

Fig. 12.2 Multimedia processor performance requirements

form factor printed circuit boards (PCBs) to reduce the total size of the final smartphone case. The same applies to the device thickness, which will need to be kept to the very minimum to allow overall thickness reduction of the phone application.

Mobile devices also need to have a very high mechanical reliability, which requires the packaged devices to pass numerous tests, including "drop tests" to ensure that they do not break or get disconnected from the PCB in case of shocks to the phone case.

Battery life is a major competitive feature for a smartphone. The capacity of the battery is very limited due to the small case volume and weight that can be allocated to it. Complementary to a traditional low power design chip implementation, power management techniques at system and device level need to be implemented to meet the talk time, audio, and video playback targets. Microprocessor-controlled use case management is used to switch entire functions, also called power islands, on the chip on or off depending on their usage for a specific application or to limit the frequency of the CPU or other critical circuitry when the full performance is not required, a technique called DVFS (dynamic voltage and frequency scaling). The silicon process needs to be optimized for device leakage to reduce the standby current if the multimedia chip is idle and for low voltage operation to reduce dynamic power. AVS (adaptive voltage scaling) methods are used to compensate for silicon wafer process variations, by slowing down fast and leaky parts and speeding up slow parts by applying transistor body biasing techniques. Chip I/O (input/output) switching power, specifically on high-speed DRAM interfaces, still account for a large amount of the overall power dissipation and can be significantly reduced with the Wide-IO 3D memory attachment technology described in this chapter. The material physics research promises the arrival of new battery technologies with higher power densities and shorter battery charging times than the currently established lithium ion technology, but none of these new technologies is yet ready for mass production.

Another strong constraint for mobile devices is thermal integrity. The ultrahigh performance of the application processor is pushing up the maximum power dissipation of the chip to a level of largely above 2 W, which cannot be managed without additional precautions at system, package, and PCB integration level. The total thermal budget of the application is defined by the thermal conduction path from the silicon to the application case. Constraints on the maximum silicon temperature (T_j) are given by the silicon process reliability and the standby leakage current, which increases exponentially at high temperatures. Common T_j maximum values are in the range of 105–125°C. The application case maximum temperature is around 40°C; above that temperature the device will be too hot for comfortable handheld use. What is done in a PC or laptop by a chip- or case-mounted fan has to be achieved in the mobile device by natural convection and thermal interface materials to spread the heat from the SOC (system on chip) to the handheld case surface.

Thermal system management needs to be implemented to ensure that the T_j and T_{case} temperatures will not exceed their specified limits. Software control can be used to avoid launching several power-hungry applications in parallel. Thermal sensors embedded close to the CPU or other high performance circuits are used to monitor the temperature and to throttle the CPU frequency down in case of imminent overheating.

12.1.3   Consumer Market Environment

The smartphone is situated at the high end of the mobile phone market and is gaining strong momentum worldwide, quickly entering the mid range with a total available market (TAM) of several hundreds of million phones per year in the next few years. This leads to a very competitive environment in terms of product cost as well as in terms of time to market pressure. To provide multimedia platform solutions in this competitive environment it is inevitable to follow a very robust and the same time flexible design process enabling the quick implementation and verification of performance, power, and thermal requirements optimized for cost, which are mainly driven by silicon area and package cost. Increasing software complexity puts the software development on the critical path for the market introduction of a new smartphone.

A high volume multimedia chipset platform production requires, owing to its complexity, a very mature and efficient supply chain with robust operations and testing to avoid quality and yield issues during production and to guarantee a high level of reliability of the final shipped good.

12.2   3D Strategies

In the following section, we will explain the basic 3D-partitioning strategies that can be considered for multimedia processors and their rationale.

12.2.1   Conventional 3D Packaging Techniques

3D device stacking has been common in the semiconductor industry for years. Examples are shown in Fig. 12.3. The first applications using this technology were DRAM and flash memory stacks, which have been in volume production for numerous years using conventional wire bonding processes. Conventional die stacking does not allow direct chip-to-chip connectivity and in the industry is very often referred to as 2.5D packaging technology.

A similar process was used in some earlier application processor packages. This die stacking technique helps to reduce the resistivity and capacitive load, which is typical for the long trace line between the SOC and an external DRAM. It also helps to reduce the overall PCB footprint area significantly. It was enabled by improved assembly methods and the KGD (known-good die) memory test strategy and business model. This packaging method presents some challenges to the supply chain and because the chipset manufacturer takes the responsibility for the memory test after assembly with the SOC.

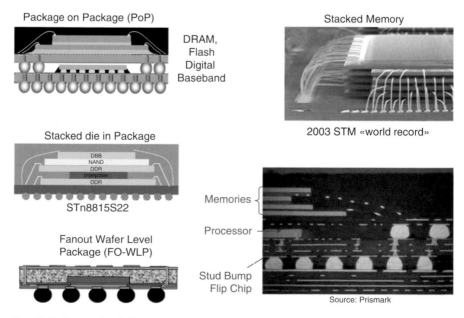

Fig. 12.3 Conventional die stacking

The package-on-package (POP) stacking method combines the advantages of the previously described stacked die in package technology with the traditional supply chain and business model where the handset manufacturer undertakes the final assembly of the product by stacking the standardized memory package, which was previously tested, on top of the SOC package. POP stacking will remain mainstream for the mid-range and high-end smartphone applications as long as the memory bandwidth requirements can be satisfied with LPDDR DRAM technology.

The fan-out wafer-level package (FO-WLP) technology, initially developed for smaller die and package areas, provides a potentially competitive solution in the near future. In the FO-WLP technology the Si die is placed into a reconstituted plastic wafer. The die bump to package ball routing is performed with a back-end metallization process, similar to that used on the silicon die. This process results in improved routing density, reduced package height, and improved thermal characteristics and provides a solution for memory stacking on the SOC die once the dual side option becomes available at a competitive cost.

12.2.2   Wide-IO DRAM Interface

In most of the current smartphone designs the memory is stacked on top of the SOC device via the POP technology described in the previous section. There are multiple limitations inherent to the POP technology that will need to be overcome for the next-generation smartphone and tablet applications. The main issue is that POP limits the access bandwidth to the DRAM. The maximum bandwidth available with LPDDR3 is limited by the interface frequency of 800 MHz, which cannot be increased further without silicon process technology changes on the memory, which would increase leakage and dynamic power. This frequency is also limited by the interface signal integrity due to the parasitic external resistance and capacitive load (RC) mainly induced by the package traces between the DRAM and SOC device and their I/O buffers. The maximum data bus width that can reasonably be implemented on a small mobile-compliant POP package with the typical size of 12×12 and 14×14 mm^2 are two 32 bit DRAM channels. This results in a maximum memory bandwidth of maximum 12.8 GBps. Previous LPDDR2 configurations without POP limit the maximum bandwidth even further to about 3.2 GBps owing to the slower maximum interface frequency dictated by the increased parasitic RC for external board traces between the memory and processor chip. Another limitation for which the LPDDR concept cannot be further scaled up is the power dissipation due to the switching of the I/O with relatively high capacitive load in the range of 5 pF at the high frequency of up to 800 MHz.

Wide-IO 3D technology is based on the concept of massively enlarged (Wide) data buses between the DRAM and the SOC. Since the total I/O count of the SOC is already the determining factor for the package size and PCB footprint, a new concept has been developed in which the memory die is directly stacked on top

Fig. 12.4 Wide-IO chip
stack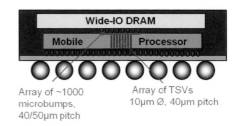

Array of ~1000
microbumps,
40/50μm pitch

Array of TSVs
10μm Ø, 40μm pitch

of the digital baseband (DBB) SOC using face-to-face or face-to-back die attachment techniques, allowing thousands of interconnects between these chips as shown in Fig. 12.4.

The Wide-IO memory interface is in the process of being standardized through the JEDEC (Joint Electron Device Engineering Council) standardization body, which was also the driver for the previously adopted memory interface standards in the PC and mobile application world. This standardization will simplify the cooperation between the chipset makers and DRAM memory suppliers because they allow common electrical, timing, and physical specifications to be incorporated in designs. This facilitates the compatibility between these devices and allows second sourcing options between multiple memory vendors and provides a healthy level of competition of memory suppliers on the same market segment, which is required for high volume products. The physical locations of the Wide-IO bumps are shown in Fig. 12.5.

The initial target of the Wide-IO standard memory bandwidth will be 12.8 GBps, which will be reached with four memory channels running at 200 MHz each over a 128 bit wide data bus at single data rate (SDR). The hardware implementation of a SDR interface is simpler, which means it will use less silicon area than the currently used dual data rate (DDR) interface in which the data toggles at the rising and falling edge of the clock signal.

In 2010, the single-die DRAM memory density available on the market is 1 Gbit. Similar to logic chips, memory chips are following Moore's law as set in [1] and will continue to grow in density, which means the next-generation 4-Gbit devices will become available in early 2012. The next-generation high-end smartphone memory requirements for multimedia and application processing are in the range of 1–8 GB. To reach this high level of memory density – which is in the same range as the PC market memory requirements today – engineers have come up with a new idea, the stacking of multiple (2–4) memories via through-silicon via (TSV) and Cu-pillar connections, as developed for the memory-to-SOC attachment. In the memory industry this concept is also called "cubing", as shown in Fig. 12.6. Data on the maturity status, i.e., the ability to manufacture these cubes in high volumes and at reasonable cost, are not yet publicly available. For that reason, system architects are exploring alternative solutions to enable higher memory densities for the next product generation. One identified solution to that problem is the combined use of Wide-IO and the traditional LPDDR technique; The Wide-IO will be used for the short latency, high bandwidth multimedia use cases, whereas

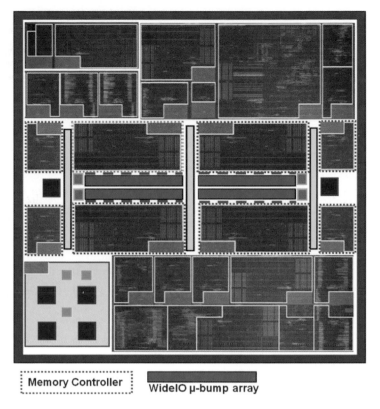

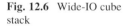 Memory Controller    WideIO μ-bump array

Fig. 12.5 Floorplan of wide-IO bottom die

Fig. 12.6 Wide-IO cube
stack

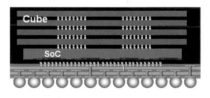

the slower and higher latency external LPDDR will be used for applications that do
not require high bandwidth. The drawback of this solution is the more complex
implementation of the application software, which needs to arbitrate the best use of
the memory resources.

Next-generation smartphones and tablet computers have memory access bandwidth requirements that can no longer be satisfied with the LPDDR-standard. Ever
more sophisticated video gaming applications are pushing the graphics performance to levels that require more than 15 GBps of memory bandwidth. These
graphic processors became available in 2010 but their full performance will only be
enabled if the memory interface can provide the necessary throughput. The rapid

market adoption of new tablet devices as well as the increased quality of TV home displays is asking for very high video display resolutions on mobile chip sets either integrated in or connected to these devices. The QXGA (Quad eXtended Graphics Array) display standard defines a 3.1 Mpixel display, which together with the requirement of an increased video frame rate of 60fps will almost saturate the current LPDDR bandwidth, which has to be shared with other parallel-running tasks on the application processor.

The Wide-IO technology is much more power-efficient than LPDDR, especially at high bandwidth-use cases, as shown in [2]. A power dissipation comparison for an external 16-bit DRAM interface with numerous Wide-IO data bus widths for a video encode use case as shown in [3] is presented in Fig. 12.7. These simulations show that even an eightfold increase in data bus width still results in lower total power dissipation of the memory, due to the dramatically reduced RC of the bus interface. The individual power management of the memory channels and banks will further increase the advantage of the Wide-IO technology in an application space where battery and thermal management is essential.

There are no immediately available alternatives on the market that can provide the required DRAM bandwidth to the next-generation products. Scaling up the LPDDR technology with POP is not an option due to pin count and power considerations as shown previously. Nevertheless there are options to enhance the standards by moving to low swing, single-ended, and terminated data on the DRAM interface. Even if improvements can be made, there will still be a remaining gap to reach the Wide-IO bandwidth at a reasonable power level. Other options on the horizon are the use of serial interfaces between SOC and external memory. These options are promising in terms of bandwidth potential, which can reach similar levels as the Wide-IO concept. The drawback of the serial interface techniques are on one hand the relatively higher switching power on the interface and on the other hand the lack of standardization. Available serial proprietary technologies are not taking off because of the business model, which creates too heavy dependencies on the intellectual property (IP) and memory provider.

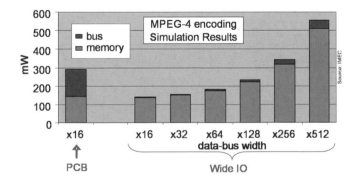

Fig. 12.7 Wide-IO power comparison

12.2.3 Technology Partitioning

Similar to the Wide-IO scenario, the idea of technology partitioning is to use the best suited silicon technology per implemented function and then stack these silicon dies together into one system. Instead of applying this concept only to the memory, it can also be applied to the functions of power management and radio-frequency (RF) transceiver technologies and others. The very expensive advanced CMOS Si process technologies are often not very well suited for analog and high voltage functions, which can be implemented at similar density in legacy CMOS process nodes, which are much cheaper. Analog IPs hardly shrink beyond 40-nm process nodes since they are very much dependent on the R, L, and C parameters, which do not change significantly over high density CMOS silicon process node generations. I/O area and electrostatic discharge (ESD) protection diode surfaces do not shrink linearly with process technology advancements. The high voltage requirements for the power management circuits and the legacy interfaces require thicker gate oxides, which come for free with legacy process nodes and which need to be paid for extra as additional gate oxide process steps in state-of-the-art CMOS processes.

Other than the potential cost optimization, there can be also some technological advantages of such a 3D partitioning concept, in which, for example, the power management unit (PMU) is closely connected via 3D to the circuitry that needs to be supplied. Figure 12.8 shows the critical RLC path from the power management chip to the processor. Shorter supply loops improve the power integrity and hence the

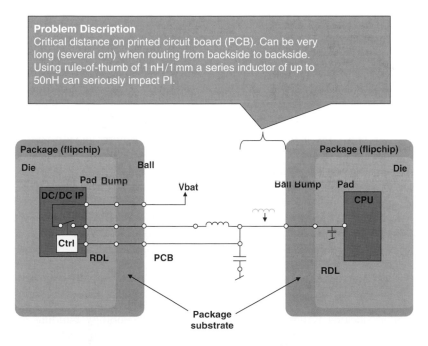

Fig. 12.8 Power integrity with conventional PMU/digital partitioning

Fig. 12.9 Silicon interposer

performance of these circuits. The implementation of decoupling capacitors to the supply nets on the analog process node will help to improve the power integrity of fast switching circuitry on the digital die. This split also allows better control of the digital voltages, which can be lowered further to gain on dynamic and leakage power.

12.2.4 Other 3D Concepts

The 3D logic circuitry split is a concept that has been discussed for some time. The underlying idea is the search for a breakthrough in further integration once Moore's law, described in [1], comes to its end at the boundary of the atomic limit, or before that time to accelerate the integration by overlaying the 2D integration with another dimension. There is currently no strong motivation to push for a logic 3D split for the multimedia application processor space since the 3D split is more expensive than the move to the next more advanced process node. Logic split could become interesting for ultra high performance CPUs. One should not forget that this will strongly increase the power density of these stacks, which will most likely require new cooling techniques.

The silicon interposer or the silicon package, another, more conventional application of 3D integration, is more likely to become interesting for the multimedia application space. Silicon interposer designs are already available as prototypes. Usually they consist of a relatively low cost substrate, which has a few levels of metallization and integrated TSV connections, and multiple silicon components placed side-by-side on top of and connected via the interposer. This concept allows the same architectures as described in the Wide-IO or technology split but has the advantage of better thermal performance with the drawback of increased package size and cost. The silicon package concept shown in Fig. 12.9 takes it one step further and replaces the organic package substrate altogether as shown. The cost of silicon compared to the available organic substrates as well as the mechanical properties of the silicon package are not yet competitive with the conventional packages for the multimedia application processor space.

12.3 3D IC Technology

3D silicon stacking requires two fundamental key technology building bricks: One for die-to-die attachment, which allows one chip to be connected vertically to a second chip, and another one that is required to connect vertically through

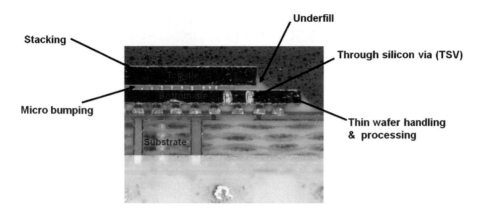

Fig. 12.10 3D stack

a chip. New techniques are required to assemble the vertical chip stack (Fig. 12.10).

12.3.1   Die-to-Die Attachment

We differentiate between two approaches: Face-to-face and face-to-back. In the face-to-face approach the active sides of the two silicon dies to be attached are facing each other. This approach has the advantage of a very low resistance and capacitive load between the two silicon dies owing to very short connections. The bottom die of the face-to-face stack can be connected to the package either via traditional wire bonding techniques or as flip chip by adding vertical connectors called through silicon vias (TSVs) through the silicon. The latter approach has the advantage of being die size independent and it works also if the bottom die is smaller than the top die. The flip chip approach also has major advantages for the power distribution network implementation of the bottom die because the supply pins of the PCB can be directly fed to the most power hungry circuitry on the bottom die (Fig. 12.11).

The face-to-back approach consists in stacking the back side of the top die on the active side of the bottom die. This is a more universal approach because it allows more than two dies to be stacked on top of each other. The Wide-IO "cube" technology uses the face-to-back approach.

Regardless of the die orientation approach, there is a need for an attachment technology between the dies. The technology most commonly used in current designs is called micro-bumping and uses copper pillars. Cu pillars are fine pitch capable down to below 40 μm, which enables a very high interconnect density and qualifies them as technology for the Wide-IO standard. Cu pillars have a high

Fig. 12.11 Cu pillar bump
cross section

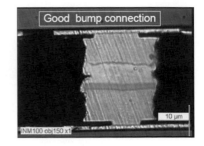

current capacity and they allow various assembly strategies that guarantee a high standoff height and thus simplify the underfill process required for the physical attachment of the two dies. The Cu pillar technology is derived from the already available die-to-package attachment methods, which are already in volume production and can be considered as a mature technology.

Solder bumping, which is still very much in use for flip chip packaging, does not allow a bump pitch below ~130 μm and as such does not qualify as high density 3D interconnect technology.

The industry is looking into new technologies that are not yet ready for volume production. Contact-less technologies, which are based either on capacitive or on inductive coupling of the signals between the dies, should be mentioned. These contact-less interconnect methods could provide an answer if high levels of vertical stacking are required, but their maturity levels are still far from potential market introduction with smartphones.

12.3.2 Through-Silicon Vias

3D silicon stacking requires a method to interconnect the dies vertically. This is performed with TSVs. One can picture a TSV as a hole through the silicon die, which is filled with metal and connected to the top and bottom metal routing layers of the silicon. There are three established approaches to implement the TSVs in silicon wafers. In the via-first process the etching is done before the FEOL (front end of line), which signifies the transistor implementation. The via-middle process implements the TSV after FEOL but before the metallization process (back end of line, BEOL). In the via-last process the TSVs are etched as the final step of the wafer process. The basic process steps for these methods are explained in Fig. 12.12.

The via-first and via-middle process options are best suited for the high-density interconnect requirements for the Wide-IO technology because they allow very high via aspect ratios, which means the best trade-off between the via diameter, which should be as small as possible, and the wafer thickness, which should be as thick as possible for better manufacturability. Typical values for via-first and via-

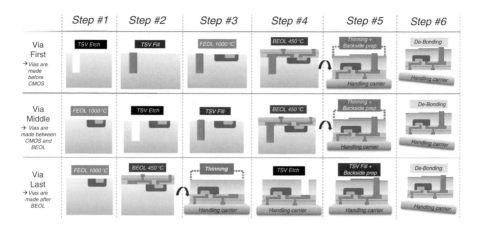

Fig. 12.12 TSV process options (Courtesy of Yole)

middle are 10 μm diameter and 100 μm die thickness. The diameters will be further reduced with increasing technology maturity. A very important parameter to monitor is the parasitic resistance and capacitance value of the TSV, which is not only linked to the aspect ratio, but also to the materials and the process used for the implementation.

The via-last process variant remains advantageous for applications where the diameter does not need to be below about 40 μm, because the process is robust and cheaper to implement. Typical implementations of the via-last process are the image sensors for camera modules shown in Fig. 12.14 used for small-form-factor digital video and still camera sensor products, which have been in mass production for several years.

12.3.3   Assembly

Depending on the application and multi-die-stacking requirements, the dies can be assembled face-to-face or face-to-back as mentioned previously. For the 3D stack assembly we also differentiate between the die-to-wafer and die-to-die attachment techniques. Die-to-wafer attachment brings the advantage of more parallel processing and the option of wafer testing of the stacked devices. The condition to allow die-to-wafer stacking is to have a significantly smaller top die than the bottom die. A more universal approach that is die size independent is the die-to-die assembly process.

Multiple assembly methods for the Cu pillars, from solder to thermocompression to ultrasonic bonding techniques, are available or under development. Most common are conventional reflow techniques, which are being replaced by

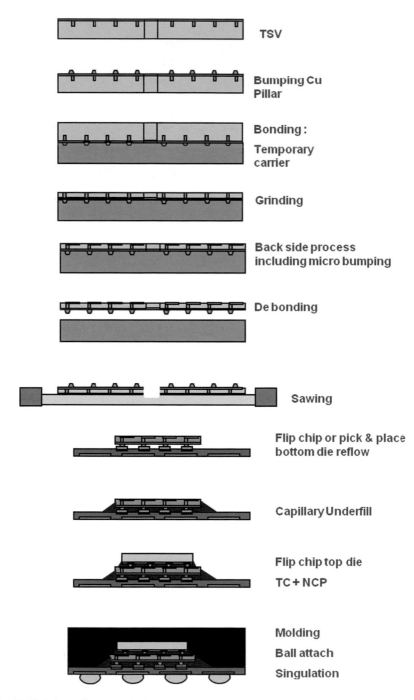

Fig. 12.13 Die-to-die assembly flow

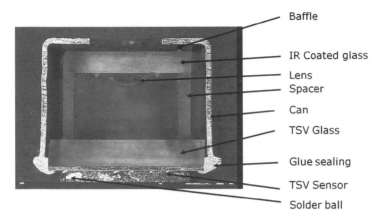

Baffle

IR Coated glass

Lens

Spacer

Can

TSV Glass

Glue sealing

TSV Sensor

Solder ball

Fig. 12.14 Camera module

thermocompression and thermosonic methods, which allow finer pitches. Figure 12.13 shows the assembly flow for a face-to-back, die-to-die scenario.

12.4   Design Automation

3D integration requires the development and use of new electronic design automation (EDA) tools. These tools need to be able to handle heterogeneous silicon technologies and include models for the interconnect micro-bumps and the TSVs.

Exploration and planning tools are required to help to partition the top and bottom die to optimize cost and performance and need to be capable of taking the different performance, power, area, and cost properties of the involved silicon process technologies into account. Electrical analysis tools are becoming more important with increased integration, higher performance, and lower silicon process voltages. Signal integrity simulation of multi-die systems for high speed interfaces and power integrity analysis taking into account the complete *RLC* chain from PCB to transistor, including the TSV and micro-bumps, are essential to guarantee the required device performances. Improved thermal analysis tools and models are necessary to ensure device reliability. A comprehensive thermal simulation for high performance use cases will need to be performed at system level. This simulation needs to include the thermal characteristics of the 3D multi-die interconnect and the environment, specifically the package, PCB, and the case of the smartphone application.

The 3D EDA tool chain is not in a very mature state yet since the R&D investments of the EDA companies have been hesitant, based on the lack of 3D industry adoption and hence the low return of investment on this type of tools. This implies that for the first generations of 3D products there will still be some lack of automation.

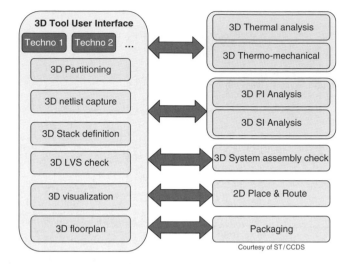

Fig. 12.15 Design automation for 3D design

Figure 12.15 shows the required elements and links for an integrated 3D design framework and user interface.

12.5 Supply Chain Considerations

3D stacking is adding more integration process steps to the already complex process of chip fabrication. This brings new challenges to meet the cost and time-to-market requirements for these devices. During the manufacturing process the package components will be processed by multiple companies in multiple locations. Before the smartphone chipset manufacturer can delivery a final packed good to the phone handset manufacturer, there will be process and test steps involving memory suppliers for the stacked memory for Wide-IO products, the silicon foundry for the SOC, the test site for the electrical wafer sort (EWS) process and outsourced assembly and test (OSAT) companies for the wafer processing, assembly, and final testing of the die stack involved. Figure 12.16 provides an overview of the process steps and possible owners. New test, supply chain, and business models will need to be worked out to cope with these new conditions.

12.5.1 Test and Failure Analysis

On top of the usual wafer test and final test steps of a system on chip, additional steps are required for a 3D stack. First, each die will need to be tested separately with a very high level of fault coverage to ensure a high yield of the separate

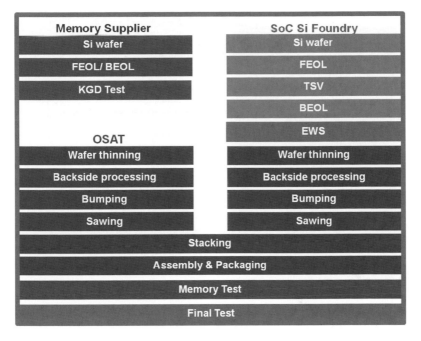

Fig. 12.16 Supply chain for Wide-IO stack

devices and to avoid the very costly aggregated yield loss after die assembly. The combined value of multiple dies in one package can become significant; a failure of one single element will cause failure of the complete system and overall impact the business if the individual yield targets are not reached.

There are also new technical challenges that arise for testing of 3D circuits. Ultra-thin-wafer handling is required since the TSV process requires wafer thinning, which is also called back grinding, of the wafer into the range of 50–100 μm. Wafers in this thickness range become very unstable and the equipment needs to be adapted to be capable of achieving still fast testing without damaging the wafers. Wafer thinning can also impact electrical parameters of the silicon and thus also the performance.

Silicon-wafer probing is done with test needles that are placed on probe pads which are on conventional silicon dies usually shared with the bumping locations for flip-chip designs or wire-bond pads for package attachment designs. These probe pads are placed in a pitch of typically 80 μm, arranged in only a few rows. High-density 3D designs will connect through more than 1,000 pads in very dense arrays of less then 50 μm. This will require the development of new more advanced test equipment to enable fine-pitch probing. An intermediate solution to that issue is the placement of probe pads with increased pitch in separate locations, which has the drawback that the bump connectivity will not be tested before assembly.

For failure analysis of customer-returned failed parts, it is important to trace back to the wafer lot, wafer number, and die location on the wafer of the failing die

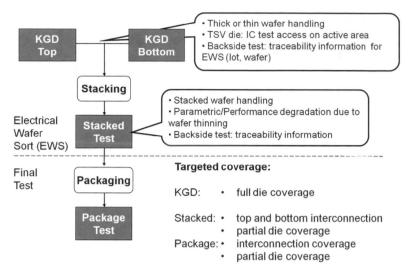

Fig. 12.17 3D test flow

in the 3D stack. For conventional designs this is done by writing the main properties of a die into a one time programmable memory (OTP), which can be read out after assembly in the case of issues. In the 3D stacked configuration, there needs to be also a way to identify the properties of the stacked devices; additional trace and debug methods need to be implemented (Fig. 12.17).

Another challenge is the integration of heterogeneous technologies into one package. Traditionally multimedia and application processing, power management, RF transceivers, and DRAM functions are placed on separate chips, which are tested separately. With the 3D technology some of these functions will be stacked into one package and need to be tested together to ensure proper functioning of the assembled system. The test equipment and test approach requirements for these different technologies are not the same, which causes additional test steps and thus costs. Improved, more versatile test equipment that is capable of running analog, digital, and memory test sequences will need to be used. Other improvement steps are to embed additional built-in-self-test circuitry (BIST) into the sSOC, which enables complex memory tests run from the CPU of the application processor through its memory interface. To enable specific direct access tests for the memory on Wide-IO devices there are a number of reserved package pins, which are directly connected to the memory through the TSV of the bottom SOC die, which can be used by the memory supplier to perform some specific test sequences for production testing as well as for failure analysis purposes.

Die-to-wafer assembly will bring an additional challenge to the wafer test since the equipment will need to be capable of handling stacked wafers in case the chipset maker would like to benefit from the possibility of an assembled wafer test prior to the die singulation and packaging.

12.5.2   Business Model

In the case of mixed suppliers for the 3D stack, which is typical for the Wide-IO systems, there is interest in new business models involving the memory suppliers, chipset makers, and smartphone handset manufacturers involved in the design, manufacturing, and testing of the chipset. Each company involved is interested in optimizing its return on investment and in increasing the profit in its core business; A memory supplier wants to sell memory; the chipset maker wants to sell the system but does not want to buy and re-sell the memory because this represents a big value that is not created inside its core business. Each process and assembly step adds cost and value to the final good and each step includes risk of yield loss due to lack of mature manufacturing processes or simply due to the complexity of the processes. The preferred model of the memory suppliers and chipset manufacturers requires the smartphone handset manufacturer to source all major components similar to the conventional discrete chipsets as shown in Fig. 12.18.

In this model the handset maker accepts responsibility for the inventory management of all components. The chipset manufacturer, who is responsible for the 3D assembly, will draft the KGD DRAM memory wafers out of a consignment cage, which can be managed either by the handset manufacturer or the memory supplier, and assembles the final goods. The value chain has to be analyzed carefully; the potential yield and resulting cost impact has to be quantified and integrated into the KGD delivery contracts. The failure-analysis flow and liability scenario in the case of customer returns of the finished goods has to be agreed by the involved suppliers.

The KGD model was established for conventional 2.5D chip stacks. Known good die means that each die on the wafer has been fully tested and failing devices have been marked on the wafer.

The manufacturing, test, and assembly process steps are likely to take place in several companies located in different continents and countries. For that reason, the legal and tax framework has to be carefully studied to avoid issues with local tax and custom regulations that could delay the assembly process significantly.

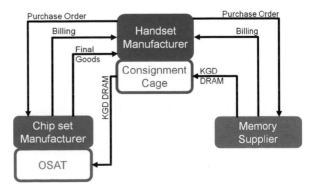

Fig. 12.18 Wide IO business model

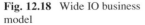

12.6   Outlook

Smartphones and tablet computers are still new on the consumer market, and many generations of multimedia handheld devices are to come with the integration of ever more features and the development countless software applications. New 4 G modem technologies will provide download speeds of 100 Mbit/s or higher and enable streaming of high quality video. Social networking software will become more and more sophisticated and location-based services will be made possible by the GPS and gyroscope circuitry integrated in smartphones. Augmented reality is one of the upcoming applications in which graphics are overlaid on live video. Head-mounted displays and virtual retinal displays (VRDs), in which the display is projected onto the retina, will enable a whole new application space as well as new advertisement techniques. Sensor and MEMS (micro-electro-mechanical systems) integration will open entirely new doors to mobile medical applications and interfacing of wireless applications with the human body. Nobody can predict where this development will end.

Very likely, the heterogeneous technology, memory, and logic stacking techniques described in this chapter will be combined with the integration of MEMS on one single silicon interposer to achieve the maximum performances of all integrated functions.

All these applications will require more CPU processing power on a very small area, which increases the power density further. The performance budget of these future devices will be defined by the ability to keep the junction and case temperatures of these devices down to guarantee lifetime reliability and user comfort. New cooling techniques will need to be developed, based on new packaging and thermal interface materials such as PCMs (phase-change materials) and new packaging and product case design concepts.

Functional, electrical, and physical interface standardization and sound business models are the key prerequisites for the high level of R&D efficiency required to enable the take-off and evolution of the 3D silicon technology. Countless software applications with as yet unimaginable functions will be available as soon as the technology maturity allows these complex hardware designs to be built in high volumes.

References

1. Moore, G.: Cramming more components onto integrated circuits. Electron. Mag. **38**(8), April 19 (1965)
2. Gu, S.Q., Marchal, P., Facchini, M., Wang, F., Suh, M., Lisk, D., Nowak, M.: Stackable memory of 3D chip integration for mobile applications. IEEE IEDM (International Electron Devices Management) 2008. doi: 10.1109/IEDM.2008.4796821
3. Facchini, M., Carlson, T., Vignon, A., Palkovic, M., Catthoor, F., Dehaene, W., Benini, L., Marchal, P.: System-level power/performance evaluation of 3D stacked DRAMs for mobile applications. Design, Automation and Test in Europe 2009, DATE'09, pp. 923–928. IEEE, Piscataway (2009)

Chapter 13
Technology for the Next-Generation-Mobile User Experience

Greg Delagi

Abstract The current mobile-handset market is a vital and growing one, being driven by technology advances, including increased bandwidth and processing performance, as well as reduced power consumption and improved screen technologies. The 3G/4G handsets of today are multimedia internet devices with increased screen size, HD video and gaming, interactive touch screens, HD camera and camcorders, as well as incredible social, entertainment, and productivity applications.

While mobile-technology advancements to date have made us more social in many ways, new advancements over the next decade will bring us to the next level, allowing mobile users to experience new types of "virtual" social interactions with all the senses. The mobile handsets of the future will be smart autonomous-lifestyle devices with a multitude of incorporated sensors, applications and display options, all designed to make your life easier and more productive!

With future display media, including 3D imaging, virtual interaction and conferencing will be possible, making every call feel like you are in the same room, providing an experience far beyond today's video conferencing technology. 3D touch-screen with integrated image-projection technologies will work in conjunction with gesturing to bring a new era of intuitive mobile device applications, interaction, and information sharing.

Looking to the future, there are many challenges to be faced in delivering a smart mobile companion device that will meet the user demands. One demand will be for the availability of new and compelling services, and features on the "mobile companion". These mobile companions will be more than just Internet devices,

Reprinted by permission from "Greg Delagi: Harnessing Technology to Advance the Next-Generation Mobile User Experience", IEEE ISSCC 2010, Dig. Tech.Papers, pp.18–24, Feb.2010. © IEEE 2010.

G. Delagi (✉)
Texas Instruments, 12,500 TI Boulevard, MS 8723, Dallas, TX 75243, USA
e-mail: t-wright@ti.com

and will function as on-the-go workstations, allowing users to function as if they were sitting in front of their computer in the office or at home.

The massive amounts of data that will be transmitted through, to and from these mobile companions will require immense improvements in system performance, including specialized circuits, highly parallel architectures, and new packaging design. Another concern of the smart-mobile-companion user will be that their device is able to deliver an always-on, always-aware environment in a way that is completely seamless and transparent. These handsets will automatically determine the best and most appropriate modem link from the multiple choices on the device, including WiFi, LTE, 5G, and mmWave, based on which link will optimize performance, battery life, and network charges to deliver the best possible user experience. In the future, adaptive connectivity will require many different solutions, including the standard modem technologies of today, as well as new machine-machine interfaces and body-area-networks.

All of the new and exciting applications and features of these mobile-companion devices are going to require additional energy due to added computational requirements. However, a gap in energy efficiency is quickly developing between the energy that can be delivered by today's battery technologies, and the energy needed to deliver all-day operation or 2-day always-on standby without a recharge. New innovations ranging from low-voltage digital and analog circuits, non-volatile memory, and adaptive power management, to energy harvesting, will be needed to further improve the battery life of these mobile companion devices.

Increased bandwidth combined with decreased latency, higher power efficiency, energy harvesting, massive multimedia processing, and new interconnect technologies will all work together to revolutionize how we interact with our smart-companion devices.

The implementation challenges in bringing these technologies to market may seem daunting and numerous at first, but with the strong collaboration in research and development from universities, government agencies, and corporations, the smart-mobile-companion devices of the future will likely become reality within 5 years!

13.1   Introduction

The mobile-handset market continues to be a dynamic and growing one, enabled by technology advances that include increased bandwidth, greater processing performance, increased power efficiency, and improved display technologies to deliver compelling user experiences. We envision a world in 5 years where mobile devices will offer new high-performance services and features, support always-on/always-aware connectivity, and deliver battery life that will provide days of active-use experience. These "smart-mobile-companion" devices of the future will be intelligent autonomous systems with a multitude of incorporated sensors and display options, all designed to make our lives easier and more productive (Table 13.1).

Table 13.1 Features of the always-on, always-connected, always-aware mobile companion

- 3D display
- 3D imaging and gesture interface
- Integrated projection technology
- Object and face recognition
- Context awareness
- Noise cancellation
- Bio-sensing applications
- Adv. GPS location awareness
- "I see what you see" HD video-conferencing
- Adaptive connectivity
- *Internet of Things*
- *The sixth Sense*
- *Brain–Machine Interface*

Ten years ago, when mobile communication was voice-centric, who could have imagined we could squeeze text and multi-media-centric mini-computers into a mobile handset of 100 cm^3?

Innovation drives new technologies in mobile handsets and is necessary to meet the challenges of bringing the smart-mobile-companion device of the future to reality. These numerous challenges include high performance, adaptive connectivity, and power efficiency.

The next generation of multimedia features, augmented with vision analytics and 3D displays, will enable users to control and interact with their mobile devices in a more natural way. Mobile companions will be more than cell phones with internet and multimedia capabilities. They will function as "work-anywhere" mobile platforms integrated with all of the needed desktop applications. The massive amounts of data processed by these mobile companions will require a 1,000× improvement in processing performance (power efficiency) from technology advancements in device density, specialised circuits, highly parallel architectures, and 3D integration.

The smart-mobile-companion device will deliver an always-on, always-aware experience in a way that is completely seamless and transparent to the user. These mobile companions will employ many different radio technologies on the same device, including WiFi, GPS, BT, 3G, 4G, and mmWave. The handset will automatically determine the best and most appropriate modem link to meet the bandwidth requirements, adapt to the current environment, and provide context-sensitive information.

Adaptive connectivity provides not only information links worldwide through standardised wireless links, but also information connectivity to the mobile companion's immediate environment (Fig. 13.1).

This may include information about the user's own body through a body-area network, or data about the surrounding establishments on the street, in the building or within the room, including other equipment and sensors. In a sense, we will be connected to a few 1000 objects surrounding each of us in our daily lives, in a world

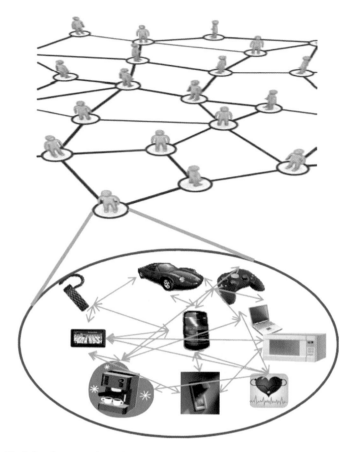

Fig. 13.1 Evolving features of mobile handsets

connected through an "Internet of Things" [1, 2]. To meet the challenge of delivering adaptive connectivity, technology advancements such as delivering seamless connectivity, providing new machine–machine interfaces, and collecting data from body-area networks will be required.

The demand of new and exciting applications and features of these smart-mobile-companion devices will quickly drain the power available with today's battery technology [3]. In order to deliver days of operation and standby [4] without recharge, technology advancements must meet the challenge of providing more energy-efficient circuits and systems. New innovations, ranging from low-voltage digital and analog circuits, zero-leakage standby, to energy harvesting, will be needed to further improve the battery life of these smart-mobile-companion devices.

Research and innovation will be needed to meet these three main challenges:

– High-performance processing,
– Adaptive connectivity,
– Power efficiency.

This chapter examines the status of each challenge, and ascertains where we need to be in the future, in order to deliver a compelling user experience within the constraints of a smart-mobile-companion device.

13.2   The High-Performance Challenge

The smart-mobile-companion device of the future is not going to be the voice- and text-centric device of the past, nor will it just be a multimedia and content-delivery device of today. Instead, it will support exponentially-increasing computational requirements for video, imaging, graphics, display controllers, and other media-processing cores. Performance requirements for these functions will go beyond one trillion (10^{12}) floating-point operations per second (TeraFLOPS) or one trillion operations per second (TeraOPS) in the next decade, as shown in Figs. 13.2 and 13.3, while power consumption for these functions must decrease.

As technology advances, a keyboard or even a touch-screen will no longer be needed; you will use gesturing in the air to interface with your device. In the film "Race to Witch Mountain" [5], 3D holographic user interfaces were shown as control devices for a space-ship, allowing the children in the movie to fly the ship without actually touching any part of it. In the future, we can look for mobile devices to have similar 3D holographic controls, allowing the user to interact in 3D space to manipulate data, play games, and communicate with others.

Yet another futuristic man–machine interface technology on the horizon is Honda's brain–machine interface (BMI) [6]. Using this technology, a computer is able to sense what a person is thinking and translate that into action. In the future, users will be able to

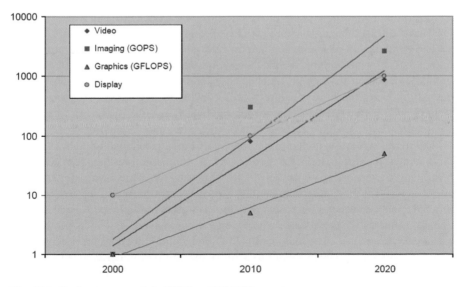

Fig. 13.2 Performance trends in GOPS or GFLOPS over time

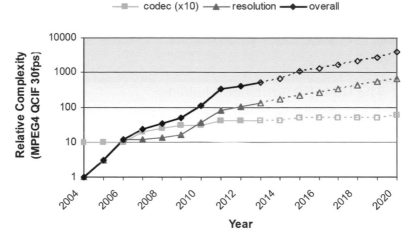

Fig. 13.3 Relative complexity of a video system (MPEG4 coder/decoder, 30 frames/s), QCIF (320 × 240 pixels) = 1

interact with their mobile companion without touch or even gesturing. Combined with 3D-holographic displays, your smart-mobile-companion device will be able to sit on a table, and you will be able to control everything about it using your brain alone.

Today, these man–machine interface concepts are moving from the silver screen to the test labs. The future of the man–machine interface is rapidly evolving to the next level with the help of technology advancements, and we expect that this interface will become a standard means of communication between a user and the various technologies, with which they interact. In particular, the BMIs discussed above allow completely motion-free control of devices with direct brain-wave detection. Voiceless communication is possible when sensors convert neuro-patterns into speech [7]. But, there are other aspects of the man–machine interface as well:

The combination of free-space human interface using gesturing and projection output onto a surface will enable the "view-and-interact" aspect of the future smart-mobile companion to become much larger than the physical dimensions of the device, allowing the always-aware mobile device in new ways. Together, these technologies better enable users to get data into and out of the smart-mobile-companion device, making it the center of their communications.

The future system-on-a-chip (SoC) must perform a wide range of computing and multimedia-processing tasks and provide a true multi-tasking environment, leveraging the concurrent operation of all the tasks associated with simultaneous applications. The best approach for tackling these diverse, concurrent tasks is to provide a range of general-purpose processors as well as multiple specialized computational elements for those tasks, where general-purpose processors cannot achieve sufficient performance. The multimedia and computation functions must be mapped onto the most appropriate processor to achieve the best performance/power ratios.

Complex-use cases will use many of these processors in parallel with data flowing appropriately between the cores. Inter-processor communications and protocols for efficiently distributing the processing among the available cores will be needed. In some cases, dedicated hardware, to enable this computation "side-loading", will be beneficial for managing the real-time nature of the concurrent-use cases, and for effectively abstracting the functions from the software operating system running on the main processors. This functionality will be provided without excessive frequency or excessive power being dissipated in any particular processor or core. In addition, the future device must provide new features, capabilities and services requiring significantly increased general-purpose computing power and support for new peripherals, memory systems and data flow through the system. With multiple input and output devices and 3D-holographic displays combined with communication-intensive 3D graphics and imaging, as well as other applications, the performance levels will continue to increase with future mobile-device services and features.

The critical architectural choices and technology enablers will be

- Multi-core application processors,
- Specialised circuits providing parallel computations,
- 3D interconnects,
- Man–machine interfaces, and
- Managing energy and thermal limits.

13.2.1 Multi-core Application Processors

Increased general-purpose computing power will enable handsets to function as our main computer at home or the office or as our main computer when we are in a mobile scenario, achieving the "workstation-on-the-go" goal. Multi-CPU processors will be required to provide the increased performance while simultaneously keeping the frequency at acceptable limits (Fig. 13.4). Today, most mobile handsets are based on two processors operating in parallel at 4 DMIPS/MHz.

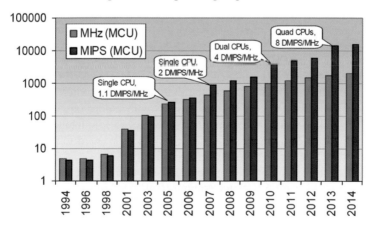

Fig. 13.4 Performance in MHz and MIPS

13.2.2 Specialized Circuits Providing Parallel Computations

For demanding multimedia and other dedicated functions, specialized circuit architectures and algorithms must be developed to perform computation in parallel so that the exponentially increasing computational demand can be met within achievable frequencies and acceptable power limits (Fig. 13.2). Figure 13.3 illustrates the relative complexity of a video core as it adapted to meet increasing demands [18]. In this particular case, the core solution evolved over multiple generations of mobile processors from a software-programmable DSP-only solution to a DSP-plus-hardware-acceleration solution and from there to a dedicated video core with many computations performed in parallel.

Recently, extreme parallelism and specialisation have been applied in a video test chip to demonstrate a video codec decoding a 720p/30fps video stream while consuming only 2 mW of power [8]. Video is used here as an example, but new services and features will require specialized circuits for other functions beyond video. These will challenge designers and researchers to create specialized circuits optimized to deliver the highest performance at the lowest possible power consumption, meeting the consumer's demand for high-performance services and features.

13.2.3 Thermal Limits

The hub of all this information exchange, compression and de-compression of information, and information intelligence on the smart mobile companion is the application processor. This device will continue to evolve to a solution that will have multiple domains for specific functions. This type of partitioning will support high-data-rate processing demands.

Within this design, the heterogeneous processors and cores will need to work concurrently while having their own memory bandwidth, creating a need for massive, low-latency, high-data-rate RAM for intermediate data and runtime storage. Today, data rates between the application processors and external SRAM and Flash are in the tens of Gb/s range. With increased demands for higher data rates, we will see a more than 10× increase in 5 years to hundreds of Gb/s (Fig. 13.5).

It is also clear that, in addition to the external memory interface, on-chip data communication among the multiple processors will require one or more interconnection fabrics that can handle Terabits/s. Specialized low-latency, asynchronous buses and low-power interconnection fabrics will be needed to achieve this 10× increase in data flow among the elements of heterogeneous systems.

A single-data-bit communication link between a processor and an external on-board memory device at 1.5 V switches about 10 pF, resulting in 10 mW/Gb/s for such a link. Similarly, using a serializer/deserializer (Ser/Des) driving a differential signal over a 90 Ohm differential wire, also consumes about 10 mW/Gb/s including all transmit (Tx) and receive (Rx) overheads in today's systems. State-of-the-art

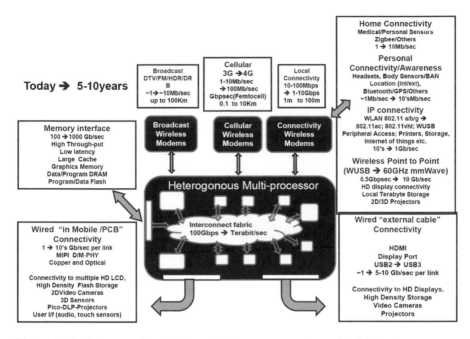

Fig. 13.5 Multiple connections to the application processor and associated data rates

published research shows that 2–3 mW/Gb/s is achievable when a number of parallel links are designed together to share PLLs and other efficient supplies [9]. Yet, the goal is to have another 10× reduction beyond today's state-of-art technology (Fig. 13.6).

As we look into the future, we can expect that data will move at speeds of a Terabit/s. Thus, the power spent per bit transmitted becomes a major challenge. Today, in a typical smartphone with a volume of $100~cm^3$, the maximum power consumption, using no mechanical air flow, is limited to about 3–5 W. Beyond that, the surface temperature becomes uncomfortable for the device to be held in a hand or carried in a pocket. With this power limit, the three key power-consumption areas are split between the processor, the display, and full-power cellular and other wireless transmission circuits. So each of these areas consumes anywhere from a fraction of 1 W to a maximum of 2 W, depending on the use case. Specifically, for the processor, the data transmission between different SoCs, memory, transceiver chips, and off-phone data communications, consume 10–20% of the processor power budget. The above calculations focus on peak-power budget, which is mainly due to temperature limitations inside the smart-mobile-companion device. Reduction in power can be achieved using circuit techniques, low-overhead efficient PLLs, and tuned processes.

Associated with the data communication on a mobile device is the complexity of EMI (electromagnetic interference) radiation in the RF bands being used for multiple wireless links. Use of optical connections for on-mobile communications is another axis, where optics versus Cu can be used to reduce EMI side effects. Supporting optical interfaces is another challenge to be met in delivering lower power well below 1 mW/Gb/s.

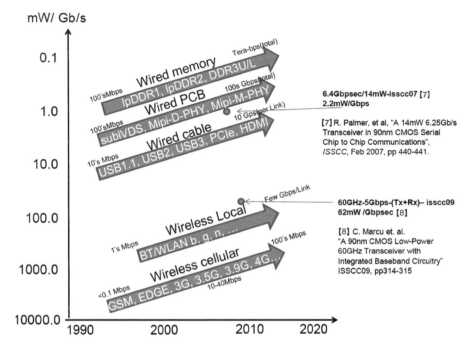

Fig. 13.6 Evolution of the communication power efficiency in mW/Gb/s. References [7] and [8] in the figure refer to [9] and [19] in the text, respectively

13.2.4   3D Interconnect

Achieving the right level of performance is not just a circuit challenge, but also a process and package challenge, amounting to a link-impedance increase and/or voltage-swing reduction by 10× or more. Bringing external memories closer to the processor will reduce the capacitance that needs to be driven, reduce noise levels, and enable lower voltage swings. One way to achieve this is by using through-Si-vias (TSVs) and die stacking for very wide data transmission between dies. Figure 13.7 illustrates the evolution of memory connectivity from wire-bonded stacked dies to POP (package-on-package), to TSV stacking. The challenge will be to move 3D interconnect from research into high-volume production. Chapter 12 provides further guidance on 3D integration.

13.3   The Adaptive-Connectivity Challenge

Besides the high-performance communication needs, the smart mobile companion of the future will be always-on and always-connected (Fig. 13.1). And, most importantly, it will not only understand the user and the environment, but it will also provide context intelligence, giving the right access to the right database at the

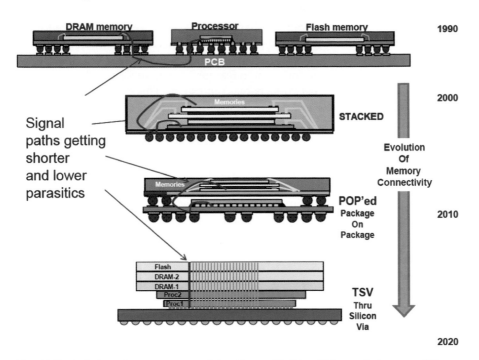

Fig. 13.7 Evolution of memory connectivity. Future 3D ICs will be a mix between different memory processes, high-performance processors, low-leakage processors, and specialty interface ICs (maybe optical) for a most optimal stack

right time. To achieve this adaptive connectivity goal, there will be at least a $10\times$ increase in the next decade in the number of bits crossing various boundaries. Such an increase in data flow poses a great challenge in terms of technologies used for such data transmission, be it wireless, wired or optical.

Figure 13.5 shows the multiple data links that can be connected to the core applications processor of the mobile companion. These links include cellular wireless, home/office LAN wireless, point-point wireless, and Bluetooth, among others. From the mobile companion to the cellular network, data rates will be increasing from today's few Mb/s to hundreds of Mb/s using LTE and other future cellular-modem technologies. For comparison, within the confines of the office or house, data rates using wireless local-area networks will increase from the tens of Mb/s to Gb/s using mm-wave links.

13.3.1 Seamless Connectivity

As technology enables mobile devices to provide increased computational capability and connectivity to the Internet, advanced-application developers will have substantial MIPS and bandwidth at their disposal. However, for advanced functionality, more than just increased performance is needed. Internet connections will be enabled by a multitude of possible modem technologies on the mobile device. The device will determine the most appropriate modem link to optimize performance, battery life, network charges, and so on. This could switch between 3G, 4G, WiFi, mm-wave, and so on, at any time without the knowledge of the handset user, making it completely seamless and autonomous. The mobile companion will make connection decisions on its own based on the user's preferences and the mobile's awareness of the context and the environment.

To enable seamless connectivity, where the mobile companion has context awareness of the user's behavior as well as environment awareness of where the mobile is physically, will require multi-processing as well as low-power technology to enable constant monitoring without draining the battery.

13.3.2   Machine–Machine Interface

For the smart-mobile-companion device to better serve its owner, it should know its location, in what situation the owner is, and which services and features should be provided automatically. Advances in the machine–machine interface, such as vision analytics, can be a solution in the future. Using the phone's cameras and sensors, the mobile companion would automatically recognize where the user is.

The always-aware environment has already begun to be shown in film and in research. For example, the sixth sense [10] was demonstrated in the film "The Minority Report" [11] when Tom Cruise's character was walking through a mall and billboards, placed on the wall, recognized who he was and tailored each advertisement specifically for him. In the future, this sixth-sense-like technology will be used to make your mobile companion always aware of where you are, what you are doing, and with whom you are interacting.

This sixth-sense system was demonstrated with real technology in March 2009 [10] with a wearable version. The demonstration showed how your handset will be able to recognize people and give you information about them, as well as providing you with a constant stream of data about things around you. While it may seem like we will be in information overload, these mobile companions will be "smart" devices that will learn about you, your likes and dislikes, so that all information that comes to your attention will be relevant and needed.

To achieve these advanced machine-machine interfaces, research is needed on different low-power sensors to measure and to communicate various body measurements, such as the brain electrical activity using electroencephalography (EEG) or a person's head or eye direction. All of these sensor/camera/3D-imaging devices can produce multiple data streams that can be interpreted into control information.

13.3.3 Body-Area Networks (BAN)

The future smart-mobile-companion device could be used with body-area networks (BAN) to interact with multiple biological sensors scattered over or within the body. Depending on the device's application, this biometric data may be used to make appropriate and necessary suggestions to improve quality of life. The smart-mobile-companion device can serve as the link between the biological sensors and a wide-area network to automatically transmit health data to doctors or emergency services. The enabling technology needed to deliver BAN will be ultra-low-power sensors, ultra-low-power, low-bandwidth communication protocols and circuits, and extreme system-security protocols.

13.4 The Power-Efficiency Challenge

As the amount and complexity of applications and features on the mobile companion increase in the future, the drain on the battery will be immense without the "right" technology advances.

Over the past 35 years, technology scaling has halved the transistor area per semiconductor-lithography node allowing the SoC designer to integrate more features without increasing chip costs. With technology scaling, SoC designers have also enjoyed the benefits of reducing the supply voltages and increasing drive currents leading to reduced power and higher performance, respectively. In turn, this has enhanced the user experience over time, with longer talk and standby times.

However, technology scaling comes with increasing process variations and leakage currents. With smaller transistors, random doping fluctuations within the channel affect transistor performance more and result in larger variations of transistor threshold voltage (Sect. 3.2) within the die as well as from die to die. Traditionally, in older process nodes, the on-chip variation is accounted for with either voltage or frequency margins, but in deep-submicron, the margins required are becoming so large that they impact the ability to advance performance or lower power. Additionally, with each process node comes an alarming relative increase in leakage (= sub-threshold = barrier-control) currents (Sect. 3.2, Fig. 3.14).

Technology-scaling benefits, coupled with architectural innovations in processor design, result in the active-power dissipation per Mio. Instructions/s (MIPS) continuing to decrease from mW/MIPS to μW/MIPS, that means by three orders of magnitude, in 20 years, as shown in Fig. 13.8 for programmable digital signal processors (DSPs).

During the period from 2000 to 2016, while the Li-battery technology is increasing its energy density (Wh/cm^3) by more than 40%, driven by package and process optimisation, the conventional Li-ion technology will likely saturate its energy density [4]. To reach the next level of energy-density gain will require a shift to new chemistries, either with new materials or new technologies or both.

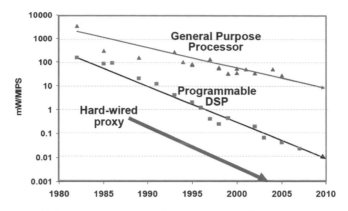

Fig. 13.8 Power efficiency in mW/MIPS (nJ/instruction) for different processor architectures

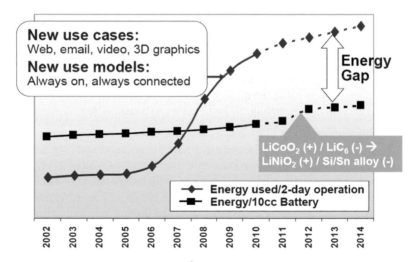

Fig. 13.9 The energy gap between a 10 cm^3 Li battery and the new mobile-companion energy

Even then, the net effect is a major energy gap (Fig. 13.9) for portable equipment. This gap is the result of the linear improvement of the Li-battery technology and the exponential growth of energy consumed in new consumer-use cases.

For mobile devices in the next 5–10 years, the goal will be to deliver at least a week of operation from a single charge with battery efficiency increasing to deliver a capacity of about 1–1.5 Wh. This will also require a more than 10× reduction in display power, allowing the viewing of an HD movie. The new always-on, always-connected use model will pose new challenges for standby operation. The energy consumed in a conventional, voice-centric mobile phone is <0.75 Wh over 2 days. But for a multimedia-centric device, it can consume 3.75 Wh of energy in the same duration for just standby operation, an increase of 5× [3]. This is primarily due to the exponentially increased leakage current as well as the complexity in logic gates

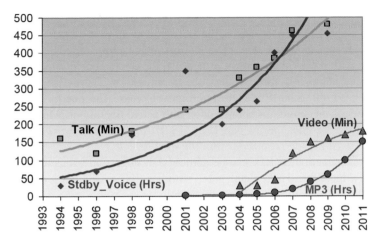

Fig. 13.10 Evolution of mobile-device capacity in minutes or hours, respectively

and on-chip memory for all the processing engines mentioned earlier. The smart-mobile-companion device must also maintain the benchmark battery life in a conventional, voice-centric talk-time and voice standby time (Fig. 13.10). With the continued improvement in the quality of video and audio technologies, playback time continues to increase, setting a new battery-life benchmark.

13.4.1   Adaptive Power Management

Among many innovations in the industry, SmartReflex™ power-and performance-management technologies have been developed and have been in high-volume production since the introduction of the 90nm OMAP2420 processor [12]. SmartReflex technologies include a broad range of intelligent and effective hardware and software techniques that dynamically control voltage, frequency and power, based on device activity, modes of operation, and temperature. SmartReflex starts with co-optimisation between process and circuit technologies to reduce memory-bit leakage by at least one order of magnitude. In the 65nm OMAP3430 multimedia-application processor [13], three major voltage domains are used, each with associated 16 power domains. Together, these enable a critical trade-off of frequency scaling as applications demand and power optimisation with fine-grain voltage- and power-domain management. The net power reduction achieved at the SoC level with autonomous software control is 1,200×. With power management turned to the SoC OFF mode, the SoC consumes only a few μA. Of the 1,200× power reduction,

– 4× comes from transistor-leakage optimisation,
– 300× comes from optimisation of circuits, SoC, and software.

Details of the second generation of SmartReflex technology have been published in [14] and [15], which focus on the active-mode power reduction. The RTA memory achieves 4× active leakage reduction. When compared to the high-activity work load with power management turned off, the sleep mode dissipates <0.1% of the energy, achieving three orders-of-magnitude of energy reduction. Both generations of the SmartReflex technologies increase the energy efficiency by three orders of magnitude in both standby and active operations. Moving forward, in order to close the energy gap (Fig. 13.9), *another three orders of energy efficiency boost will be required within the next decade.*

A power reduction of 10× can be achieved through transitions to new process technology including Hi-k metal gate. Another 100× will have to come from additional research and innovations, including

– Ultra-low-voltage and -power circuits, logic, SRAM
– Non-volatile memory,
– High-efficiency DC–DC and A/D converters, and
– Parallel architectures.

13.4.2   Ultra-Low-Voltage Circuits

For the user-centric mobile companion of the future, we need to develop ultra-low-voltage circuits in order to achieve orders-of-magnitude reduction in the power of the key components. Operating at low voltage ($V_{DD} < 0.5$ V) impacts both the dynamic power and the leakage (DC) power. A hypothetical chip might include low-voltage SRAM, NV(non-volatile)RAM (Chap. 11), ultra-low-voltage, ultra-low power ADCs (analog–digital converters, Chap. 4), and distributed high-efficiency DC–DC converters for each voltage domain on the chip. Properly designed CMOS logic continues to operate, even when V_{DD} is below threshold (Sect. 3.3). However, the logic speed is very slow. $V_{DD} = 0.5$ V is a good compromise between speed and ultra-low power. For 6T SRAM, operation at 0.5 V is also possible (Sect. 11.1). For lower-voltage operation, 8T is an option.

On every interface to the world, there is a sensor or an antenna followed by a signal path ending with either an ADC or a DAC (digital–analog converter). Converters are fundamental blocks. Their resolution and power efficiency determines the power and cost efficiency of that new sensor or wireless solution. Fortunately, research is showing that, with the "right" converter architecture and process advancements, ADCs are on a path for energy efficiency. As wireless bandwidths extend to support increasing bit rates and more complex modulation schemes, the power per conversion-step, an ADC figure-of-merit (FOM), must continue on the reduction path to maintain a constant power per signal path, even though the bandwidths and rates are increasing. The path shown in Fig. 13.11 indicates that, with the "right" choice of architecture, the FOM improves with the process node, where today ADCs are well below 100 fJ per conversion step (Chap. 4), while a 110db audio-quality DAC consumes <0.5 mW [16].

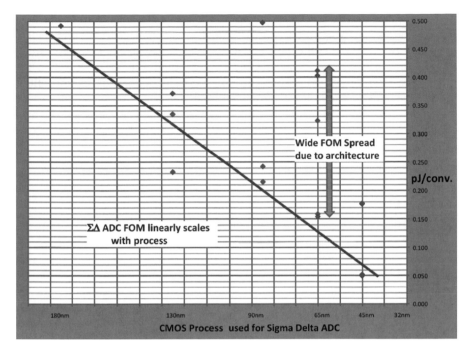

Fig. 13.11 Energy figure-of-merit (*FOM*) for analog–digital converters (*ADCs*) in pJ per conversion step

In order to realise the savings in power that is enabled by 0.5 V operation, a DC–DC converter is required to deliver ~10 μW of power with high efficiency. LDOs, which are typically used for voltage conversion today, have poor efficiency at low-output voltage. Switching regulators require an external inductor and can have poor efficiency at low output power as well. Micropower management technology needs to be developed to deliver ~10 μA at an output voltage of ~0.5 V from an input voltage of ~1.8 V with ~80% efficiency [17].

But low-voltage circuits by themselves are not enough to achieve the 100× power reduction that is needed for mobile devices of the future. Hardware accelerators can also be used to perform high-throughput computation at low voltage (Sect. 3.6). For mobile monitoring of medical diagnostics such as ECG (electro-cardiogram), we need hardware accelerators for FT (Fourier transformation), FIR filtering, and CORDIC functions. Together, these have the potential to reduce power by 100×.

13.4.3 Zero-Leakage Standby

In many systems, SRAM-leakage power is a large fraction of the total power. For many years, researchers have been searching for a practical non-volatile RAM

(NVRAM) that dissipates zero-standby power because it retains its state when the power is turned off. Magnetic RAM, Ovonic or phase-change RAM, or ferroelectric RAM (FeRAM) are all candidates for NVRAM. As of today, none of these technologies has proven to be scalable to very high densities. However, for many ultra-low-power applications, ultra-high density is not required. For ultra-low-power RAM in the range of 1–100 Mb, FeRAM has great potential.

Today's newest technology is wireless power or inductive charging, which allows mobile devices to be charged without plugging them into a wall. New technologies in the future will allow true wireless power under certain conditions. When the phone is in standby or monitoring mode, the power demands will decrease significantly, and future devices will not depend on the battery, but will use energy harvesting techniques (Chap. 19) to power the device. At these very low power levels, it is possible to "harvest" energy from the environment to provide power to ULP (ultra-low-power) circuits. For example, with energy harvesting, no battery is required to make life-critical medical monitoring possible. There are several sources, from which useful electrical power can be extracted, including mechanical vibration, heat-flow from a temperature differential, RF energy in the environment, and light energy. Figure 13.12 shows the practical power levels that are achievable from these sources.

Energy levels can be realised from the μW range using thermal or piezoelectric sources to the centiwatt cW level with electromagnetic sources (Chap. 19). With the combination of energy harvesting and ULP ICs, future mobile devices could have a near-limitless standby mode.

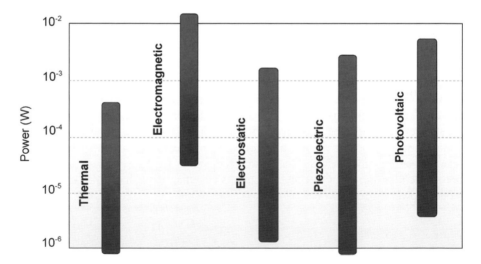

Fig. 13.12 Energy-harvesting power levels achieved by 2010

13.5 Conclusion

Users of future mobile-companion devices expect their devices to be always-on, always-aware, to have extended battery-life, and to provide a highly adaptive experience at very high performance levels of communication. This requires broad innovation in technology. By using heterogeneous multi-processing, multi-core processors, specialised circuits and parallel architectures along with 3D interconnect and multi-chip partitioning, the high performance of these smart-mobile-companion devices will be achievable while maintaining acceptable thermal limits. An always-on, always-aware use case will require adaptive connectivity that will be solved with machine-machine interfaces, seamless connectivity, and body-area networks. While a $10\times$ improvement in energy efficiency can be provided by technology, another $100\times$ improvement will have to come from research and innovations in areas such as

- Ultra-low-voltage and –power circuits, logic, and SRAM (Chaps. 3 and 11),
- Non-volatile memory (Chap. 11),
- High-efficiency DC–DC and analog-digital converters (Chap. 4), and
- Parallel architectures.

New technologies, like TSV, optical interfaces, and efficient PLLs, will have to be investigated to reduce the power level to well below 1 mW/Gb/s (Fig. 13.6), in order to move around the massive amounts of data these new mobile companions will require. The implementation challenges to bringing ultra-low-power and high-throughput technologies to market may seem daunting and numerous at first, but with the strong collaboration in research and development from universities, government agencies, and corporations (Chap. 22), these smart-mobile-companion devices will become reality within the next decade.

References

1. Santucci, G.: Internet of things – when your fridge orders your groceries. In: International Conference on Future Trends of the Internet, Luxemburg, 28 Jan 2009
2. ITU Report on "Internet of Things" (2005)
3. Samms.: Evolving Li-Ion technology and implications for mobile phones, IWPC (2008)
4. Öfversten, J.: Mobile internet impact on battery life. In: IWPC (2008)
5. Walt Disney Pictures.: "Race to Witch Mountain", 13 Mar 2009
6. Honda Research Institute Japan Co., Ltd.: Honda, ATR and Shimadzu jointly develop brain-machine interface technology enabling control of a robot by human thought alone, News Release, 31 Mar 2009
7. Texas Instruments Inc.: World's first, live voiceless phone call made today at Texas instruments developer conference, News Release, 26 Feb 2008
8. Sze, V., et al.: A low-power 0.7V H.264 720p video decoder. In: Asian Solid-State Circuits Conference (A-SSCC), Fukuoka, Apr 2008
9. Palmer, R., et al.: A 14mW 6.25Gb/s transceiver in 90nm CMOS serial chip to chip communication, ISSCC Digest, pp. 440–441, Feb 2007

10. Maes, P., Mistry, P.: Unveiling the "Sixth Sense", Game-Changing Wearable Tech. In: TED 2009, Long Beach, Mar 2009
11. Dream Works/20th Century Fox Pictures.: "Minority Report", 21 June 2002
12. Royannez, P., et al.: 90nm low leakage SoC design techniques for wireless applications, ISSCC Digest, Feb 2005
13. Mair, H., et al.: A 65-nm mobile multimedia applications processor with an adaptive power management scheme to compensate for variations, VLSI, June 2007
14. Gammie, G., et al.: A 45-nm 3.5G baseband and multimedia application processor using adaptive body-bias and ultra-low power techniques, ISSCC Digest, Feb 2008
15. Carlson, B.: SmartReflex power and performance management technologies: reduced power consumption, optimized performance, http://www.techonline.com/learning/techpaper/ 209100225. Accessed Apr 2008.
16. Hezar., et al.: A 110db SNR and 0.5mW current-steering audio DAC implemented in 45nm CMOS, ISSCC Digest, Feb 2010
17. Ramadass, Y., et al.: A 0.16mm^2 completely on-chip switched-capacitor DC-DC converter using digital capacitance modulation for LDO replacement in 45nm CMOS, ISSCC Digest, Feb 2010
18. Meehan, J., et al.: Media processor architecture for video and imaging on camera phones. In: ICASSP (2008), Las Vegas
19. Marcu, C., et al.: A 90nm CMOS low power 60GHz transceiver with integrated baseband circuitry, ISSCC Digest, pp. 314–315, Feb 2010

Chapter 14
MEMS (Micro-Electro-Mechanical Systems) for Automotive and Consumer Electronics

Jiri Marek and Udo-Martin Gómez

Abstract MEMS sensors gained over the last two decades an impressive width of applications:

(a) ESP: A car is skidding and stabilizes itself without driver intervention
(b) Free-fall detection: A laptop falls to the floor and protects the hard drive by parking the read/write drive head automatically before impact.
(c) Airbag: An airbag fires before the driver/occupant involved in an impending automotive crash impacts the steering wheel, thereby significantly reducing physical injury risk.

MEMS sensors are sensing the environmental conditions and are giving input to electronic control systems. These crucial MEMS sensors are making system reactions to human needs more intelligent, precise, and at much faster reaction rates than humanly possible.

Important prerequisites for the success of sensors are their size, functionality, power consumption, and costs. This technical progress in sensor development is realized by micro-machining. The development of these processes was the breakthrough to industrial mass-production for micro-electro-mechanical systems (MEMS). Besides leading-edge micromechanical processes, innovative and robust ASIC designs, thorough simulations of the electrical and mechanical behaviour, a deep understanding of the interactions (mainly over temperature and lifetime) of the package and the mechanical structures are needed. This was achieved over the last 20 years by intense and successful development activities combined with the experience of volume production of billions of sensors.

This chapter gives an overview of current MEMS technology, its applications and the market share. The MEMS processes are described, and the challenges of MEMS, compared to standard IC fabrication, are discussed. The evolution of

J. Marek (✉) • U.-M. Gómez
Robert Bosch GmbH, Automotive Electronics (AE/NE4), Postfach, 1,342, 72703 Reutlingen, Germany
e-mail: Jiri.Marek@de.bosch.com; Udo-Martin.Gomez@de.bosch.com

B. Hoefflinger (ed.), *CHIPS 2020*, The Frontiers Collection,
DOI 10.1007/978-3-642-23096-7_14, © Springer-Verlag Berlin Heidelberg 2012

MEMS requirements is presented, and a short survey of MEMS applications is shown. Concepts of newest inertial sensors for ESP-systems are given with an emphasis on the design concepts of the sensing element and the evaluation circuit for achieving excellent noise performance. The chapter concludes with an outlook on arising new MEMS applications such as energy harvester and micro fuel cells.

14.1   Introduction

Silicon is arguably the best-studied material in the history of the world to date. Millions of man years of research have created a material with superior quality and excellent control of electronic properties – this know-how is the basis of the electronics industry, which drives the innovation of today's world.

Besides the excellent electronic properties, silicon provides us with mechanical properties barely found elsewhere:

– Three times the density of steel
– Four times higher yield strength
– Three times lower thermal expansion

Silicon is brittle; plastic deformation does not occur in monocrystalline silicon. As a result, high-magnitude physical shocks less than the yield strength can be applied to mechanical structures fabricated in silicon with no observable changes in material properties. Overstress will destroy micro-electro-mechanical (MEMS) structures; the destruction can be detected by a self-test.

As a result, silicon is the material of choice for critical safety applications prone to high physical shock magnitudes, such as MEMS accelerometers used for crash detection and subsequent automotive airbag deployment.

The combination of silicon's superior electronic and mechanical properties enabled the creation of completely new devices: micro-electro-mechanical systems, or MEMS for short. Examples of MEMS devices (Fig. 14.1) include inkjet nozzles, arrays of millions of gimballed micro-mirrors used for "digital light processing" (DLP) projection systems, and the suspended spring-mass with comb-like capacitive finger structures used to create an accelerometer. These seismic mass structures are capable of measuring the rapid acceleration change experienced during an automotive crash or by a laptop computer during a 1 g free-fall event.

Fig. 14.1 Acceleration sensor for airbag systems

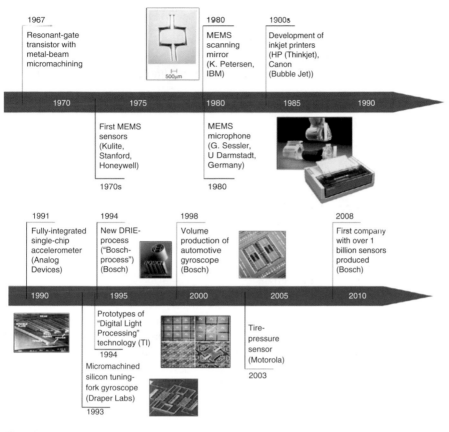

Fig. 14.2 History of MEMS

Initial MEMS micro-machining work was demonstrated in 1967 with singly clamped metal cantilever beams serving as both an electrostatically excited beam resonator and the gate electrode of an underlying silicon field-effect transistor [1]. Subsequently, the mechanical properties of silicon were intensely studied during the 1970s, and researchers started to realize mechanical devices directly in silicon [2, 3]. Figure 14.2 shows a short overview of MEMS history.

14.2   Today's MEMS Market and Applications

The growth of the MEMS market and the market segmentation is shown in Fig. 14.3. Today's market size is around $6 billion with three main segments:

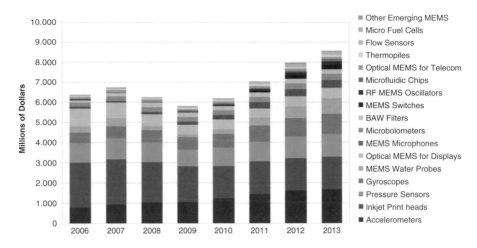

Fig. 14.3 MEMS market overview (iSuppli Corporation). BAW: bulk acoustic wave

Fig. 14.4 Detail of a digital light processing (DLP) chip (©Texas Instruments)

- Displays; dominated by TI's DLP chip (Fig. 14.4)
- Ink jet nozzles
- Sensors (Fig. 14.1)

After the crisis in 2008/2009, the MEMS market is recovering very rapidly. The driving force for sensors is the measurement of mechanical quantities for (a) acceleration, (b) yaw rate/angular rate, (c) pressure, and (d) air flow.

The first wave of MEMS commercialization came in the 1980s and 1990s with sensor applications in automotive electronic systems and with inkjet printers. Around 2000, DLP chips joined in. Starting in 2005, the next wave arose with new volume applications in the consumer-electronics market such as acceleration sensors or microphones in mobile phones, laptops and game controllers.

According to iSuppli, the market segment of mobile and consumer electronics is growing with a CAGR (compound annual growth rate) of 16.5% and will be the largest one by 2013, accounting for 30% of the total MEMS market [4]. Main applications are orientation/motion-sensitive user interfaces in mobile phones (e.g.

in Apple's iPhone or Samsung's Omnia) or in gaming stations (e.g. in Nintendo's Wii), vibration compensation in digital still-cameras and MEMS microphones.

14.3   The Challenges of MEMS Devices

14.3.1   Bulk Micromachining

MEMS technology is based on processes of the semiconductor industry. However, new processes have to be added to produce mechanical structures in silicon. At the same time, mechanical properties of silicon and the whole system have to be taken into account, which are normally neglected by the semiconductor industry.

The first MEMS devices used bulk micromachining. Using anisotropic wet-etching cavities, membranes and bridges can be etched into silicon (Fig. 14.5). There are many different etchants with different etch rates along the various crystallographic orientations of silicon. KOH is most widely used in the industry due to the high etch rate as well as high anisotropy between the (100) and (111) directions.

Owing to the wafer thickness the device miniaturization is limited in bulk micromachining since the complete thickness of the wafer has to be etched through.

14.3.2   Surface Micromachining

In order to shrink the MEMS devices even further, surface micromachining was developed in the 1990s. The mechanically active layer is deposited onto the silicon wafer covered with a sacrificial layer. The material is structured with high mechanical precision; the deep reactive-ion etching (DRIE) process developed by Bosch (also called the "Bosch process") is capable of etching trenches with highly perpendicular walls at a high etch rate. The sacrificial layer is removed, and the mechanical structure is released (Figs. 14.6 and 14.7). For high reliability, these structures are protected from particles using a second silicon wafer bonded as a cap-wafer or by depositing a thin-film encapsulation.

Recently, we have invented the APSM (advanced porous silicon membrane) process to create monocrystalline membranes for pressure sensors in a fully CMOS-

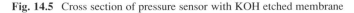

Fig. 14.5 Cross section of pressure sensor with KOH etched membrane

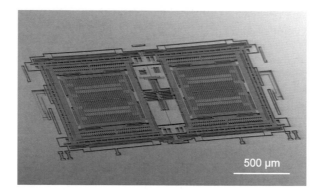

Fig. 14.6 Gyroscope fabricated with surface micromachining

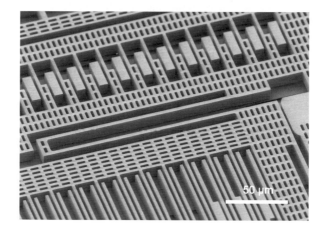

Fig. 14.7 Details of the gyroscope structure

compatible process [5]. Monocrystalline silicon pressure sensor membranes have been previously formed in most cases by anisotropic etching with an electrochemical etch-stop [6]. High-quality monocrystalline silicon membranes are produced using the APSM process using the following process sequence: (a) anodic etching of porous silicon in concentrated hydrofluoric (HF) acid, (b) sintering of porous silicon at temperatures in excess of 400°C in a non-oxidizing environment [7], and (c) subsequent epitaxial growth using the underlying rearranged porous silicon as a single-crystal seed-layer.

14.3.3 Challenges of MEMS Production

MEMS requires the extension of the semiconductor technology into the world of mechanical devices. Therefore several challenges arise:

(a) Special MEMS-processes: For the fabrication of MEMS additional dedicated equipment is needed. Examples are tools for vapor etching with HF or ClF$_3$ or for DRIE.

(b) Contamination: Some MEMS processes constitute contamination risks to standard ICs, e.g. potassium originating from KOH can lead to severe drifts in analog circuits. A stringent separation of standard CMOS production and MEMS-specific production is therefore inevitable.

(c) Electromechanical system simulation: The MEMS devices need a completely new design environment in order to simulate the mechanical structures in silicon. Silicon has a low thermal expansion coefficient compared to the materials used for assembly such as die-attach glue, mold plastics and die pads. The simulation has to identify the influence of mechanical strain of the package onto the MEMS device and optimize the device performance.

(d) Testing for mechanical parameters: A MEMS device has to be tested for its mechanical performance: New test methods for pressure, acceleration or yaw rate have to be developed for high volume manufacturing.

(e) Media protection: The devices must get in contact with various, partly aggressive media in order to measure properties like pressure or mass flow. Despite this exposure they have to work properly for years at high accuracy with failure rates below the ppm range.

14.4 The Development of MEMS

Similar to the semiconductor industry, MEMS technology makes possible a continuous shrinking of the devices. This development is demonstrated using accelerometers for airbag application: Acceleration sensors decreased in size during the last two decades by a factor of 5, enabling us to produce devices of smaller and smaller package size; see Fig. 14.8.

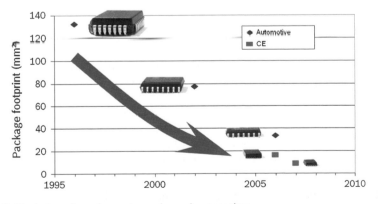

Fig. 14.8 Evolution of accelerometer package size over time

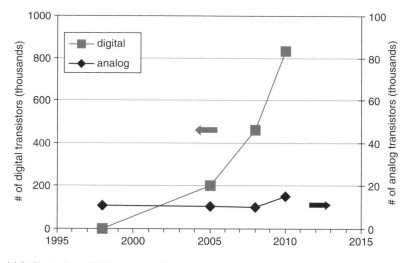

Fig. 14.9 Evolution of ESP sensor ASICs (yaw rate + acceleration)

The evaluation circuitry evolved from bipolar to CMOS; nowadays more and more digital signal processing is applied. The typical sensor ASIC contains only an analog front end; the remaining signal evaluation is performed on digital level. Figure 14.9 shows the evolution of the evaluation ASICs for an ESP inertial sensor (yaw rate plus acceleration): the number of transistors in the analog part stayed nearly constant at 10,000–15,000 but the number of transistors in the digital part increased from 0 to ~1 million over a period of 10 years.

In the next two sections, we will describe two typical MEMS devices used currently in automobiles at high equipment rates: yaw rate and pressure sensors.

14.5 Sensor Cluster (Yaw Rate and Acceleration) for Electronic Stability Program

The vehicle dynamics control (VDC) system or Electronic Stability Program (ESP) is one of the most effective active safety systems in the automotive domain, assisting the driver to keep the vehicle on the intended path, thereby helping to prevent accidents. A current description can be found in [8]. Since its introduction in 1995, it has displayed a steadily growing installation rate in new vehicle platforms, especially in Europe. Numerous studies show that this technology drastically reduces the number of fatalities. In the USA, for example, a comprehensive installation rate of ESP systems – as has been made mandatory by NHTSA (National Highway Traffic Safety Administration) from 2012 on – could save up to 10,000 human lives and avoid up to 250,000 injuries every single year [9].

The first ESP systems used a macro-mechanical yaw rate sensor, which was based on a piezoelectrically actuated, vibrating brass cylinder with piezoelectric

Fig. 14.10 First mechanical ESP yaw rate sensor

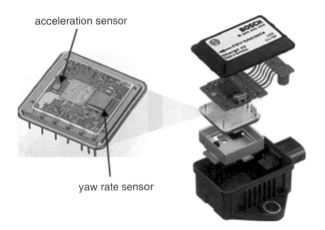

Fig. 14.11 First micromechanical ESP yaw and acceleration sensor

material as a sensing element of the Coriolis force [10], for detection of the car's rotation along its vertical axis (Fig. 14.10). In addition, a mechanical single-axis low-g accelerometer was applied to detect the vehicle's dynamic state and for plausibilisation of the yaw rate signal.

In 1998, as ESP systems were starting to gain a broader market share, Bosch introduced its first silicon micromachined yaw rate sensor [11]. The sensing elements were manufactured using a mixed bulk and surface micromachining technology and were packaged in a metal can housing (Fig. 14.11).

Growing demand for new additional functions of ESP and of future vehicle dynamics systems – such as Hill Hold Control (HHC), Roll Over Mitigation (ROM), Electronic Active Steering and others – required the development of improved inertial sensors with higher precision at lower manufacturing costs. These goals have been achieved by the third-generation ESP sensors [12], a digital inertial sensor platform based on cost-effective surface micromachining technology (Fig. 14.12). Electromagnetic drive of the first generation has been replaced by a simplified electrostatic actuation and sensing elements have been packaged in

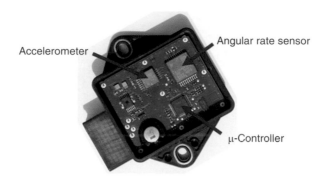

Fig. 14.12 The ESP sensor cluster MM3

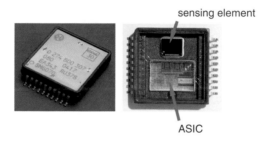

Fig. 14.13 Yaw-rate sensor for sensor cluster MM3

a premolded, plastic package (Fig. 14.13). However, at that time the size of the readout circuit prevented the combination of angular rate and acceleration sensors in a small-size package.

Recent development at Bosch resulted in the first integrated inertial sensor modules, combining different sensors (angular rate and low-g acceleration sensors) and various sensing axes (x, y, and z) into one single standard mould package at small size and footprint. In detail, the sensor consists of a combination of two surface micromachined MEMS sensing chips – one for angular rate, one for two-axis acceleration – stacked onto an application-specific readout circuit (ASIC) in a SOIC16w package (Fig. 14.14).

14.5.1 Angular Rate Sensing Element

The angular rate sensing element is an electrostatically driven and sensed vibratory gyroscope, which is manufactured in pure silicon surface micromachining, using a slightly modified Bosch foundry process [13], with an 11-μm thick polysilicon functional layer to form the movable parts of the micromachined structure (Fig. 14.6).

Fig. 14.14 SMI540: The world's first combined ESP inertial sensor (yaw rate and acceleration sensor) in a mold package

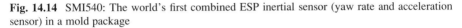

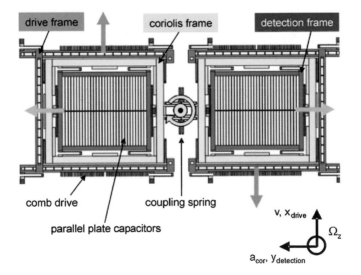

Fig. 14.15 Operating principle of a yaw rate sensing element

The operating principle of the sensing element is depicted in Fig. 14.15. It consists of two almost identical masses, connected by a coupling spring to ensure common mode vibration shapes of the whole structure. Each part has a "drive" frame on the outer circumference, followed by a "Coriolis" frame und ending at a "detection" frame on the innermost position. All frames of each part are connected via U-shaped springs, and the outer and innermost frames are also anchored via U-shaped springs onto the substrate.

The drive frames are excited to a resonant vibration at approximately 15 kHz by electrostatic comb drives anti-parallel along the x-axis. This drive motion is

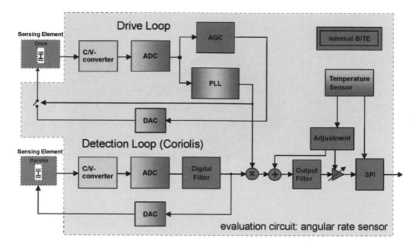

Fig. 14.16 Block diagram of evaluation circuit of yaw-rate sensor

translated to the Coriolis frames by appropriately attached U-shaped springs, whereas the detection frames are hardly affected by the drive motion, leading to a decoupling of drive and detection motion at the detection frames. When an external angular rate is applied around the z-axis, the Coriolis frames are driven into a vibrating motion by the occurring Coriolis forces. This motion is translated to the detection frames via U-shaped springs and picked up using parallel-plate capacitor structures by their change in capacitance.

A scanning electron microscopy (SEM) image of the sensing element is shown in Fig. 14.6. The application of U-shaped springs results in low mechanical nonlinearities. The particular design with the feature of decoupled detection structures reduces mechanical and electrical crosstalk emerging from drive motion. Mechanical crosstalk is additionally reduced by tightly controlled lithography and trench etching, which are the critical MEMS manufacturing processes. Differential signal evaluation and the high operating frequency of approximately 15 kHz lead to sufficient insensitivity to external mechanical perturbation (vibration, shock) usually present in automotive environments. High Q-factors of drive and detection modes, due to the enclosed vacuum, lead to high bias stability.

14.5.2 Signal Evaluation Circuit

A block diagram of the signal evaluation circuit is shown in Fig. 14.16. It consists of drive-control loop, detection loop, synchronous demodulator, signal conditioning and self-test circuits. Except for the front ends, all of the signal processing is performed digitally. The drive-control loop maintains a stable oscillation of the sensing element by means of a phase-locked loop (PLL) to establish the phase condition and an automatic gain control (AGC) to achieve constant magnitude.

Measurement of the Coriolis acceleration is carried out by electromechanical $\Delta\Sigma$-modulators including force-balanced feedback loops. The analog front ends are differential capacitance-to-voltage and flash A/D-converters. Early transition to the digital domain provides for excellent offset stability. The micromachined sensing element itself is part of the $\Delta\Sigma$-modulators, giving a huge improvement to signal-to-noise ratio (SNR) by applying noise-shaping techniques. Post-packaging trimming of the angular-rate sensors is enabled by on-chip PROM and temperature measurement. Angular-rate output is provided by a digital serial peripheral interface (SPI).

Due to the demand of advanced vehicle-control-system safety and availability of sensor signals, extensive internal safety and monitoring techniques have been implemented in the sensor-system design. They consist of self-test and monitoring internal features during start-up as well as operation; their status is presented at the SPI interface. In addition, several internal parameters, such as temperature, quadrature and current-control-loop quantities, are also output at the interface, thus enabling the integration of safety and monitoring concepts into higher-level safety and monitoring systems. In addition, the availability of the sensor's internal signals enables additional high-precision trimming.

14.5.3   Experimental Results: Sensor Performance

The sensor modules are designed to operate under usual automotive conditions. This involves a temperature range from $-40°C$ to $+120°C$ and external mechanical and electrical perturbations (vibration and EMI).

The most important properties of inertial sensors and in particular of angular-rate sensors are

- Noise-power spectral density, determining SNR and the limits of resolution,
- Offset drift of the sensor over operating temperature range and lifetime,
- Sensitivity of the sensor system to external mechanical perturbations (e.g. vibrations).

Some typical measurements concerning these properties are discussed in the following.

14.5.3.1   Noise Performance

The output-noise characteristic is shown in Fig. 14.17, measured on a typical production sample. The inset displays the original time-domain data during a measurement interval of 10 s at a sample rate of 1 kHz. Note that the noise peaks are within approximately $\pm0.1°/s$. The main plot depicts the corresponding

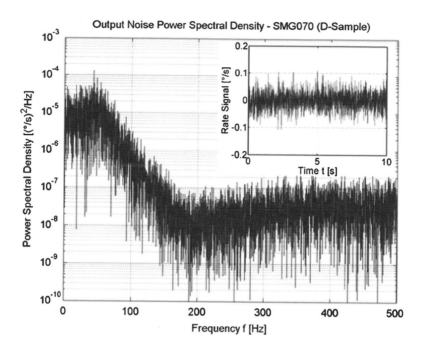

Fig. 14.17 Output-noise characteristic of yaw-rate sensor

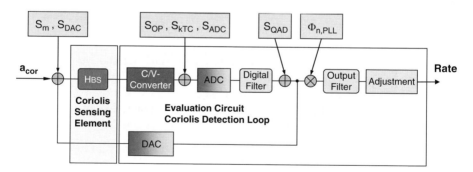

Fig. 14.18 Noise simulation model

noise-power spectral density. The characteristics of the sensor's output low-pass filter can easily be identified. The angular-rate signal bandwidth is from DC to approximately 60 Hz. Within this bandwidth, the output-noise density is typically about 0.004 $(°/s)/\sqrt{(Hz)}$. At higher frequencies, the noise is attenuated by the output filter.

In order to determine the contribution of the particular noise sources, the design of the ASIC and the sensing element, including parasitic effects and different noise sources, is implemented in a simulation model according to Fig. 14.18. Using this model, good agreement between experimental and simulated output noise could be obtained, and the contributions of the different noise sources to the output noise were evaluated as shown in Fig. 14.19.

Symbol	Quantity	Calculated relative contribution to the output noise
S_{OP}	Voltage noise of the C/V converter	~35%
S_m	Thermal noise of the measuring element	~20%
S_{ADC}	Quantisation noise of the AD converter	~20%
S_{KTC}	KT/C noise of the C/V converter	~10%
S_{QAD}	Quantisation noise of the DAC converter	~10%
S_{DAC}	Noise of feedback force due to time-jitter and amplitude variations	<1%
$\Phi_{n, PLL}$	Phase noise of the PLL	< 1%

Fig. 14.19 Contribution of different noise sources

The simulation reveals three main noise sources:

S_{OP} voltage noise of the C/V converter,
S_m thermal noise of the measuring element,
S_{ADC} quantization noise of the AD converter.

14.5.3.2 Offset Drift

Beside low noise and large signal bandwidth, high offset stability (over temperature and time) is a prerequisite for the application of angular-rate sensors in advanced vehicle-stabilization systems. Therefore, excellent offset stability is another key feature of the sensor. A usual measure for the offset (in)stability over time is the root Allan variance [14], which has been calculated from the measured rate-offset data (Fig. 14.20, sample time 10 ms, measurement time 6×10^4 s, room temperature). The slope at short cluster times corresponds to angle random walk (white noise). The resolution noise floor is reached after ~600 s integration time with bias instability of ~1.35°/h. Typical values are outstanding: <3°/h @ ~200–600 s. The increase at larger cluster times (>1,000 s) indicates rate random walk ($1/f$ noise contribution) or an exponentially correlated process with a time-constant much larger than 1,000 s. Furthermore, the rate-offset deviation over the whole temperature range is typically ±0.4°/s.

The resolution limit <3°/h corresponds to amplitude changes in the micromechanical structure as small as ~4 fm (10^{-15} m, about the size of the atomic nucleus) or <2 zF (10^{-21} F). These outstanding numbers can only be achieved by carefully controlling tolerances during the fabrication processes of the

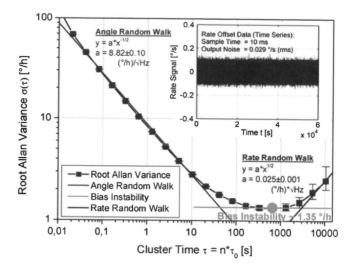

Fig. 14.20 Root Allan variance of yaw rate sensor

micromechanical sensing element and by validating the capability and maturity of the applied design concept.

The excellent noise and offset performance of these angular-rate sensors makes them ideal choices, even for the highest demands of current and future VDC systems.

14.6   Pressure Sensors Based on APSM Technology

New pressure sensor generations are based on a novel surface-micromachining technology. Using porous silicon, the membrane fabrication can be monolithically integrated with high synergy in an analog/digital semiconductor process suited for high-volume production in an IC fab. Only two mask layers and one electro-chemical etching step are added at the beginning of a standard IC process to transform the epitaxial silicon layer used for IC circuits into a monocrystalline membrane with a vacuum cavity underneath (Fig. 14.21).

14.6.1   Fabrication

The epitaxial layer is deposited on a porous silicon layer (Fig. 14.22). The cavity is formed by subsequent thermal rearrangement of this porous silicon during the epitaxial growth process and the following high-temperature-diffusion processes [5]. The hydrogen enclosed inside the cavity during the epitaxial deposition diffuses

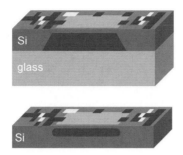

Fig. 14.21 Membrane fabrication technology: *Top*: Conventional bulk micromachining with KOH etching and glass bonding to fabricate a membrane over a reference vacuum. *Bottom*: Porous silicon technology: a monocrystalline membrane is formed by epitaxial-layer growth. The porous silicon is converted into a vacuum cavity by sintering in H_2 atmosphere

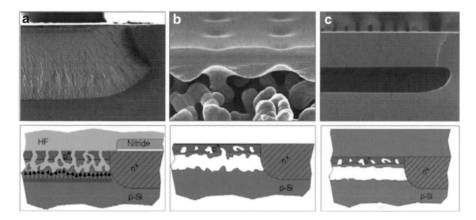

Fig. 14.22 Schematic fabrication steps: (**a**) substrate after anodic etching, (**b**) during sintering in hydrogen atmosphere, and (**c**) after epitaxial growth

out, but no other gases can diffuse in. This leads to a good reference vacuum inside the cavity (Fig. 14.23). The square membrane is deflected by 0.7 μm/bar. It is mechanically robust against overload because of the "floor stopper" when it touches the bottom of the cavity (Fig. 14.24).

A 1 bar sensor is overload-proof to more than 60 bar. The signal is sensed by a piezoresistive Wheatstone bridge, which is a measurement principle with better linearity than capacitive sensors. The piezoresistors are fabricated using the same diffusion layers as the electronic circuit to achieve good matching properties. The bridge signal of 55 mV/bar is amplified by the surrounding electronics. The mixed-signal semiconductor process is suited for the automotive temperature range of −40°C to +140°C. The circuit is designed to be electrostatic discharge (ESD)-proof and for electromagnetic compatibility (EMC).

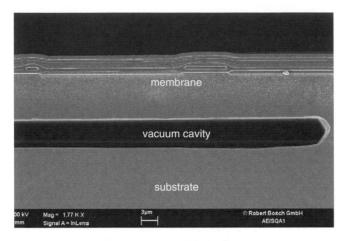

Fig. 14.23 Pressure-sensor membrane with vacuum cavity under it and implanted piezoresistors on top

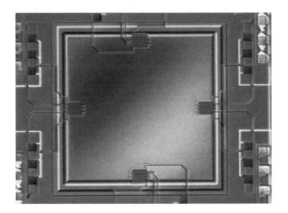

Fig. 14.24 Sensor membrane with piezo resistors. The differential interference contrast of the optical microscope makes the bending visible

14.7 Results

The sensor is digitally calibrated in four parameters: offset, sensitivity, temperature coefficient of offset, and temperature coefficient of sensitivity. The calibration is done in the packaged state to compensate for packaging stress (Fig. 14.25). The digital calibration does not need openings in the chip-passivation layer (in contrast to laser trimming), which makes it compatible with operation in a harsh environment. Results based on first engineering samples show that the sensor meets a total accuracy of ±1.0% over the pressure range of 200–1,200 hPa and from 0°C to

Fig. 14.25 Sensor chip wire-bonded in a premold package

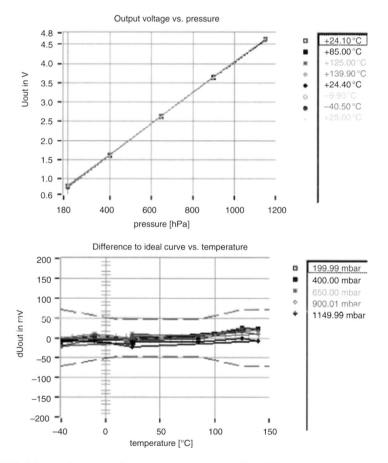

Fig. 14.26 Measured output voltage versus pressure (*top*) and absolute error from −40°C to 140°C (*bottom*). The *dashed line* shows the 1.0% error band from 10°C to 90°C

110°C (Fig. 14.26). In the selected window for altimeter applications, the linearity is significantly better.

14.8 Conclusion

The sensor chip is thin compared to KOH-etched pressure sensors because no anodic bonding on the glass substrate is required. The piezoresistive pressure sensor reported here looks like a standard IC chip. The easy packaging makes it suitable for new consumer applications and tire pressure measurement systems, besides the known automotive applications manifold-air-pressure and barometric-pressure sensors.

14.9 The Future: MEMS Everywhere!

After automotive applications, MEMS is conquering currently the area of consumer electronics. Acceleration sensors and MEMS microphones are contained in many laptops, mobile phones and game controllers; gyros are following soon. In the near future, we will see magnetic sensors for navigation, small projectors with MEMS mirrors embedded in mobile phones or cameras (Fig. 14.27), or inertial units containing acceleration sensors and gyros, including the associated algorithms. Further on, MEMS energy harvesters, (Fig. 14.28) and micro-fuel cells for energy storage, to provide wireless autonomous sensor nodes, will be launched. MEMS sensors will enable applications never imagined before.

Fig. 14.27 Microprojector (©Microvision)

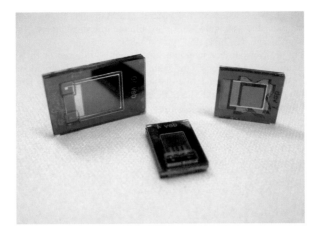

Fig. 14.28   Energy harvester (©IMEC)

Acknowledgments   Many thanks to Gary O'Brien and Frank Schäfer who contributed to the preparation of this chapter.

References

1. Nathanson, H.C., Newell, W.E., Wickstrom, R.A., Davis Jr., J.R.: The resonant gate transistor. IEEE Trans. Elec. Dev. **14**(3), 117–133 (1967)
2. Angell, J.B., et al.: Silicon micromechanical devices. Sci. Am. **248**(4), 44 (1983)
3. Petersen, K.E.: Silicon as a mechanical material. Proc. IEEE. **70**, 420 (1982)
4. Global MEMS market suffers first-ever decline in 2008. iSuppli Corporation, 2009 (http://www.isuppli.com/News/Pages/Global-MEMS-Market-Suffers-First-Ever-Decline-in-2008.aspx)
5. Armbruster, S. et al.: A novel micromachining process for the fabrication of monocrystalline Si-membranes using porous silicon. TRANSDUCERS 2003. In: 12th International Conference on Solid-State Sensors, Actuators and Microsystems, Vol. 1, pp. 246–249. DOI: 10.1109/SENSOR.2003.1215299
6. Kress, H.J., Bantien, F., Marek, J., Willmann, M.: Silicon pressure sensor with integrated CMOS signal conditioning circuit and compensation of temperature coefficient. Sens. Actuators A **25**(1–3), 21–26 (1990)
7. Herino, R., Perio, A., Barla, K., Bomchil, G.: Microstructure of porous silicon and its evolution with temperature. Mater. Lett. **2**, 519–523 (1984)
8. E.K. Liebemann, K. Meder, J. Schuh, G. Nenninger: Safety and performance enhancement: The Bosch Electronic Stability Control (ESP). SAE Convergence on Transportation Electronics 2004, Detroit, MI, October 2004, Technical Paper 2004-21-0060. http://papers.sae.org/2004-21-0060/
9. Electronic Stability Control (ESC) and Electronic Stability Program (ESP) automotive applications. http://en.wikipedia.org/wiki/Electronic_stability_control. Accessed April 2011
10. Reppich, A., Willig, R.: Yaw rate sensor for vehicle dynamics control system. SAE Technical Paper 950537. http://papers.sae.org/950537 (1995)

11. Lutz, M., Golderer, W., Gerstenmeier, J., Marek, J., Maihöfer, B., Mahler, S., Münzel, H., Bischof, U.: A precision yaw rate sensor in silicon micromachining. In: Proceedings of Transducers'97, Chicago, IL, June 1997, pp. 847–850
12. U.-M. Gómez et al.: New surface micromachined angular rate sensor for vehicle stabilizing systems in automotive applications". In: Transducers'05: 13th International Conference on Solid-State Sensors, Actuators and Microsystems, Seoul, June 2005, Digest of Technical Papers, pp. 184–187. DOI: 10.1109/SENSOR.2005.1496389
13. Offenberg, M., Lärmer, F., Elsner, B., Münzel, H., Riethmüller, W.: Novel process for a monolithic integrated accelerometer. Technical Digest Transducers'95 and Eurosensors IX, Stockholm, June 25–29, 1995, Vol. 1, pp. 589–592. DOI: 10.1109/SENSOR.1995.717293
14. IEEE Std 952–1997, IEEE Standard Specification Format Guide and Test Procedure for Single-Axis Interferometric Fiber Optic Gyros

Chapter 15
Vision Sensors and Cameras

Bernd Hoefflinger

Abstract Silicon charge-coupled-device (CCD) imagers have been and are a specialty market ruled by a few companies for decades. Based on CMOS technologies, active-pixel sensors (APS) began to appear in 1990 at the 1 μm technology node. These pixels allow random access, global shutters, and they are compatible with focal-plane imaging systems combining sensing and first-level image processing. The progress towards smaller features and towards ultra-low leakage currents has provided reduced dark currents and μm-size pixels. All chips offer Mega-pixel resolution, and many have very high sensitivities equivalent to ASA 12.800. As a result, HDTV video cameras will become a commodity. Because charge-integration sensors suffer from a limited dynamic range, significant processing effort is spent on multiple exposure and piece-wise analog-digital conversion to reach ranges >10,000:1. The fundamental alternative is log-converting pixels with an eye-like response. This offers a range of almost a million to 1, constant contrast sensitivity and constant colors, important features in professional, technical and medical applications.

3D retino-morphic stacking of sensing and processing on top of each other is being revisited with sub-100 nm CMOS circuits and with TSV technology. With sensor outputs directly on top of *neurons,* neural focal-plane processing will regain momentum, and new levels of intelligent vision will be achieved. The industry push towards thinned wafers and TSV enables backside-illuminated and other pixels with a 100% fill-factor. 3D vision, which relies on stereo or on time-of-flight, high-speed circuitry, will also benefit from scaled-down CMOS technologies both because of their size as well as their higher speed.

B. Hoefflinger (✉)
Leonberger Strasse 5, 71063 Sindelfingen, Germany
e-mail: bhoefflinger@t-online.de

B. Hoefflinger (ed.), *CHIPS 2020*, The Frontiers Collection,
DOI 10.1007/978-3-642-23096-7_15, © Springer-Verlag Berlin Heidelberg 2012

15.1   Vision and Imaging Fundamentals

Visual information provided by images and video has advanced the most due to the ever increasing performance of nanoelectronics. The first decade of the new century offered a jump by three orders of magnitude in the product of pixels × frame-rate to the mobile user of camera-phones or digital cameras: Resolution went from 100 k pixels to 10 M pixels and the frame rate from 3 fps (frames/s) to 30 fps.

It is interesting to note that the pixel size on camera chips followed a roadmap with the same slope as that of its associated technology node with a magnifier of about 20× (Fig. 15.1). We will see what enabled this progress, which subtle trade-offs in expected pixel performance have been accepted to realize this scaling strategy, and what the future challenges and opportunities will be.

As Dr. Theuwissen states convincingly in [1], there are attractive scaling rules associated with the pixel size (pixel edge) p:

1. Pixel area.	$\sim p^2$
2. Chip area	$\sim p^2$
3. Chip cost	$\sim p^2$
4. Read-out energy/pixel	$\sim p^2$
5. Lens volume	$\sim p^3$
6. Camera volume	$\sim p^3$
7. Camera weight	$\sim p^3$
8. Sensitivity	$\sim p^2$
9. Dynamic range (bright–dark)	$\sim p^{-2}$

As is typical for a chip-dominated electronic product, these economic factors have led to more pixels/camera, increasing market opportunities to the incredible volume of *almost one billion camera-phones sold in 2010*. However, there have been subtle physical penalties regarding sensitivity and dynamic range.

The weaker performance of recent phone-cameras and compact cameras in darker scenes can be noticed by more frequent activation of flash and/or double exposure to compensate motion blur. Is this indicative of an inflection point in the

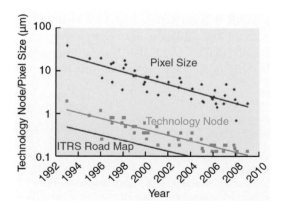

Fig. 15.1 The pixel size of CMOS image-sensor chips, their associated technology node, and the minimum half-pitch according to the ITRS [1] (© 2010 IEEE)

scaling of digital imagers? We study some basics to find out where an inflection point for simple scaling may occur, and we discuss the key future opportunities for advanced imaging and vision systems for everyone.

The basis for the success story of solid-state image sensors, replacing earlier vacuum-tube diodes, is the Si p-n junction diode and its near-perfect photosensitivity (Fig. 15.2). We see that in this photodiode, in spite of the heavy processing overhead for an advanced CMOS process, every second impinging photon generates an electron – hole pair, which can be recorded as an electronic charge or current (50% quantum efficiency) over most of the *visible* spectrum as well as into the near infrared (NIR) and into the ultraviolet (UV). How many electrons does the photometric flux density, the illuminance, generate? To be quantitative:

$$1 \text{ green}(555 \text{ nm})1x = 4 \times 10^{11} \text{photons cm}^{-2}s^{-1}$$
$$= 2 \times 10^3 e \ \mu m^{-2}s^{-1}(\text{quantum efficiency } 50\%).$$

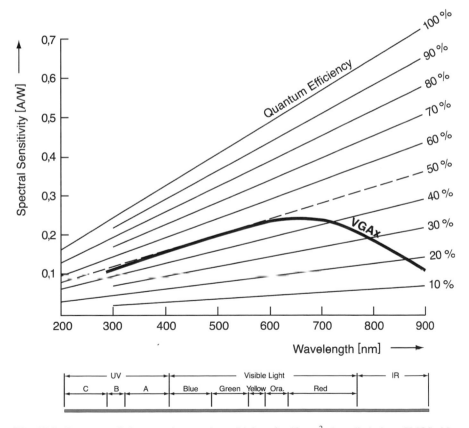

Fig. 15.2 Quantum efficiency and spectral sensitivity of a 40-μm^2 photodiode in a CMOS chip versus photon wavelength [2] (©Springer)

Table 15.1 Sensitivity standards and photoelectrons in a Si photodiode (quantum efficiency 50%, exposure time 25 ms)

ASA	100	200	400	800	1,600	3,200	6,400	12,800
DIN	21	24	27	30	33	36	39	42
mlx s	8	4	2	1	0.5	0.25	0.12	0.06
Signal S_{min} [e μm^{-2}]	16	8	4	2	1	0.5	0.25	0.12
Signal [e $\mu m^{-2} s^{-1}$)	640	320	160	80	40	20	10	5

This sensitivity of Si basically can result in a minimum detectable illuminance better than that of the human eye and considerably better than film. We can calibrate a photodiode against the well-known sensitivity standards ASA and DIN (Table 15.1). To gauge ourselves: Standard film is 100ASA with high-sensitivity film reaching 800ASA.

Table 15.1 tells us what the minimum detectable signal S_{min} would be for a given sensitivity. This signal has to be as large as the noise level, so that any extra electron would make a just noticeable difference (JND). The dark limit is the leakage current of the p-n junction, which has a dominant peripheral component around the edges of the metallurgical junction against the environment and an area component, which only shows up at large reverse bias voltages close to the breakdown of the diode. In the context of photodiodes, this diode reverse current is called the *dark current*. It is obviously the most critical parasitic component of a photodiode and the real treasure in the proprietary know-how of a manufacturer.

The fundamental noise component in the photodiode is that of the variance of the dark-current electrons N_D [e μm^{-1}]. As in any observation of small numbers, the standard deviation of N_D is $(N_D)^{1/2}$, also called the dark-current shot noise. We can determine the size $L \times L$ of a photodiode to reach a certain sensitivity as follows:

$$S_{min}L^2 = (N_D \cdot 4L)^{1/2},$$

$$L = \left(\frac{2}{S_{min}}\right)^{2/3} N_D^{1/3}.$$

This is the photodiode size L required for a given sensitivity S_{min} in a technology with dark charge N_D. The inverse is the sensitivity achievable with a given diode size:

$$1/S_{min} = 2(N_D)^{1/2}L^{3/2}.$$

We show this relationship in Fig. 15.3 for three levels of dark electrons: 50, 5, and 0.5 e/μm^2.

Several lessons can be learned from Fig. 15.3:

- Minimize the dark current.
- Since dark current depends on perimeter, the penalty for scaling down the diode size L is severe.

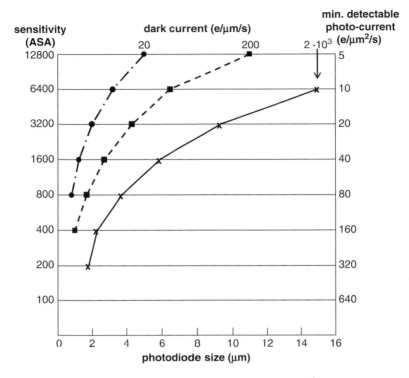

Fig. 15.3 Sensitivity levels in ASA and minimum signal charge [e μm^{-2}] versus photodiode sizes L and three levels of dark electrons

- Sensitivity can be increased with larger diode sizes, because the signal increases with the diode area $\sim L^2$, while the dark noise increases as only $L^{1/2}$.
- The binning of n pixels, i.e., combining their signals into one, improves the sensitivity only by $n^{1/2}$.
- The binning of 4 pixels improves the sensitivity only by $2\times$ (1 f-stop or f-number), while a photodiode $4\times$ larger improves the sensitivity $8\times$ (3 f-stops).

The sheer drive for more megapixels in the product sheet by scaling down pixel size severely reduces the pixel sensitivity. The customer should be told: *Ask for sensitivity first, then for the number of pixels!* Furthermore, the resolution of lenses poses a lower limit on useful pixel sizes. This resolution can be identified either by the diameter of the image dot resulting from an original spot, the so-called Airy disk, or the minimum resolvable pitch of line-pairs [3]. These characteristics are listed in Table 15.2 for the f-stops of lenses.

How do we sense the electrons generated in the photodiode? In most sensors, we integrate the charges for a certain integration time onto a capacitor, and we measure the resulting voltage change on this capacitor. This is accomplished basically with a *source-follower* transistor, serving as a voltage-to-current (or impedance) converter, as shown in Fig. 15.4.

Table 15.2 Image resolution of lenses (line resolution for 9% contrast)

Lens	Airy disk diam. [μm]	Line pairs per mm	Line-pair pitch [μm]
f/1.0	1.3	1,490	0.65
f/1.4	1.9	1,065	0.95
f/2.0	2.7	745	1.35
f/2.8	3.8	532	1.88
f/4.0	5.4	373	2.68

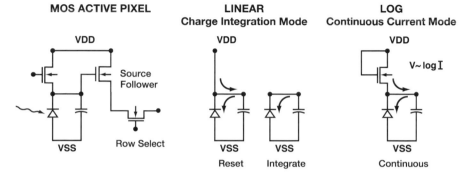

Fig. 15.4 The fundamental CMOS active pixel [2] (©Springer)

This *active* pixel basically consists of the photodiode, three transistors, and a capacitor. The so-called linear charge-integration mode (Fig. 15.4 center) has two phases: In phase 1, the reset, the first transistor is turned on, and the capacitor is charged to V_{DD}, irrespective of the photocurrent. In phase 2, the first transistor is turned off, the capacitor is discharged for a certain time by the (photo-plus dark-) current of the diode, and the resulting voltage is detected as the source voltage on the source-follower through the third transistor, the select transistor. The dynamic range of signals, which can be detected in this mode, is given by

(a) The dark-charge voltage loss in the absence of photo charge and
(b) The maximum charge, CV_{DD}, on the capacitor.

It is on the order of 800:1.

In CMOS area-array sensors with *rolling shutters*, all pixels in a selected row are reset simultaneously, integrated for the same (exposure) time, and, with all row-select transistors turned on, the column-sense lines are charged by the source-followers to their individual signal levels. All recent sensors have per-column analog/digital (A/D) converters so that digital outputs are available for the whole row at the same time. We illustrate the characteristics of this linear-type CMOS sensor with an advanced 2010 state-of-the-art sensor [4] in Table 15.3.

If a 100% fill-factor is assumed, the photodiode area of this sensor is 2.7 μm². The sensitivity reported by Wakabayashi et al.then means that 1 photoelectron/μm² is generated by 0.27 mlx s, a signal just exceeding the random noise of 0.6 e/μm² so that we would rate this sensor close to ASA3200. A SEM cross-section of this sensor is shown in Fig. 15.5.

Table 15.3 2010 advanced CMOS sensor for imaging and video [4]

Pixel-array size	3,720 (H) × 2,780 (V)
Technology	140 nm
Pixel size	1.65 μm × 1.65 μm
4-pixel clusters, 1.5Tr./pixel, back-illuminated	
Supply voltage	2.7 V/1.8 V
Saturation signal	9130e
Dark current	3e/s 60 °C
RMS random noise	1.7e
Sensitivity	9.9e/(mlx s)
Dynamic range	71 dB
Output:	10b serial LVDS
Clock:	72 MHz
Pixel rates	10.3 Mpix at 22–50 fps
	6.3 Mpix at 60 fps
	3.2 Mpix at 120 fps
Power consumption	375 mW HD video

Fig. 15.5 Cross-section of an advanced CMOS video sensor with backside-illuminated (*BI*) photodiode (*PD*) and microlenses (*MLs*) [4] (© 2010 IEEE)

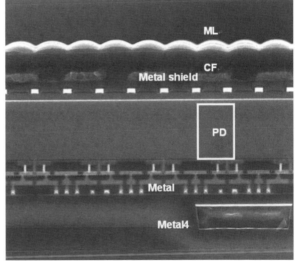

Pinpointing an important future direction is the formation of the photodiode deep inside the Si active layer and removing the backside of the wafer after completion of the full CMOS process. This way, the photodiode can fill the whole pixel area (~100% fill-factor), and the optical components such as color filters (CFs) and microlenses (MLs) become most effective. The viewing angle of the photodiode also is much larger than that for diodes inside the deep canyons of a multi-level interconnect topography (Fig. 15.6). The other advantage of this structure is that the back-end metal structures (at the bottom of Fig. 15.5) are ideal for the 3D interconnect to chip stacks with further processing and memory capacity.

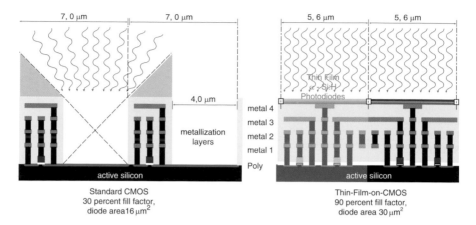

Fig. 15.6 Cross section of a standard CMOS active-pixel sensor (APS, *left*) and a 3D configuration (*right*) with an amorphous Si photodiode on top of a finished CMOS structure [5]

An alternative that offers 100% fill-factor and the optimization of the diode is the 3D integration of an amorphous-Si photodiode layer on top of a CMOS wafer as shown on the right in Fig. 15.6 [5]. Both topographies, back-illuminated and 3D-photodiode-on-CMOS, are indicative of future imager strategies.

15.2 Natural-Response and High-Dynamic-Range Sensors

The charge-integration-based sensors have a linear optoelectronic conversion function (OECF), as shown in Fig. 2.2 of [2]. Their fundamental problems are

– Limited dynamic range
– Poor contrast resolution in the dark
– White saturation in bright scenes
– Complex image processing because of luminance dependence of shapes
– Color distortion because of local luminance dependence
– Unnecessarily large bit rates

Many of these problems have been solved in still-cameras, and even in their Video Graphics Array (VGA)-resolution video modes, with truly high-speed, high-performance image processing on-board. A key example is dual or multiple exposure to increase the dynamic range, to improve resolution in the dark, to avoid white saturation, and to estimate correct colors. The adoption of certain routines and standards has helped this a posteriori image improvement to impressive results and to a well-established science [6]. The sophisticated approach is to take seven exposures in order to cover a dynamic range of 24 f-stops. In the case of motion, two or three exposures have to suffice, while the steady progress in motion

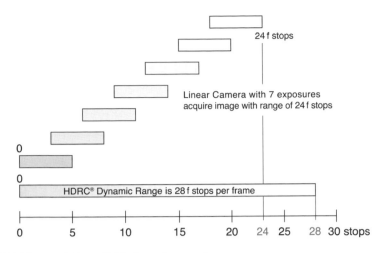

Fig. 15.7 The acquisition of high-dynamic-range images

compensation and multi-resolution recording/processing promises further improvements along this avenue. Figure 15.7 illustrates this approach.

Natural and technical scenes can have a dynamic range of eight orders of magnitude or 28 f-stops from a clear night sky at 1 mlx to direct sunlight at 100 klx and beyond, and our eyes manage this with a logarithmic response to illuminance, like many other natural senses. This natural logarithmic-response function has been adopted in digital voice codecs since the 1950s with the so-called A-law or μ-law for coders, which have several fundamental features:

- Constant contrast resolution independent of voice level, which means a constant signal-to-noise ratio from very low voice-levels to very high levels,
- Natural bit-rate compression by compressing a 12-bit dynamic range to an 8-bit signal.

The human eye has this logarithmic response (like our ears), with a high range of over five orders of magnitude in its spontaneous response and another three orders of magnitude in long-time adaptation (Fig. 15.8). *A HDRC (High-Dynamic-Range CMOS) log-response sensor with an input dynamic range of 24 bit and a 10 bit output provides a natural bit-rate compression of 2.4:1.*

In a comparison with the multi-exposure approach, which requires at least 7 × 8 = 56 bit, the advantage of the natural (log)-response sensor is significant for future developments, particularly video.

The inner pixel of the HDRC sensor is shown in Fig. 15.4 on the right. Gate and drain of the transistor are connected, $V_{GS} = V_{DS}$, so that this two-terminal transistor works in the sub-threshold (barrier-controlled) mode with the current–voltage characteristic (3.20)

$$I_D = I_0 \exp\{[(A - B)V_{GS} - AV_T]/nV_t\},$$

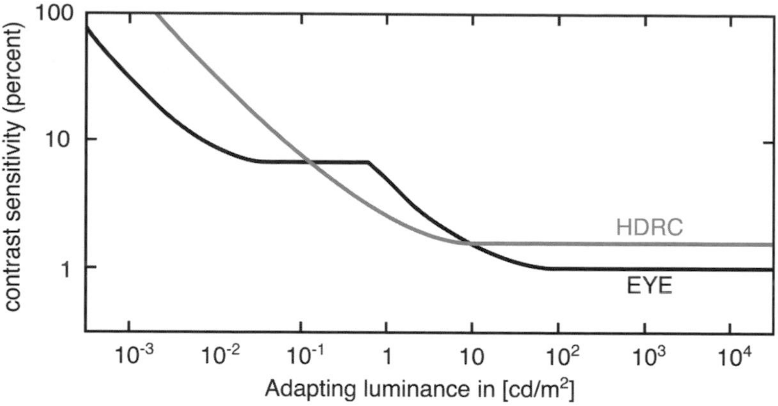

Fig. 15.8 The contrast-sensitivity function (CSF) of the human eye, indicating that a difference in a local gray level of 1% is just noticeable. The contrast sensitivity of a HDRC sensor with a dynamic range of 7 orders of magnitude (or 24 f-stops) and a digital 10-bit output is also shown to have minimum detectable differences of 1.6% [2] (©Springer)

where I_D is the sum of the photocurrent and dark current of the photodiode. The voltage V on the capacitor follows the current over more than five decades as

$$V \propto \ln I_D,$$

except at very low levels, where it settles at the dark-current level so that the optoelectronic conversion function (OECF) has the form

$$y[\text{DN}] = a \ln(1 + x),$$

and the contrast sensitivity function (CSF), the minimum resolvable difference in percent, is

$$\Delta C = \frac{1 - x}{a \cdot x}.$$

Here, y is given as a digital number (DN), and x is the illuminance relative to its 3-dB value, the illuminance where its CSF has increased to twice its optimum value. Figure 15.9 presents the OECF and the CSF for $a = 100$.

The contrast sensitivity, as a measure of the minimum detectable difference in gray levels, establishes the fundamental quality of how uniform a field of pixel images can become under a homogeneous illuminance signal S, and it enters as quantization noise into the total flat-field noise. Other noise contributions are the dark-current shot noise, discussed before, the signal shot noise, the fixed-pattern noise, and the reset noise in charge-integration sensors (for a detailed treatment see [3] and Chap.3 in [2]). The resulting signal-to-noise ratio is plotted in Fig. 15.10.

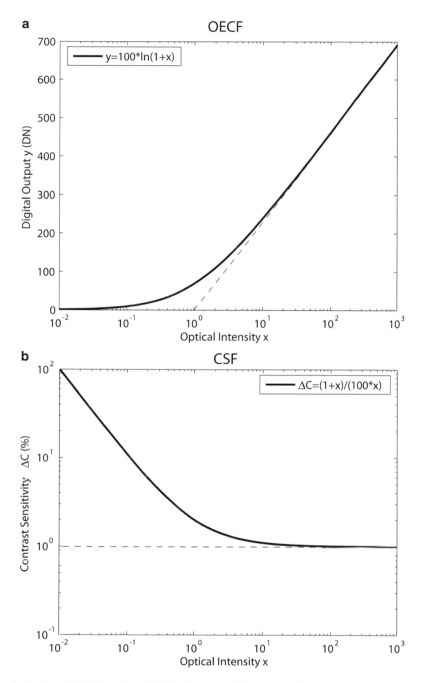

Fig. 15.9 The OECF (**a**) and the CSF (**b**) for an eye-like sensor with $a = 100$

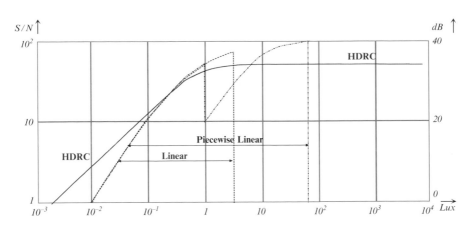

Fig. 15.10 Flat-field signal-to-noise ratio for a linear, a piecewise linear, and a HDRC sensor as a function of the illuminance

The linear sensor at low levels is plagued by quantization and reset noise and at higher levels by white saturation, resulting in a very limited dynamic range (300:1 in our example). For this reason, many linear sensors operate with a two-(our example) or multi-step gain [1] in order to achieve a piecewise linear approximation to a log-response OECF. Even then, they remain limited in dynamic range by white saturation (*full-well* electron capacity), 2,000:1 in our example. The HDRC sensor starts with the best contrast resolution and, as a continuous-current sensor, it is purely shot-noise-limited at low levels, free of reset noise. It maintains a *natural*, constant S/N ratio of 70:1 with a 10-bit output over more than four orders of magnitude.

The HDRC sensor appeared as a radical innovation in 1992 [7] with the fundamental features of a natural, biomorphic vision system:

- Logarithmic response with unparalleled dynamic range (>140 dB at room temperature), unlimited on the bright side except for lens limitations.
- A photometric continuous sensor, which means:
- No aperture adjustment needed,
- No shutter and no integration-time adjustment needed,
- Separation of luminance and (object) reflectance, making object and pattern recognition very simple,
- Color constancy, superb for contrast and color management as well as tone mapping,
- Gain control and white balance are offsets rather than multiplies,
- Very-high-speed read-out of regions of interest.

All these capabilities remain as challenges for any alternative future development either as a continuing refinement of the HDRC sensors and cameras [2] or as a radical alternative. A critical-size European research effort in this direction was started in 2008 with the project HiDRaLoN [8].

For future high-performance focal-plane imaging and, even more for video processing and compression, logarithmic sensor data offer significant advantages (see Chaps. 11–14 in [2]), and the reduction of digital multiplies to the addition of logarithms remains an all-time goal.

15.3 Stereo and 3D Sensors

The formidable repertory of image registration programs and of low-cost sensors has helped in the design of digital stereo cameras both for consumer and professional use so that such systems including polarization glasses have appeared on the market for the production of 3D movies as well as for photographers. Highly compact consumer cameras are being designed using neighboring chip-pairs directly off the wafer, which are automatically perfectly aligned, and prism-optics to establish the proper stereo basis.

For the precise determination of depths in a scene, a base plate with three cameras is being used [9]. A triangular arrangement with sensor distances of ~25 cm allows ranges of ~8 m with a depth accuracy of <10 cm. 400kpixel HDRC sensors with global shutters are used because of their photometric, high-dynamic-range capabilities, including *disregard of luminance* and safe edge detection as well as straightforward global synchronization, because there are no concerns for individual integration times.

The generation of 3D depth or range maps is achieved with artificial lighting and array sensors, in which the pixels resolve the delays of local reflections. Clearly, the gains in switching speed achieved by scaling down CMOS to sub-nanosecond responses have made various such concepts practical since the late 1990s. We describe two such systems, one because it demonstrates the versatility of Si technology, and the other because of its high-speed, high-performance capabilities.

The first system is based on a Si photodiode with a robust response to extremely low signals, namely the arrival of single photons. The principle, which has been used for a long time, is the multiplication of a primary photoelectron by secondary electron – hole pairs resulting from the impact ionization of Si atoms in high-field regions close to the avalanche breakdown of the diode. The naturally large variance of the multiplication factors can be averaged in the case of pulsed lighting by correlated detection and repetition of the measurement $100\times$ or $10^4\times$. A 32×32 pixel CMOS 3D camera with a 25-V single-photon avalanche diode (SPAD) in each 5 T pixel was reported 2004 [10]. Owing to the high multiplication factor, the detected signals could be routed directly to an external time-to-digital converter, which had a resolution of 120 ps with a standard deviation of 110 ps. An uncollimated laser source provided an illumination cone with a rate of 50 MHz, a pulse length of 150 ps, a peak power of 100 mW, and an average optical power of 750 µW. For a life-size object at 3 m, and 10^4 repetitions, the range uncertainty was

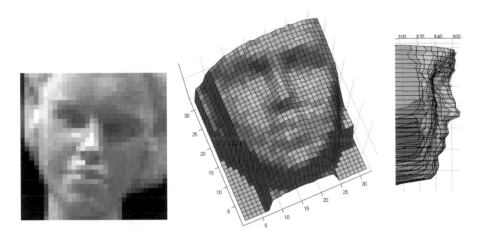

Fig. 15.11 A 2D image and the 3D depth map and profile (in millimeters), taken with a 32×32 pixel time-of-flight (TOF) camera in 2004 [10] (©2004 IEEE)

1.8 mm or 0.06%. This mode would offer 5 frames/s. Figure 15.11 shows a 2D gray image as well as a 3D depth map and profile taken with the same camera.

These 3D imagers are very important for the understanding and generation of high-fidelity 3D scenarios as well as for intelligent operations in such scenarios, be it people, health care, traffic, and, in particular, the interaction between man and machines. The pressure for higher map resolution and higher frame rates will continue, and our second system [11] is an example for the rate of progress: With a resolution of 256×256 pixels, it generates 14,000 range maps/s using a sheet-beam illumination and the so-called light-section method. Together with a row-parallel search architecture, the high speed and a high accuracy was achieved. Taking 10 samples/map, a maximum range error of 0.5 mm was obtained for an object distance of 40 cm, a relative range accuracy of 0.12%.

The field of 3D vision will see exceptional growth for several reasons:

– Ever more powerful sensors
– High-density, power-efficient focal plane coding and processing
– 3D integration of optics, sensing, and processing
– The eternal fascination with the most powerful sense of our world

15.4   Integrated Camera Systems

The sensor is only a means to an end, which is perfect vision encompassing the camera (or several cameras) as well as encoding, storing, and transmitting the needed image information in the most effective way. At the one end of these system requirements is the effective manufacturing of miniature cameras in a batch process

Fig. 15.12 3D wafer-level
integration miniature cameras
[12] (©2010 IEEE)

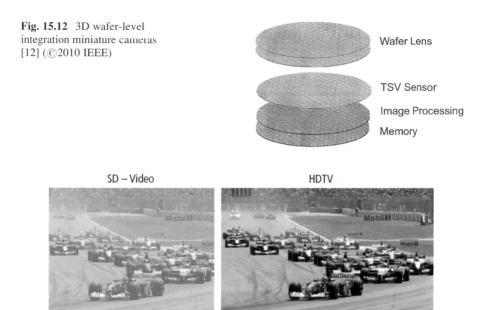

Wafer Lens

TSV Sensor

Image Processing

Memory

SD – Video   HDTV

270 MBit / sec   1485 MBit / sec

Fig. 15.13 Comparison of a video frame in standard definition video and in HDTV, and their bit rates [13] (©KAPPA)

Fig. 15.14 One frame in a
high-speed video of a fire
breather [2] (©Springer)

including lens, sensor, and all necessary focal-plane operations. 3D wafer-level integration of these elements is illustrated in Fig. 15.12, certainly the *next big thing* for the cameras in the billions of mobile phones, but also for all minimally invasive cameras in medical, technical-inspection, and surveillance applications.

At the other end, in the world of high-definition video (Fig. 15.13), we face the big challenge of compressing the large bit rates, 1.5 Gb/s in our example or 2× to

$3\times$ higher in 3D high-definition, to rates that are compatible with future energy and bandwidth limitations.

15.5 Conclusion

The availability and quality of visual information and sensation is the single biggest roadblock or driver of our future digital world, both in wireless and in the internet backbone (Chaps. 5, 13, and 20). The figure-of-merit will be video quality/bandwidth. An attractive direction to pursue is photometric-log-compression sensing and processing, producing frames as shown in Fig. 15.14.

References

1. Theuwissen, A.J.P.: Better pictures through physics. IEEE Solid-St. Circ. **2**(2), 22 (2010)
2. Hoefflinger, B. (ed.): High-Dynamic-Range (HDR) Vision. Springer, Berlin/Heidelberg (2007)
3. Janesick, J.R.: SPIE course "Introduction to CCD and CMOS imaging sensors and applications", SC504. SPIE Education Services, San Diego (2003)
4. Wakabayashi, H., et al.: A 1/2.3-inch 10.3Mpixel 50frame/s back-illuminated CMOS image sensor. IEEE ISSCC (International Solid-State Circuits Conference) 2010, Digest Technical Papers, pp. 410–411
5. www.ims-chips.de
6. Debevec, P., Patanaik, S., Ward, G., Reinhard, E.: High Dynamic Range Imaging. Elsevier, San Francisco (2006)
7. Hoefflinger, B., Seger, U., Landgraf, M.E.: Image cell for an image recorder chip, US patent 5608204, filed 03-23-1993, issued 03-04-1997
8. www.hidralon.eu
9. www.pilz.com/safetyeye
10. Niclass, C., Rochas, A., Besse, P.A., Charbon, E.: A CMOS 3D camera with millimetric depth resolution. In: Proceedings IEEE Custom Integrated Circuits Conference (CICC), pp. 705–708. (2004)
11. Mandai, S., Ikeda, M., Asada, K.: A 256 × 256 14 k range maps/s 3-D range-finding image sensor using row-parallel embedded binary search tree and address encoder. IEEE ISSCC (International Solid-State Circuits Conference) 2010, Digest Technical Papers, pp. 404–405 (2010)
12. Jaffard, J.L.: Chip scale camera module using through silicon via technology. IEEE ISSCC (International Solid-State Circuits Conference), Forum F1: Silicon 3D-Integration Technology and Systems (2010)
13. www.kappa.de

Chapter 16
Digital Neural Networks for New Media

Lambert Spaanenburg and Suleyman Malki

Abstract Neural Networks perform computationally intensive tasks offering smart solutions for many new media applications. A number of analog and mixed digital/analog implementations have been proposed to smooth the algorithmic gap. But gradually, the digital implementation has become feasible, and the dedicated neural processor is on the horizon. A notable example is the Cellular Neural Network (CNN). The analog direction has matured for low-power, smart vision sensors; the digital direction is gradually being shaped into an IP-core for algorithm acceleration, especially for use in FPGA-based high-performance systems. The chapter discusses the next step towards a flexible and scalable multi-core engine using Application-Specific Integrated Processors (ASIP). This topographic engine can serve many new media tasks, as illustrated by novel applications in Homeland Security. We conclude with a view on the CNN kaleidoscope for the year 2020.

16.1 Introduction

Computer architecture has been ruled for decades by instruction pipelining to serve the art of scientific computing [1]. Software was accelerated by decreasing the number of clocks per instruction, while microelectronic advances raised the speed of the clocks (the "Moore" effect). However, algorithms became also more complex, and eventually the "processor-to-memory" bottleneck was only solved by revealing a more severe "algorithm-to-processor" bottleneck, a "more than Moore" effect [2]. The algorithmic complexity was first raised in communications, and recently we see a further need in new media tasks. As the microelectronic advance has bumped into

L. Spaanenburg (✉) • S. Malki
Department of Electrical & Information Technology, Lund University, P.O. Box 118, 22100 Lund, Sweden
e-mail: Lambert.Spaanenburg@eit.lth.se; Suleyman.Malki@eit.lth.se

B. Hoefflinger (ed.), *CHIPS 2020*, The Frontiers Collection,
DOI 10.1007/978-3-642-23096-7_16, © Springer-Verlag Berlin Heidelberg 2012

the "power wall," where merely speeding up the clock makes little sense, new ways to accommodate the ever-growing algorithmic needs are demanded.

A typical example of a complex algorithm is the neural network. Though originally inspiring von Neumann to computing structures, the neural network had little life for decades [3]. It saw rejuvenation in the mid 1980s to demonstrate the power of the upcoming supercomputers. Large-scale correlation on physical datasets demanded a "more than Moore" acceleration to overcome the principle problems in the highly interconnected computations. Going from the urge to partition the software for parallelization on the underlying computational platform to the inherent parallelism of ASICs was then just a small step.

A cellular neural network (CNN) is a mathematical template with interesting properties. Especially, it allows nonlinear dynamic behavior to be modeled. A well-known example is in the Game of Life (GoL) [4]. But the apparent simplicity is deceptive for such examples where the simple binary cellular GoL topology is sufficient. Where realistic examples with reaction/diffusion behavior come in, one sees that the local connectivity brings a tide of data hazards, producing a surplus of processor stalls and greatly reducing the potential speed of a classical central processing unit (CPU). Specialized fine-grain hardware seems to be the remedy.

However, in [5] it is argued that the performance deficit remains due to the presence of global control. The computing is limited to the single cell, and many cells will make the task faster in proportion. Global control will destroy this perfect scalability by the need for long control lines. Long lines are slow and system performance is dominated by the longest lines. It seems that the cellular network will have a close-to-constant performance at best, when scaling.

As a typical media application, vision offers interesting applications because the basic neural equation fits nicely with low-level pixel operations. Analog realizations allow a focal-plane processor: a tight integration of light sensors with on-chip low-level operations to reduce the off-chip bandwidth. Parallelism between many small nodes is a given, but programmability is still needed. This led first to analog [6] and then to digital realizations. With scrutiny of number representations and on-chip network communication, the digital realization has slowly evolved from a simple pipelined CPU to a tiled application-specific integrated processor (ASIP) [4].

The many-core architecture with thousands of tiles was developed, but it showed quickly that the classical compilers are failing by a margin. In the early days, the sequential programming paradigm was expanded with partition and allocation methods, but gradually the awareness is growing that other programming paradigms on different abstraction levels are required. It is not even clear what algorithms are to be preferred, and how such algorithms should be implemented. Therefore, we will look here into the scaling potential of cellular-neural-network applications to establish a better understanding of the new phenomena.

The chapter discusses the history of CNNs as a gradual development to a many-core implementation. From a review of the basic neural design equation, it first reviews some of the analog and acceleration systems. Hence we show the communication needs in a digital implementation. Subsequently, we introduce the

application-specific CNN processor and treat the need for new technology to
create the CNN platform for new media applications.

16.2   Some Basic Theory

A neural network is, in its most general format, a system of connected elementary
processing units (PUs), which for historical reasons are called neurons. Each neuron
is associated with a real numerical value, which is the state of the neuron.
Connections between neurons exist so that the state of one neuron can influence
the state of others. The system operation can be programmed by weighting each
connection to set the degree of mutual influence. Each processing unit performs the
following operations: (a) the incoming signals x are collected after weighting, (b) an
offset θ is added, and (c) a nonlinear function φ is applied to compute the state [7].

A single neuron is depicted in Fig. 16.1. It states the series of computations when
going from left to right. The offset θ is modeled as the weighted contribution of
a source with value -1 and therefore not explicitly shown in the remainder of
this chapter. Creating a network of neurons can be done in numerous ways, with
different architecture for different applications. The system gets a specific function
from the structure and the weight settings: the structure allows for general functionality, while the weights set the specific function.

The neural architecture represents the network function globally and graphi-
cally. The network function (network transfer function, mapping function, or
"answer of the network" are equivalent terms) accomplishes the main task of the
network: to adapt to associate the questions posed to the network with their

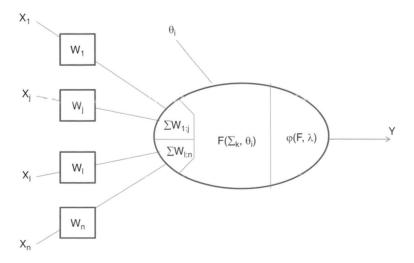

Fig. 16.1 The operation of a single neuron

answers. "Questions" and "answers" are terms borrowed from the human association process. As widely used alternatives, the terms input and output examples are
used: examples are shown to the network of how the problem to be solved behaves;
the input example (the stimulus) is shown together with the adequate reaction to it –
the response or the output example. This process is performed in the forward pass
through the network: the outputs are obtained as a response of the neural system to
the input (question) stimulation. Thus the forward pass through the network
evaluates the equation that expresses the outputs as a function of the inputs, the
network architecture, the nodal transfer, and the parameters so that, during the
backward pass, the learning algorithm can adapt the connection strengths.

For the multilayer feed-forward neural network (Fig. 16.2), the input vector x
and the scalar output y are connected via the network function f, as described by

$$f(\vec{x}, \vec{w}) = \varphi\left(\sum_i w_{ji}\varphi\left(\cdots\varphi\left(\sum_k w_{ik}x_k\right)\cdots\right)\right) \qquad (16.1)$$

where φ is the nodal transfer and w_{mn} denote the different weight connections
within the network architecture with indices according to the direction of information transport. The nested structure of the formula represents the steps that the feed-
forward track of the learning algorithm has to pass.

The nature of the nonlinear neural transfer function has not been specified so far.
Basically it could be anything, from a straight threshold to a complicated expression. As (16.1) already indicates, it is even possible to wrap an entire neural
network to be used in another network as a single neuron with a complex transfer
[8]. We will refrain from that and apply only an S-shaped type of transfer,
especially a sigmoid function. More specifically, we will use the logistic and the
zero-centered variety: two sigmoid transfer functions with an exemplary symmetry,
the effects of which on the overall learning performance are nicely suited to display
the behavior as studied here. Their practical importance relies on the fact that the
logistic sigmoid has found major application in digital circuitry, while the zero-
centered sigmoid appears foremost in analog circuitry. The specification of these
functions is shown in Table 16.1.

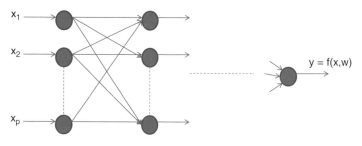

Fig. 16.2 A multi-layer feed-forward network

Table 16.1 Specification of two sigmoid transfer functions

	Logistic	Zero-centered
Function φ	$1/(1 + e^{-x})$	$(1-e^{-x})/(1 + e^{-x})$
Output range	$(0, +1)$	$(-1, +1)$
First-order derivative	$e^{-x}/(1 + e^{-x})^2$	$2e^{-2x}/(1 + e^{-x})^2$
$\varphi' = f(\varphi)$	$\varphi' = \varphi(1-\varphi)$	$\varphi' = 0.5(1-\varphi)^2$

The artificial neural networks (ANNs) adapt themselves to map the event–consequence, evidence of behavior of the problem to solve, on the known problem areas. In human terms, it can be said that the network learns the problem. This learning is an optimization process, accomplished in many of the networks by a gradient optimization method. The gradient methods can be described as a hill-climbing (descending) procedure, which ends by finding an optimum (extreme) state in which the network models the problem best. At this state it is expected that every question from the problem area that is presented to the network will bring the right reaction (answer) at the output of the network.

Any optimization problem can be defined by a set of inequalities and a single cost function. In neural networks, this cost function is usually based on a measure of the difference between the actual network response and the desired one. Its values define the landscape in which an extremum must be found. By deliberately adapting the weights in the network, both responses are brought together, which defines the desired extremum. In other words, a measure of the difference in responses for one or more inputs is used in an algorithm to let the network learn the desired weight settings: the learning algorithm. It is a fundamental network design issue to determine just the right value [9].

16.2.1   The Cellular Neural Network

Cellular automata have always created a scientific attraction by displaying complex behavior on the basis of a regular computing structure with simple rules [40]. From the observation that a cellular automaton can both imitate biology and work as a Turing machine, the understanding has grown that they can also create autonomous behavior. Many cellular automata implementations have been made in software, but this approach will focus on a specific variation, the cellular neural network, as it has a number of promising hardware realizations [10].

After the introduction of the Chua and Yang network [11], a large number of CNN models appeared in literature. Like a cellular automaton, a CNN is made of a regularly spaced grid of processing units (cells) that only communicate directly with cells in the immediate neighborhood. Cellular neural networks [12] are widely used with real-time image processing applications. Such systems can be efficiently realized in analog techniques. Early attempts to realize the CNN functionality in macro-enriched field programmable gate-arrays (FPGAs) have shown impressive potential [13].

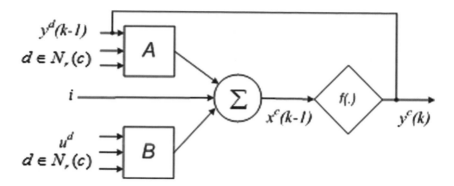

Fig. 16.3 Block diagram of a DT-CNN cell

A DT-CNN (discrete-time CNN), introduced by Harrer and Nossek [14], is a regular multi-dimensional grid of locally connected cells. Each cell c communicates directly with its r neighbors, i.e., a set of cells within a certain distance r to c, where $r \geq 0$. For example, if $r = 1$ we have a 3×3 neighborhood and if $r = 2$ we have a 5×5 neighborhood. A cell can still communicate with other cells outside its neighborhood owing to the network propagation effect.

The state of a cell c, denoted x^c, depends mainly on two factors: the time-independent input u^d to its neighbors d and the time-variant output $y^d(k)$ of its neighborhood. The neighborhood always includes c itself. This dependency in a discrete time k is described by

$$x^c(k) = \sum_{d \in N_r(c)} a_d^c y^d(k) + \sum_{d \in N_r(c)} b_d^c u^d + i^c, \tag{16.2}$$

while Fig. 16.3 gives an illustration.

Spatially invariant CNNs are specified by a control template A containing a_d^c, a feedback template B containing b_d^c, and the cell bias $i = i^c$. Node template, $T = <A, B, i>$, determines together with input image u and an initial output $y(0)$ completely the dynamic behavior of the DT-CNN. Convergence to a stable state will be achieved after a number of iterations n, i.e., at time $k = n$.

Although neighborhoods of any size are allowed in DT-CNNs, it is almost impossible to realize large templates. Limited interconnectivity imposed by current VLSI technology demands that communication between cells is only local. Let us restrict ourselves to the use of 3×3 neighborhoods, where templates A and B are 3×3 matrices of real-valued coefficients. Additionally, the input range of a DT-CNN cell is restricted to $[-1, +1]$.

16.2.2 The Analog-Mostly Network ACE

Since the introduction in 1992, the concept of CNN-UM [41] (universal machine) has been considered to realize the most complex CNN applications electronically, owing to its universality and implementation capability. Several realizations have seen the light of day, some focusing on analog-only or mixed-signal implementation in CMOS while others follow the footprints of the predecessor emulators CNN-HAC [42] (hardware accelerator) and ACE. In this section, the mixed-signal type is considered with focus on a specific chip series mainly developed by a group at Centro Nacional de Microelectrónica at the University of Seville in Spain [6, 15–20].

These CNN implementations have a number of drawbacks (Table 16.2). One has to do with the difficulty of electrical cell design due to the various ranges for the internal voltages and currents. These ranges have to be considered to reduce the MOS transistor nonlinearities. Another issue is that input signals are

Table 16.2 Comparison of mixed-signal full-custom CNN universal chips. All chips use a modified CNN model, i.e., the FSR model

		CNNUC1[a] [6]	ACE400 [15]	ACE4k [17, 18]	ACE16k [19]
CMOS technology		1 μm	0.8 μm	0.5 μm	0.35 μm
Density [cells/mm²]		33	27.5	82	180
Array size		32 × 32	20 × 22	64 × 64	128 × 128
Input	Type	Binary	Binary	Binary and Gray scale	Binary and Gray scale
	Optical	✔	✔	✔	✔
	Electrical	✔	✔		✔
Output	Type	Binary	Binary		
	Electrical	✔	✔		
Dig. external control				✔	
Global instr. memory		Static	Dynamic	Static	Dynamic
# Ana. instructions		8	8	32	32
# Dig. instructions			0	64	64 × 64
Local memory	Type	Digital	Digital		Dig.&Ana.
	Dynamic		✔		✔
	Amount	4 Binary (1-bit)	4 Binary	4 Binary	2 Binary
				4 Gray	8 Gray
Ana. Acc.	Input			8 bits	8 bits
	$A \& B$	7 bits + sign	7 bits + sign	7 bits	7 bits + sign
	Bias	7 bits + sign	8 bits + sign	N/A	7 bits + sign
Ana. circuit area/cell		N/A	70%	N/A	
Cell array area/chip		N/A	53%	58%	
Cell area		180 × 170 μm²	190 × 190 μm²	120 × 102.2 μm²	73.3 × 75.7 μm²
Power	Entire chip	N/A	1.1W @ 5V	1.2W @ 3.3V	< 4W@ 3.3V [20]
	Per cell	N/A	N/A	370 μW	180 μW

[a]This architecture was not given a name in [6], but is called CNNUC1 here to emphasize that it was the first universal chip of the series

always voltages while internal signals may be voltages or currents. This is crucial in focal plane architectures where sensors provide the signals in the form of currents. Incorporation of the sensory and processing circuitry on the same semiconductor substrate is pretty common [21] as CMOS technologies offer good photo-transduction devices [22]. A conversion into voltages is then needed, which complicates CNN interface design. Finally, the combination of internal voltage and current signals leads to internal high-impedance nodes and, hence, large time constants. This results in a lower operation speed than desired.

The full signal range (FSR) model has been introduced to overcome these limitations [23, 24]. Here, all variables are in the form of currents, thus eliminating the need of current-to-voltage conversion. The main difference compared to continuous-time (CT) and DT-CNNs is found in the way state variables evolve. State variables have the same variation range as input and output variables, i.e., independently of the application (16.3).

$$\tau \frac{dx_c}{dt} = -g(x_c(t)) + d_c + \sum_{d \in N_r(c)} \{a_{cd}y_d + b_{cd}y_d\},$$

$$\text{where} \quad g(x) = \lim_{m \to \infty} \begin{cases} m(x+1) - 1 & x < -1 \\ x & |x| \le 1 \\ m(x-1) + 1 & x > 1 \end{cases} \tag{16.3}$$

This results in a reduced cell complexity for both CT and DT cases and thus reduces area and power consumption in VLSI implementations. Stability and convergence properties are guaranteed and proven to be similar to the original models. It is further shown that uniform variations of the coefficients of the cloning template affect only the time constant of the network.

The flexibility and generality of the CNN-UM lies in the ability to freely reprogram the system using distinct analogic parameters, i.e., different cloning templates and logic functions. This is guaranteed in [6] through a synergy of analog and digital programmability. Internally, all cells are equipped with an analog-programmable multiplier, while digital control signals are provided externally, i.e., outside of the cell array. A specific interface circuitry is required to generate the internal weights from the corresponding external digital signals. The interface is located at the periphery of the cell array and behaves as a nonlinear D/A converter. The analog weights are gradually adapted to the desired level and then used to control the analog multiplier within the cells in the array. Each peripheral weight tuning stage consists of an analog-controlled multiplier and a digital-controlled multiplier connected in a feedback loop through an integrator. The conceptual architecture of the ACE16k, also called VSoC as being a clear advance in a roadmap towards flexible vision systems on chips, is shown in Fig. 16.4.

Notable improvements in performance are further reached by pre-processing the pixels locally to decrease the communication load [25]. By integrating more than only low-level pixel processing, it aims to reduce the communication load with the other system parts. This is clearly a next step in the development of smart sensors.

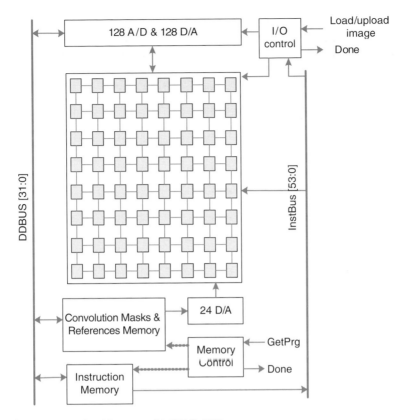

Fig. 16.4 Conceptual architecture of ACE16k [19]

The digital feature bus brings the advantage of intelligent networking of parts that mask their analog nature but retain the low power and small area.

16.2.3 Pipelined CNN Architecture

In contrast to the analog focal-plane processor, the digital version will assume a separate device to extract the image and to make the pixels available in a distinct digital memory. This allows any image size, but cannot handle all pixels simultaneously. Large images are divided into sub-frames that are handled in sequence. Consequently the operation of a DT-CNN on images covers many dimensions. The local operation is performed in a two-dimensional plane (width and length) and iterates in time (Fig. 16.5). Owing to the limited capacity of the CNN implementation, this has to be repeated over image slices and iterates over the surface to handle potential wave propagation. Finally, the *state-flow architecture* is performed on

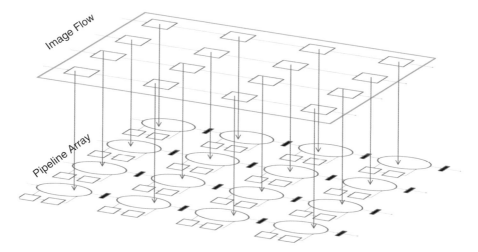

Fig. 16.5 The ILVA pipeline

sequences of images. This can be implemented in many ways, compromising between speed and resource usage.

The CASTLE architecture [26] is a representative of the class of fully digital emulators. The architecture is capable of performing 500 CNN iterations using 3×3 templates on a video stream with frequency of 25 fps taking 240×320 pixels each. This is valid for a system with 24 processing units (PEs) with a precision of 12 bits. CASTLE makes use of the FSR model where the absolute value of the state variable is never allowed to exceed the value of $+1$.

Loading input pixels on-the-fly from an external memory into the processing array constitutes a performance bottleneck. On the other hand, storing the entire image on chip is impossible due to the limited resources. Instead, the image is divided into a number of belts with a height of $2r$ pixels, where r represents the neighborhood. Each belt is then fed to a PE (Fig. 16.6, right). In this case, the I/O requirements of the PE, i.e., the cell, are reduced to two inputs and two outputs per cell update. Each pair consists of one state value and one constant value, corresponding to the combined contribution of control template together with the bias (Fig. 16.6, left). The main memory unit in the PE consists of three layers of equally sized circular shift-register arrays for the state input and two layers for each of the constant and template selection inputs. Inputs from left and right neighboring PEs are directly connected to the corresponding ends of the shift-register arrays.

In line with the proposed approach in the CNN-UM, the functionality of the CASTLE architecture is ruled by means of a global control unit. One of the most important features of this unit is the selection of the employed precision. Data precision is variable and can be set to 1, 6, or 12 bits. The lower the accuracy, the faster the system.

An important issue is the amount of logic occupied by the register arrays constituting the internal memory units (IMUs, Fig. 16.6, left). In the first

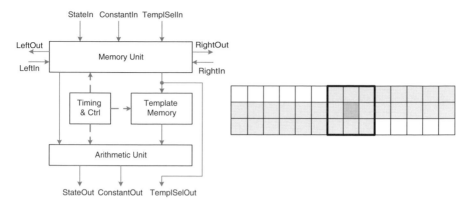

Fig. 16.6 *Left*: A schematic view of the processing unit in CASTLE, where dashed lines represent control signals and continuous lines show the data path. *Right*: The belt of pixels stored on-chip for 1-neighborhood, where the black-outlined square indicates the current position of the convolution operation [27]

experimental chip that has been fabricated in 0.35 μm CMOS [27] technology, about 50% of the total area of a single PE is allocated to register arrays, while the arithmetic block occupies not more than 21% of the area. Furthermore, experiments show that a CASTLE emulator with 24 processors outperforms the digital signal processor (DSP)-based ACE engine (Sect. 16.2.2) only when the rate of logic operations is high enough [27].

The CASTLE architecture suffers from a number of drawbacks. One has to do with the inability of emulating complex dynamic systems where operative parallelism is a key feature. The single-layered architecture handles only one operation at time. Other drawbacks include the limited template size, cell array size, and accuracy. Hence, a new architecture called FALCON has been developed to provide higher flexibility and to allow for multi-layer accommodation [28]. The implementation is based on the FSR model with discretized state equations. In contrast to all CNN-UM-inspired implementation discussed so far, the design is hosted on a Xilinx Virtex series FPGA. This increases the ability of reconfiguration, brings down developing time cycle, and decreases the overall cost.

The arithmetic uses a fixed-point representation, where possible value widths are 2–64 bits. Configurability is essential to allow accommodation of flexible precision when needed. But for the highest possible precision the cell array will consist of not more than four processing cells! The configuration is unfortunately not dynamic but the entire design has to be re-synthesized and loaded on the FPGA every time a new configuration is required! Apparently, for algorithms with alternating operations of low and high precision the FPGA has to be reconfigured several times in order to provide accurate results.

To overcome this drawback, spatial and temporal elements are mixed by interweaving three pipelines corresponding to a row of three pixels (Fig. 16.7). This reduces the latency and makes better utilization of the available resources. The nodes are grouped in columns where each column holds a scan-line in the image

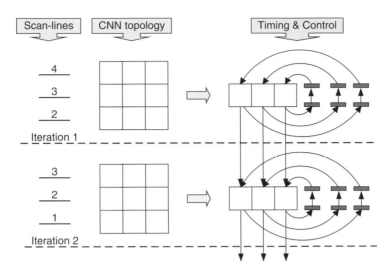

Fig. 16.7 Mixed spatial-temporal state-flow architecture operating directly on the pixel pipeline

stripe. The columns will then form iterations performed on the image. In this way, one dimension (width or length) of the image frame together with the number of iterations are implemented as columns of nodes while the other dimension of the frame is handled by slicing as illustrated in Fig. 16.7. One of the resulting realizations is a design called ILVA [4].

The underlying idea is that a two-dimensional computation of the local cell is flattened into a series of one-dimensional computations by dropping intermediate results on the computational path. In this way, the requirement of each node to have data from eight neighbors for finding the output is met. In other words, we let every node in the network contain image data from three pixels, i.e., pixel values for the cell itself and for its left and right neighbors are stored in each node. A direct connection with the two nodes above and below completes the communication between a node and its neighborhood. In short, one node contains three pixels and calculates the new value for *one* pixel and *one* iteration.

The prescheduled broadcasting in ILVA keeps the communication interface at minimum, which allows for a large number of nodes on chip. The performance is high as the system directly follows the line accessing speed, but the design suffers from a number of weaknesses. The iterations are flattened on the pipeline, one iteration per pipeline stage, making the number of possible iterations not only restricted due to the availability of logic, but also fixed. Operations that require a single iteration only still have to go through all pipeline stages. Output data has to be fed back to the pipelined system in order to perform additional iterations, making it far from trivial to handle larger iterations without accessing the external image memory. This requires additional logic for loading and uploading pixel data and therefore adds overhead for timing control and thereby severely slows down the system.

16.2.4   In Short

Neural networks are programmed by templates that modify the connectivity of many small but identical processing elements. A suitable technology realizes such a PE by keeping it small. Analog design does this by suitable circuit design to answer some reasonable precision requirements. Digital design has struggled with the size of the multiplication operator. Therefore a number of specialized algorithm implementations have been developed to limit the dependency on multiplier-size by going for bit-level implementations [29].

Where size does not matter anymore, speed becomes of interest. But a faster element does not make a fast system. Getting the data in and out of the system is a major concern. This already plays a role in the matrix interconnect of the focal-plane architecture, where early-on the light sensors are buffered to separate the image extraction from the pixel reading. The concern about system-level communication dominates the further developments towards the current Eye-RIS vision system [30].

In the digital domain, the state-flow architecture, in its desire to exploit pipelining for compromising speed and resource usage, gets a streaming character-istic that loses simultaneously the efficiency of two-dimensional diffusion-reaction behavior. This limits the attraction of CNN algorithms to provide more than just low-level pixel processing.

Moreover, the FALCON architecture comes with no possibility of algorithmic control on chip. All algorithmic steps, as well as local logical operations and programs, are executed on a host PC. This reveals that the system cannot stand alone but is always dependent on the host PC! Obviously, all the benefits of performing complex tasks on the CNNs are lost.

To remedy these problems, the architecture is extended with a global control unit GAPU (global analogic programming unit) [31] in line with the conceptual CNN-UM. In addition to on-chip memories and some peripheral blocks, the GAPU is built up using an embedded MicroBlaze processor core with 32-bit RISC architecture. Most modern FPGAs provide at least one of these processor cores on chip. The extended FALCON architecture is implemented on a Xilinx Virtex-II 3000 FPGA. Apart from the embedded processor core, the GAPU occupies about 10% of the available logic, which can be compared to the area of a single CNN processor that requires about 2.8% of the logic. It is worth mentioning that the GAPU runs on lower clock frequency than the processing units (PEs), thus setting a higher limit of the overall speed.

16.3   Many-Core CNN Architecture

The sequential evaluation order of CNN equations is closely related to the software evaluation. Pixel lines are consumed immediately and subsequently the iterations are pipelined, the time-ordered style. The alternative is the spatial approach, based

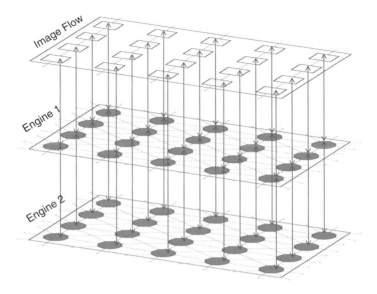

Fig. 16.8 The n-dimensional Caballero architecture

on nodes which each relate to a CNN cell. In this situation, the pixel lines come into the FIFO (first-in-first-out) until it is fully filled. Then these values are copied into the CNN nodes, which subsequently start computing and communicating. Meanwhile new pixel lines come in over the FIFO. When the FIFO is filled again and the CNN nodes have completed all local iterations, the results are exchanged with the new inputs (Fig. 16.8). This leaves the CNN nodes with fresh information to work on and the FIFO can take new pixel lines while moving the results out.

16.3.1 Architectural Variety

Another communication technique is involved with the network itself. The CNN equation is unrolled not in time but in space, and the nodes retain the result of the equation evaluation so that next iterations do not involve access to the external data memory. In this *state-scan architecture*, the neighborhood is actively scanned for the input values, known as Caballero [4]. The schedule is predetermined, but computation and communication needs are decoupled by splitting the simple node into a processor and a router. The nodes can theoretically transfer their values within the neighborhood in parallel.

The need for explicit scheduling on nodal activities works out differently for different CNN-to-network mappings. Two main categories can be distinguished:

– The *consumer node* is fully in accordance with (16.1). The discriminated output of a cell is also the cell output and broadcasted to all connected nodes, where it is

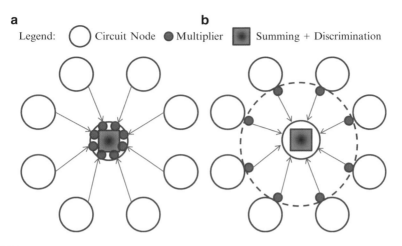

Fig. 16.9 Consumer (**a**) and producer (**b**) cell-to-node mapping

weighted with the coefficients of the applied template before the combined
effect is determined through summation (Fig. 16.9a).
– The *producer node* discriminates the already weighted inputs and passes to each
connected node a separate value that corresponds to the cell output but weighted
according to the applied template (Fig. 16.9b).

Ideally all nodes are directly coupled and therefore bandwidth is maximal. In
practice, the space is limited and the value transfer has to be sequenced over a more
limited bandwidth. This problem kicks in first with the producer-type of network,
where we have $2n$ connections for n neighbors. The network-on-chip (NoC)
approach is meant to solve such problems. However, as the CNN is a special case
for such networks, being fully symmetric in the structure and identical in the nodal
function, such a NoC comes in various disguises.

In the consumer architecture, scheduling is needed to more optimally use the
limited communication bandwidth. Switches are inserted to handle the incoming
values one-by-one. To identify the origin of each value, one can either schedule this
hard to local controllers that simply assume the origins from the local state of the
scheduler (circuit switching, Fig. 16.10a), or provide the source address as part
of the message (packet switching, Fig. 16.10b). The former technique is simple.
It gives a guaranteed performance as the symmetry of the system allows for
an analytical solution of the scheduling mechanism. The latter is more complicated,
but allows also for best effort.

The counterpart of consumption is production. Every node produces values that
have to broadcast to all the neighbors. Again, where the communication has
a limited bandwidth, we need to sequence the broadcast and this can be done in
the same way as for the value consumption (Fig. 16.11). A typical example, where
each node can only handle one value at a time, is discussed in [4]. Such a scheme
suffers from a large latency overhead.

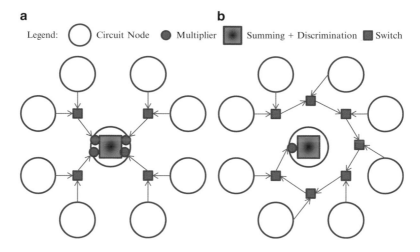

Fig. 16.10 Value routing by multiplexing in space (**a**) and in time (**b**)

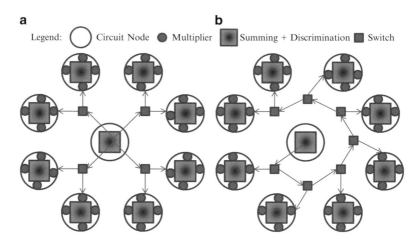

Fig. 16.11 Value routing by multiplexing in space (**a**) and in time (**b**)

In the case of producer architectures, the nodal output is already differentiated for the different target nodes. Each target node will combine such signals to a single contribution. This combining network is an adder tree that will reduce the *n* values to 1 in a pipeline fashion. Consequently, this tree can also be distributed, allowing for a spatial reduction in bandwidth.

The overall processing scheme as shown in Fig. 16.12 is then similar to what has been discussed for the consumer architecture. The main difference is that the communicated values will be larger than the represented products and are therefore of double length. Whereas the consumer architecture is characterized by "transfer and calculate," the producer architecture is more "calculate and transfer". Furthermore,

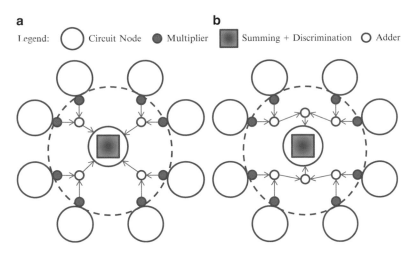

Fig. 16.12 Adder trees combine the network in the production architecture

they both rely on a strict sequencing of the communication, simultaneously losing much of the principle advantage of having a cellular structure.

16.3.2   Numbers and Operators

Digital neural networks are like digital filters in the sense that they take a lot of additions and multiplications. Actually they take mostly only multiplications. As digital multiplications take a lot of space, one has always had an open eye for serial multiplication or even "multiplier-less" algorithms.

16.3.2.1   Precision Needs

Moving from floating-point to fixed-point numbers is not merely a question of reducing the value scope, but a careful arbitration between precision and accuracy. Precision is addressed by the smallest step in the value space, such that small variations in the value space have little significance in the problem space. Accuracy on the other hand has to do with the degree by which the computation can achieve the desired result. Figure 16.13 visualizes the difference between these two notions.

We see the difference between precision and accuracy back at the moment when we move from floating- to fixed-point notation. The fixed-point notation requires an explicit scaling, which is implicit in the floating-point counterpart. Let us assume the decimal point in the right-hand location of the number. When we move the decimal point to the left, the absolute size of the number decreases while we add a fractional part. What happens when we move the decimal point to the left is that we make the smallest step in the value space smaller, such that variations in the

Fig. 16.13 Precision versus accuracy

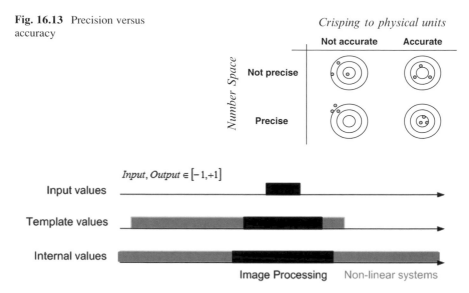

Fig. 16.14 Precision needs in number representation

lower-order bits will have a reduced meaning. In other words, we gain precision, but the absolute meaning of the value has decreased, giving a smaller range for the number, and therefore chances are that we lose accuracy at the same time.

In the case of a DT-CNN, input and output values of every node are scaled between -1 and $+1$ by definition. However, this is not true for the inputs to the system. In the case of template values for image processing, such inputs have a dynamic range that is close to that of the internal values. For other nonlinear systems, this is not the case: they have a larger dynamic value range and/or that range lies far away from the internal values (Fig. 16.14).

When scaling is properly performed, pruning can be used for further optimization. In earlier work [32] it has already been shown that for 8-bit inputs the internal representation does not have to allow for the full arithmetic precision of 21 bits because the CNN structure is largely permissive for statistically random rounding errors. The compensatory behavior is principally due to the feedback structure. The approach performs gradual adjustment of template values, which allows pruning of internal numbers down to the minimum with retained accuracy.

Throughout the experiment, the same fixed boundary condition has been used. However, the choice of boundary conditions has a direct impact on the complete stability of a certain category of CNNs, i.e., those with opposite-sign templates [33]. The instability of these CNNs depends far more on the boundary conditions than on the template. Simple tests of, e.g., hole filling, have shown that gradual refinement of the boundary conditions leads to noticeable deviations in convergence in terms of functionality and speed [4]. Furthermore, robustness is also sensitive to input data [34]. Hence, derived templates must be checked for robustness for the "most difficult input image," as this guarantees correct functionality. Further modification in the pruning approach shows that the number of robust templates is about 15% less for an image containing two holes, a complete hole positioned in the center of the frame, while an incomplete hole is placed on one of the edges [4]. It is therefore mandatory to have as many nodes as possible on a single chip.

16.3.2.2   The Case of the Multiplier

The high throughput of the system is due to the accommodation of three parallel multipliers performing, in parallel, three multiplications that use pixels and corresponding template coefficients as operands. Multiplication results are shifted 1-bit in the least significant bit (LSB) direction before they are forwarded to a tree of adders to accumulate the results with the previous intermediate result. In order to improve the accuracy, rounding units are introduced between the shifters and the following adders. A limiter unit brings the final sum into the operational region. Figure 16.15 depicts the structural architecture of the arithmetic unit. It is obvious that the reduction of communication demands comes at the cost of larger arithmetic units with more functional blocks.

In the serial approach, we pass all values simultaneously but in bit sequence. For every step, the coefficient bits multiply the single-bit input from each neighbor; the results are added and accumulated. As usual in bit-serial logic, the data-path

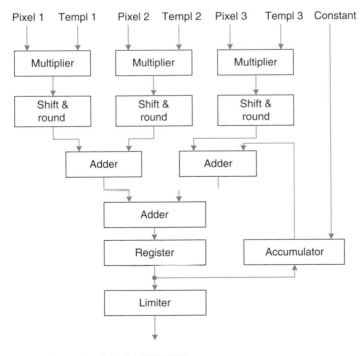

Fig. 16.15 The arithmetic unit in CASTLE [27]

becomes small but at the expense of a more complicated control. Furthermore we can expect a higher latency, as more clock ticks are needed to get to the result. But that is only true for the single node. The basic ten clock cycles for a single node have to be repeated for five neighboring nodes for reasons of bus contention. It does not seem likely that a serial solution that eliminates such bus contention problems will need more. As the small serial node allows for a larger network to be implemented on a single chip, it is worthwhile evaluating its potential.

A fully serial node computation is shown in Fig. 16.16. It requires the coefficients to be locally available in a ring-buffer. This is not as bad as it seems, because a serial shifter can be implemented in a single look-up table (LUT) per 4 bits. For longer coefficients one may consider building the ring-buffer in the BlockRAM, but usually coefficients are not long. Together with the bit-register for the input and the bit multiplier, a 4-bit base unit takes just a slice. The outputs are tree-wise added and give a 4-bit result to be added to the shifting accumulator. This final addition has to be in parallel because long carry propagation may occur. Also the result has to be available in parallel, because a final table lookup is needed for the output discrimination.

All values are notated in 2-complement numbers and range over [1...0]. For such 2:*x* numbers we take *x* as 7 for the inputs and for the template coefficients. This brings an internal representation of 2:21. Because all the values are identically

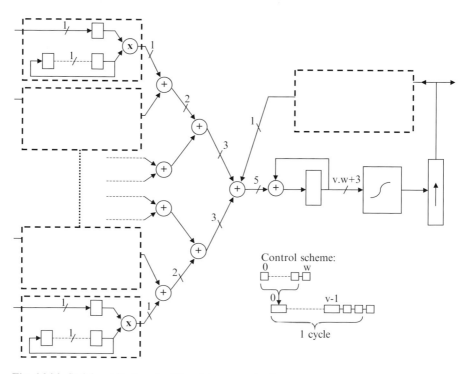

Fig. 16.16 Serial architecture for bit-serial communication

aligned, there is no real problem in finding the binary point in the final results. Addressing the discrimination function by looking up the table implies trimming the internal representation to the size of the external one.

16.3.2.3 The Series/Parallel Approach

This overview of alternative designs shows a rich variety of compromises between speed and area. Starting from the bit-serial structure, even more alternatives can be created by logic transformation. A typical example of such a derivative implementation is in series/parallel computation (Fig. 16.17). Every single input bit multiplies the entire coefficient. The outputs are tree-wise added and give a $w + 3$-bit result to be added to the shifting accumulator. This reduces the latency and the control significantly, but at the expense of wider adders. Connected to this comes the implementation of buffering. Where in the pure bit-serial approach, the buffers are directly created in hardware; in the derivatives it becomes worthwhile implementing the buffers in the local memory.

We introduce here a very little known third alternative that has been introduced in the past to achieve the best of both worlds: good addition and good multiplication. It is based on fixed-point arithmetic; hence addition and subtraction deserve no further discussion. Multiplication and division are accomplished by an in-line

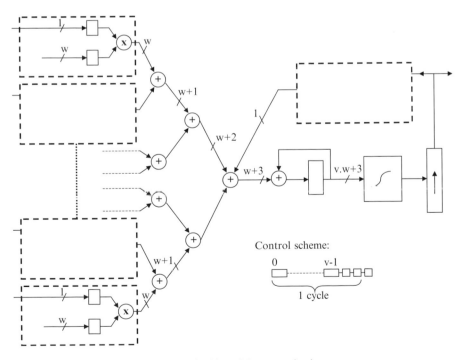

Fig. 16.17 Series/parallel architecture for bit-serial communication

interpretation (rather than off-line conversion) into the logarithmic domain in a seemingly similar way as performed in FP.

The basic idea is that DIGILOG multiplication (Sect. 3.6 and [35–37]) can be transformed to addition in the logarithmic domain. A binary logarithmic number, $A = 2^n a_n + 2^{n-1} a_{n1}^- + \ldots + 2^1 a_1 + 2^0 a_0$ can be rewritten as $\log^2 A = j + \log^2(1 + A_r/2^j)$ in the binary logarithmic system; here j is the leading "1"s position of A, and defined by $j|a_j = 1$ and $a_i = 0$ for all $i > j$. A_r is the remainder, $2^{j-1} a_{j1}^- + \ldots + 2^1 a_1 + 2^0 a_0$. Therefore, in the DIGILOG system, two n-bit binary numbers, A and B, can be written as

$$A = 2^n a_n + 2^{n-1} a_{n-1} + \cdots + 2^1 a_1 + 2^0 a_0$$

and

$$B = 2^n b_n + 2^{n-1} b_{n-1} + \cdots + 2^1 b_1 + 2^0 b_0.$$

These two numbers have a leading "1" at the j and k positions, respectively. If all the leading "0"s bits higher than the j in A or k in B are neglected, these two n-bit binary numbers also can be written as a combination of the leading "1" and a remainder:

$$A = 2^j a_j + 2^{j-1} a_{j-1} + \cdots + 2^1 a_1 + 2^0 a_0 = 2^j + A_r$$

and

$$B = 2^k b_k + 2^{k-1} b_{k-1} + \cdots + 2^1 b_1 + 2^0 b_0 = 2^k + B_r.$$

Then the multiplication of the two binary numbers A and B becomes

$$A \times B = 2^j 2^k + 2^j B_r + 2^k A_r + A_r \times B_r.$$

In this formula, the three terms 2^j, 2^k, $2^j B_r$, and $2^k A_r$ can be easily implemented as shift-and-add on the hardware level. In the DIGILOG multiplication, the rest term $A_r \times B_r$ can be realized through iteration until A_r or B_r becomes zero, rather than that B_r becomes zero in the conventional FP (floating point) multiplication. Therefore the iteration time in DIGILOG only depends on the total number of "1"s in A or B, whichever has the smaller number. In the FP, the iteration time of multiplication depends on the total numbers of "1"s in the argument of choice.

16.3.3 Broadcast Mechanism

Also here, we have to look at the way values are broadcast. In contrast to the consumer architecture, we have as many output values as there are neighbors. This makes for an identical situation and no additional measures are needed, except for the fact that we are not be able to generate all the different products at the same and the sequencing issue pops up again.

The elimination of the global controller will speed up the operation of a CNN. It has to synchronize all the local nodal controllers to ensure that no value transfer is missed. The presence of such a lock step mechanism will undoubtedly decrease the operational speed of the network. Therefore it has been investigated whether such a lock step method is needed. It is proposed here to replace the local controller that actively scans the inputs by an administration that is not dependent on the order of arrival. This allows the operation of individual nodes to be speeded up without harming the overall behavior.

Because of the regularity of the CNN network model we can combine the switch and the network interface into a single unit that administrates per node the status of all local communication per nodal computation. For each virtual connection, we have (Fig. 16.18)

– A status field, showing whether the value has been rewritten;
– A redirection field showing whether the value has reached the final destination or should still be forwarded;
– A value field, where the content of the last transmission is stored.

If a node receives a new value, this will be passed to the virtual switch and this unit will in turn try to transmit this value to its neighbors. As all nodes may be thus activated at the same time, infinite bandwidth will be needed. As mentioned earlier, the bandwidth will be limited to handle one or two values at a time for most practical applications. An arbitration mechanism is required to handle multiple requests that can potentially create deadlocks. In the existing implementations, this is done by enforcing a strict sequencing of all nodes, which can be easily shown to be correct.

Local control is supported by the use of the forward field, which is meant to pass values between the switches until eventually all final destinations are reached. We have limited the forwarding to such switches that are also final destinations. By stretching the communication cycles of a 1-neighborhood so that it overlaps with the sequence of operations, routing demands are reduced and controller is simplified.

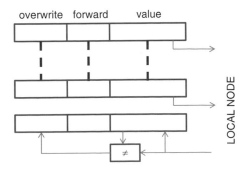

Fig. 16.18 The virtual switch model

From this basic concept, we can see an easy simplification. Where the values come in, they can be immediately absorbed by the node. Performing the CNN equation evaluation on-the-fly only the most current input value has to be saved if

- All values arrive in some order and
- Within this period no duplication occurs.

What remains is therefore an input register, an output register, and flags for all virtual connections.

The moral of the serialized broadcasting approach is that the transfer needs to be sequenced when the communication bandwidth is limited due to area restrictions. We have also found that doing so by state machines leads not only to architectural rigidity but also to degraded performance. For instance, 15% of the utilized area in the serial broadcasting scheme is occupied by the state machine (Fig. 16.19). One way to eliminate the need for the nodal controller, at least partially, is by transferring all values in a source-addressed packet. The original data-only packet used previously is padded with a small header containing the position of the source node in the grid. Hence, the packets carry their addressing information in the header, which can be exploited in two different ways.

A better approach will make use of the intrinsic positioning information carried in the header to address the local template memory of the current node. The nodal equation is then performed in the manner the packets are received. The logic required for the addressing of the value/coefficient pairs is greatly reduced through the use of mirrored binary numbers of both the rows and the columns. In the case of a 1-neighborhood only 2 bits for the row and 2 bits for the column address are required. In general we need only $2(r + 1)$ bits, where r is the neighborhood.

A second remark is on the need for full buffering in the network interface. In serial broadcasting, we can consume every incoming value immediately, as long as we can guarantee monotonicity. For this, however, we need only the flags, not the values themselves. Hence, the presence of the overwrite and forward flags allows us to consume and pass on-the fly, reducing the need for buffering to only a read and a write register.

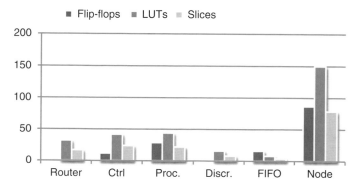

Fig. 16.19 Area utilization of the different components with the serial broadcasting scheme

We are now left with a simple and scalable network interface, taking:

- Overwrite/forward flags for all cells in the neighborhood;
- A single node communication register for temporary storage of the incoming values;
- At least one CNN state value register
- At least one CNN input register

As the status of each realized node takes only two registers, it can implement a number of CNN cells. This raises the capacity for the multi-layer and multi-frame situation. Add to this the elimination of the "control & forward bottleneck," which not only savesg clock cycles but also increases clock speed.

16.3.4   In Short

The many-core architecture provides a complete implementation including the reaction/diffusion behavior. This enables the occurrence of nonlinear dynamics, which uses more accurate template values. This seems to suggest the use of a floating-point number representation. On a closer look, the precision has not changed, mostly because of the precision of the nonlinear smashing function. Consequently, a block floating-point representation is usually sufficient. This can be accommodated simply on fixed-point numbers by the use of a scaling factor that comes to each individual template.

A major architectural issue is the balance between the data-flow rate and the computational complexity. The many-core architecture has special advantages where each template requires much iteration, as is usually the case in nonlinear dynamics. It has less of an advantage for low-level pixel operations for image processing, where a pipelined architecture seems superior. In that light, the seeming disadvantage of serial processing (on word-level or on bit-level) is not severe, where otherwise the computation needs to be halted to wait for the data to flow in.

16.4   The CNN Processor

The CNN Instruction-Set Architecture (ISA) defines the exterior of the CNN Image Processor in terms of signals and visible memory locations. The overall CNN ISA is depicted in Fig. 16.20. Overall, we find four modes of operation and their respective instructions using two separate bus systems: the Image-Memory Bus (IMB) and the Host-Interface Bus (HIB), both with a R/W signal and strobed address and data bus.

Fig. 16.20 External view of
the CNN architecture

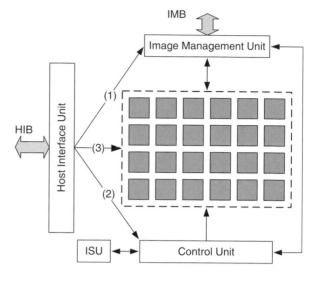

16.4.1 Instruction Set

The Instruction Set covers four domains of activity: (a) the window on the image,
(b) the configuration of the network, (c) the running of the network, and (d) the
debug of the CNN software.

The window operations influence the image-management unit only. It converts
physical into virtual pixels and will autonomously fill the CNN with pixel information with respect to the designated Region of Interest (RoI) for any frame format
using the IMB. Using the window settings it is possible to repeat the CNN program
on a steadily smaller part of the image while increasing the resolution.

- Frame Size: the width and height of a frame in pixels.
- Center coordinate: the non-sampled center of the first frame to be handled.

The internal operation is governed by a number of tables, downloaded over the
HIB. They all start with a preamble that gives the general table information and then
subsequently provides the table entries. The template and discrimination table will
be distributed to all nodes, while the program table is saved in the Instruction Store
Unit (ISU).

- Discrimination: table for discrimination function
- Program: Instruction Store (opt.)
- Template: label and content of a template

The discrimination function lists the transformation from internal node status to
external data result. The length of the table is therefore given by the table size
divided by the table step. The program tells the successive applications of pixel
operations that can be either templates or hard-coded linear instructions. It implic-
itly relates the use of various layers and how they are combined either in time or in

space. A template gives each CNN function. Templates can be downloaded and stored in every CNN node for use later on. The pixel operations can be selected from a number of linear (hardwired) and nonlinear (downloadable) options. The instructions will be placed into a separate ISU.

- Logical: NOT, AND, OR, EXOR
- Arithmetic: Sum, Minus per pixel or horizontal or vertical
- CNN: refers to downloaded templates

The program execution is controlled by:

- Run: none, per clock, per iteration, per template till a specified breakpoint in the program.
- Boundary: the boundary conditions as stated in the templates can be overwritten for debug purposes.
- Sample Size: the amount of physical pixels represented by one virtual (CNN internal) pixel as implied by the window can be overwritten for debug purposes.
- Mode: only this window, or a stripe of the entire image.

The ISA makes the CNN network architecture invisible to the host program and therefore allows a late binding of the actual structure to an application at hand. More often than not, the development network is different from the production network. Starting from a MATLAB model with values represented in a double floating-point format, a gradual conversion into fixed-point numbers is needed [6]. The length of the internal words is application dependent, though accuracy can be easily guaranteed by block-based scaling with a factor derived by inspection of the templates. In practice we have not seen the need for more than 8-bit precision, but for simple templates a smaller length can be accepted.

In line with this, we have inserted a number of in-line debug facilities. The system can be run in various time step sizes, inspected for network data, while allowing the network status to be overwritten and to continue from the existing status. In our reference system we assume that the network is configured separately from the rest. Consequently we have to ensure that the system components can handle appropriate network architectures.

16.4.2 Blocks

The processor largely comprises three blocks: (a) the Host Interface Unit, (b) the Image Management Unit, and (c) the Control Unit.

A host must be able to control the overall functionality of the system by sending instructions and cloning templates and by setting a number of configuration parameters. The communication is handled by the HIU, which receives the requests from the host over the HIB and forwards them to the system using a wishbone bus. The HIU is also responsible for data delivery to the host. Two different FIFOs are used, one for acquiring host requests and one for putting out data to the host.

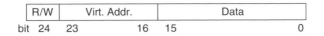

R/W	Virt. Addr.	Data

bit 24   23   16   15   0

Fig. 16.21 A host request is subdivided into flag, address, and data fields

A host request is 25 bits long and is divided into three fields: a Read/Write flag that determines the type of the request, a virtual address field, and a data field that is of interest only in write-requests (Fig. 16.21). Once a request is captured by the FIFO, the virtual address is translated into a system memory address by the I/O memory mapping unit (IOMMU). This address will serve as a base address for all incoming data as long as the virtual address field in the subsequently received requests remains unchanged. The bus master acts partially as a direct memory access (DMA); it generates the proper addresses from the base address and puts it on the address port of the wishbone bus. In the case of a read request, once data are available, the wishbone bus raises an ACK(nowledgment) signal notifying the bus master that reads the data and puts it on the output FIFO. Write requests are handled similarly. Here the ACK signal notifies the bus master that the writing of data is accomplished so the next pair of address/data can be handled.

The camera captures images and stores them in an external memory. The 8-bit gray-scale pixels are then retrieved and converted by the IMU to a signed fixed-point notation with a precision of 7 bits for the fractional part. One of the main operations of the IMU is the windowing operation. As the size of the network is far smaller than the processed image frame, a gradual zooming toward the RoI is required. At the beginning the RoI covers the entire frame, where each CNN node on the chip is mapped onto a virtual pixel that corresponds to a group of real pixels in the image. The virtual pixel is suitably obtained through a conventional averaging of all pixels in the corresponding group. In the next round the RoI covers a smaller part of the frame, depending on the output of the previous round.

The unit has direct communication to the CNN core and the HIU through wishbone buses. It is built with the concept of pipelining in mind and consists of two main components: Instruction Fetch and a Controller (acts as instruction decoder). The naming convention is somehow misleading as the former pipelining stage generates two additional signals; control (used by the Controller pipeline stage) and iteration; in addition to the instruction that is fetched from a dual-port RAM. The controller consists of two major components. One is the actual instruction decoder and provides the proper template, while the other generates CNN-enable and instruction-done signals, depending on the number of iterations and whether equilibrium is reached or not.

The instruction memory (ISU) is arranged as shown in Fig. 16.22 with a space for 64 instructions. Taking a Xilinx Virtex-II 6000 as reference, which accommodates 34000 slices, we find that the overhead incurred by turning a CNN network into a system ranges from 1% for a limited edition to 5% for a complete one with large buffers.

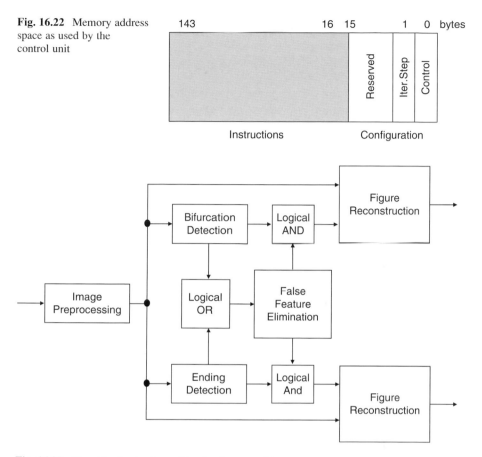

Fig. 16.22 Memory address space as used by the control unit

Fig. 16.23 Algorithmic structure of hand vein recognition

16.4.3   Multi-level Implementation

The CNN computations can be formally identified within morphological algebra. The primitives of low-level pixel operations are the three set-operations "not", "or", and "and", supplemented with directions by iterative operations on "dilation" and "erosion." It has been demonstrated how any morphological expression can be written out into a flat list using only primitives, such that this flat list can be procedurally assembled into an optimal CNN structure [12].

The dataflow graph of the CNN computation displays a number of parallel paths. Each parallelism runs on each instance of the basic CNN structure, which can be either a separate level or a recursive structure over a separate memory domain. A typical example is in hand vein recognition, where the same image is analyzed for different features (bifurcation and ending) to achieve a collective annotation of the reconstructed overall image (Fig. 16.23).

Fig. 16.24 Motion extraction

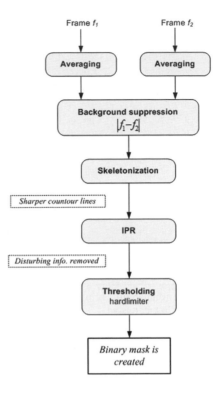

A similar situation is where the same CNN computation operates on subsequent images. In Fig. 16.24, it is shown how motion can be extracted from subsequent images. Where the expression is evaluated before the next frame is available, it is feasible to use the memory domain without loss of system speed. Nevertheless, the fundamental issue involves state explosion, that is strictly reduced by time-multiplexing, where image extraction is less accelerated than by microelectronic speed.

The integration of multiple images helps not only to analyze in time, but also to analyze in space. The objects on a single image, but at a different distance from the sensor, are nonlinearly and therefore non-constantly scaled. Finding the scaling is important for 3D reconstruction. A similar technique can be used for stitching together images of a single object from sensors at different angles [38].

16.4.4   In Short

Branch handling is not only a concern for a conventional processor, but also for a CNN. In the pipelined architecture, a branch condition will stop the operation until the current cycles are performed while new input is halted, similar to flushing. For the many-core architecture, this is not an issue, but the question arises how the instruction fetch is realized. In our CNN processor, the CNN area has the CNN program inclusive branch conditions, and it will pre-fetch. This shows the control is local, while the attached processor is not involved with the details of the CNN operation but simply performs the additional administration. In other words, the CNN part autonomously handles a complete algorithm: no additional DSP is required.

The ingredients of the CNN architecture are close to current tiled architectures of the many-core type. This similarity can be exploited by the evaluation of many-core compilers. The (optimal) CNN many-core solution is at least understood, even if not solved. Therefore, it provides a benchmark for compilers by evaluating how well the compiled code exploits the symmetry issues in the CNN algorithm. The similarity is not complete, as the CNN allows a cellular network without global control, whereas the compiler technology will usually have problems taking the Wave Computing effect of nonlinear dynamics into account.

Clock synchronization is not part of the mathematical concept of cellular networks. On closer inspection, the always "natural" addition of clock synchronization makes it easier to understand but at the same time takes the performance profit away. The analogic realization does not need a clock, because it reaches, by nature, the more speed- and power-efficient solution. Boxing the CNN as a programmable digital unit is a consequence of the need for a generally applicable software development environment, but the obvious detrimental effects of achieving generality in loosing specificity may be worth to spend new thoughts.

For the digital realization, the clock synchronization paradigm has always been taken for granted. As a consequence, a number of architectural experiments have broken down in scaling as the global control lines became dominant for the system speed. Architectures that can run for many cycles without being controlled have clearly provided a far better operation.

Apparently, the actual problem is not in the clock, but in the global control that is usually been designed in a synchronized way. Having a small instruction stack in each node will provide some remedy, but the effect of the small control spread will still not become visible if the nodes are globally synchronized. A more attractive solution comes from the area of asynchronous design. If a node can autonomously determine that it has no activity, it can go into a "no operation" mode. If a node sees that all surrounding nodes are in "no operation," it can fetch the next nodes.

This simple solution is not sufficient, as it cannot be guaranteed that all nodes in the network are inactive by simply looking at the neighbours. Where the CNN creates system activity by doing only local interaction, a mirror mechanism is clearly needed.

16.5 New Media Applications

The naive image application of CNNs has largely been involved with low-level pixel computations, such as edge detection. It takes the image from the camera and makes it a better one. But that is drastically changed with the advent of smart cameras. Adding high-level image computations, the camera starts to determine the parts of the scene. Then it is just a small step towards an understanding of what appears in the scene. This comes together with a further complication of the camera. To handle the algorithmic complexity, the platform requires a processor with a high processing power, typically a graphics processor with the potential of a supercomputing node.

This is invoking a counter movement in an algorithmic swap. The camera is not the encapsulation of the functionality, but the function will encapsulate vision. And even more! The overall system will operate on information that is real-time extracted from the embedding system. As the embedding system reacts to this, it will enforce the sensor, and the time-sequenced interaction creates the desired knowledge. This vision-in-the-loop is the area of New Media. In a similar fashion, audio and video can be handled. The hardware platform discussed here is only restricted to any topological map.

The cellular hardware is especially efficient in video extraction. It is known from the continuous MPEG improvements that 90% of the resources are exploited by movement extraction. There is an interesting alternative, when each object with interesting movement has a small signal added to it outside the natural light spectrum. Pulsation brings a further identification. Consequently, the signal samples have to be stored.

It is only a small step from the pulsating signal to audio. And again, we have to consider the increased order of the state explosion. The window on an audio signal is small from a suitable selected pulse, but in general the window will be much larger. Using modularity the window size is still limited, but it will be too large for comfort [39].

The multi-level CNN is handled by time-multiplexing: the same cells are run but by new parameters from the local memory. For the Fourier transform, we cannot use the same cells, but we can time-multiplex new cells that reconfigure a number of CNN cells. This is a straightforward extension of the previous concept, as illustrated in Fig. 16.25.

So far we have discussed exploiting CNN operation to its theoretical potential. The added problem is to raise that potential. For the simple one-level CNN, all the computations will be local and therefore short and fast. For more complex systems, and notably for new media applications, we need multi-level and multi-mode. As the planar microelectronic technology is essentially two-dimensional, we have used multi-levels through memory structures, but this technique is limited.

The use of 3D technology is an interesting concept from the 1980s, but has recently become more viable. Especially by looking at the adder tree structures in the nodes, it seems that stacking transistors can provide a clear decrease of size and therefore a structured method to shorten the signal path in computational logic. With new vigor in stacking CMOS technology, smaller cells could be realized within the coming decade, but it does not solve all of the problems.

The other technology need is illustrated by the figures that raised the increasing modes. Instead of time-multiplexing over the memory structures, it seems of advantage to stack wafers. Again, the idea is not new. In the 1990s, research was proposed to stack an analog, a mixed-signal, and a fully digital wafer for complex, new media applications. Most of these projects come from a desire to imitate

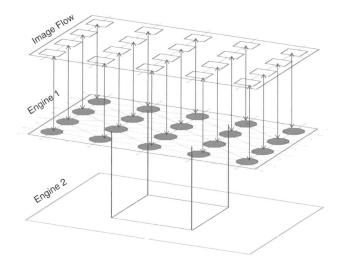

Fig. 16.25   The heterogeneous CNN-based system

biology. Recently, the idea has been renewed, though initially as an accelerator for bio-morphic systems.

Overall, we see that the CNN was inspired in the late 1980s, and it has gradually matured in theory and application, with both analog and digital architectures. The new interest in scaling this paradigm to larger systems, where few or no other technologies can solve the complex problems, matches nicely with recent technological progress in different areas. This makes the basic ingredients of a disruptive, promising development.

References

1. Hennessy, J.L., Patterson, D.A.: Computer Architecture: A Quantitative Approach, 4th edn. Morgan Kaufmann, San Francisco (2007)
2. Tummala, R.R.: Moore's law meets its match. IEEE Spectr, **43**(6), 44 (2006)
3. von Neumann, J.: The general and logical theory of automata. In: Newman, J.R. (ed.) The World of Mathematics, pp. 2070–2089. Simon and Schuster, New York (1954). Reprinted by Dover Publications (2000)
4. Malki, S.: On hardware implementation of discrete-time cellular neural networks. Ph.D. Thesis, Lund University, Sweden (2008)
5. Zhirnov, V., Cavin, R., Leeming, G., Galatsis, K.: An assessment of integrated digital cellular automata architectures. IEEE Comput. **41**(1), 38 (2008)
6. Domínguez-Castro, R., Espejo, S., Rodríguez-Vázquez, A., Carmona, R.: A CNN universal chip in CMOS technology. In: Proceedings 3rd IEEE International Workshop on Cellular Neural Networks and their Applications (CNNA-94), Rome, pp. 91–96, December 1994
7. Haykin, S.: Neural Networks: A Comprehensive Foundation. Macmillan, New York (1994)
8. Keegstra, H., Jansen, W.J., Nijhuis, J.A.G., Spaanenburg, L., Stevens, J.H., Udding, J.T.: Exploiting network redundancy for lowest–cost neural network realizations. In: Proceedings

IEEE International Conference on Neural Networks, ICNN '96, vol. 2, Washington, DC, pp. 951–955 (1996)

9. Hornik, K., Stinchcombe, M., White, H.: Multilayer feedforward networks are universal approximators. Neural Netw. **2**, 359 (1989)

10. Spaanenburg, L., Malki, S.: Artificial life goes 'in-silico'. In: Proceedings 2005 IEEE International Conference on Computational Intelligence for Measurement Systems and Applications, CIMSA 2005, Giardini Naxos, Taormina, Sicily, Italy, pp. 267–272, 20–22 July 2005

11. Chua, L.O., Yang, L.: Cellular neural networks: theory. IEEE Trans. Circuits Syst. **35**, 1257–1273 (1988)

12. ter Brugge, M.H.: Morphological design of discrete-time cellular neural networks. Ph.D. Thesis, Rijksuniversiteit Groningen, Netherlands (2004)

13. Uchimoto, T., Hjime, H., Tanji, Y., Tanaka, M.: Design of DTCNN image processing (Japanese). Transactions of the Institute of Electronics, Information and Communication Engineers **J84-D-2**, 1464–1471 (2001)

14. Harrer, H., Nossek, J.A.: Discrete-time cellular neural networks. Int. J. Circuit Theory Appl. **20**, 453 (1992)

15. Domínguez-Castro, R., et al.: A 0.8 μm CMOS two-dimensional programmable mixed-signal focal-plane array processor with on-chip binary imaging and instructions storage. IEEE J. Solid-St. Circ. **32**, 1013 (1997)

16. Espejo, S., Domínguez-Castro, R., Liñán, G., Rodríguez-Vázquez, A.: A 64 × 64 CNN universal chip with analog and digital I/O. In: Proceedings 1998 IEEE International Conference on Electronics, Circuits and Systems (ICECS98), Lisbon, pp. 203–206 (1998)

17. Liñán, G., Espejo, S., Domínguez-Castro, R., Roca, E., Rodríguez-Vázquez, A.: CNNUC3: a mixed-signal 64 × 64 CNN universal chip. In: Proceedings 7th International Conference on Microelectronics for Neural, Fuzzy and Bio-Inspired Systems (MicroNeuro-99), IEEE, Piscataway, NJ, pp. 61–68 (1999)

18. Liñán, G., Espejo, S., Domínguez-Castro, R., Rodríguez-Vázquez, A.: ACE4k: an analog I/O 64 × 64 visual microprocessor chip with 7-bit analog accuracy. Int. J. Circuit Theory Appl. **30**, 89 (2002)

19. Rodríguez-Vázquez, A., Liñán-Cembrano, G., Carranza, L., Roca-Moreno, E., Carmona-Galan, R., Jimenez-Garrido, F., Domínguez-Castro, R., Espejo Meana, S.: ACE16k: the third generation of mixed-signal SIMD-CNN ACE chips toward VSoCs. IEEE Trans. Circuits Syst., I: Regul. Pap. **51**, 851 (2004)

20. Liñán, G., Rodríguez-Vázquez, A., Espejo, S., Domínguez-Castro, R.: ACE16k: A 128 × 128 focal plane analog processor with digital I/O. In: Proceedings of the 7th IEEE International Workshop on Cellular Neural Networks and their Applications (CNNA 2002), Frankfurt, Germany, pp.132–139, July 2002

21. Liñán, G., Foldesy, P., Espejo, S., Domínguez-Castro, R., Rodríguez-Vázquez, A.: A 0.5 μm CMOS 10^6 transistors analog programmable array processor for real-time image processing. In: Proceedings 25th European Solid-State Circuits Conference ESSCIRC '99, Stresa (Italy), pp. 358–361

22. Delbrück, T., Mead, C.A.: Analog VLSI phototransduction by continuous-time, adaptive, logarithmic photoreceptor circuits. California Institute Technology, Computation and Neural Systems Program, Tech. Rep. CNS Memo no. 30, Pasadena, May (1994)

23. Rodríguez-Vázquez, A., Espejo, S., Domínguez-Castro, R., Huertas, J.L., Sanchez-Sinencio, E.: Current-mode techniques for the implementation of continuous- and discrete-time cellular neural networks. IEEE Trans. Circuits Syst. II: Analog Dig. Signal Process. **40**, 132 (1993)

24. Espejo, S., Rodríguez-Vázquez, A., Domínguez-Castro, R., Carmona, R.: Convergence and stability of the FSR CNN model. In: Proceedings 3rd IEEE International Workshop on Cellular Neural Networks and their Applications (CNNA-94), Rome, pp. 411–416 (1994)

25. Jimenez-Marrufo, A., Mendizdbal, A., Morillas-Castillo, S., Dominguez-Castro, R., Espejo, S., Romay-Judrez, R., Rodriguez-Vazquez, A.: Data matrix code recognition using the Eye-RIS

vision system. In: Proceedings IEEE International Symposium on Circuits and Systems, ISCAS 2007, New Orleans, p. 1214

26. Zarándy, Á., Keresztes, P., Roska, T., Szolgay, P.: An emulated digital architecture implementing the CNN Universal Machine. In: Proceedings 5th IEEE International Workshop on Cellular Neural Networks and their Applications (CNNA-98), London, pp. 249–252 (1998)

27. Keresztes, P., et al.: An emulated digital CNN implementation. J. VLSI Signal Process **23**, 291–303 (1999)

28. Nagy, Z., Szolgay, P.: Configurable multilayer CNN-UM emulator on FPGA. IEEE Trans. Circuits Syst. I: Fundam. Theory Appl. **50**, 774 (2003)

29. Lopich, A., Dudek, P.: An 80×80 general-purpose digital vision chip in 0.18 μm CMOS technology. In: Proceedings IEEE International Symposium on Circuits and Systems, ISCAS 2010, Paris, pp. 4257–4260, May 2010

30. Alba, L., et al.: New visual sensors and processors. In: Arena, P., Patanè, L. (eds.) Spatial Temporal Patterns for Action-Oriented Perception in Roving Robots. Cognitive Systems Monographs, pp. 351–370. Springer, Berlin/Heidelberg (2009)

31. Vörösházi, Z., Nagy, Z., Kiss, A., Szolgay, P.: An embedded CNN-UM global analogic programming unit implementation on FPGA. In: Proceedings 10th IEEE International Workshop on Cellular Neural Networks and Their Applications, CNNA '06, Istanbul, Turkey, Aug 2006. doi: 10.1109/CNNA.2006.341652

32. Fang, W., Wang, C., Spaanenburg, L.: In search of a robust digital CNN system. In: Proceedings 10th IEEE International Workshop on Cellular Neural Networks and their Applications, CNNA '06, Istanbul, Turkey, Aug 2006. doi: 10.1109/CNNA.2006.341654

33. Thiran, P.: Influence of boundary conditions on the behavior of cellular neural networks. IEEE Trans. Circuits and Syst.– I: Fundam. Theory Appl. **40**, 207 (1993)

34. Mirzai, B., Lím, D., Moschytz, G.S.: Robust CNN templates: theory and simulations. In: 4th IEEE International Workshop on Cellular Neural Networks and their Applications, CNNA-96, Seville, Spain, pp. 393–398 (1996)

35. Grube, R., Dudek, V., Hoefflinger, B., Schau, M.: 0.5 Volt CMOS Logic Delivering 25 Million 16×16 Multiplications/s at 400 fJ on a 100 nm T-Gate SOI Technology. Best-Paper Award. IEEE Computer Elements Workshop, Mesa, (2000)

36. van Drunen, R., Spaanenburg, L., Lucassen, P.G., Nijuis, J.A.G., Udding, J.T.: Arithmetic for relative accuracy. In: Proceedings IEEE Symposium on Computer Arithmetic, ARITH 1995, Bath, England, pp. 239–250 (July 1995)

37. Spaanenburg, L., Hoefflinger, B, Neusser, S., Nijhuis, J.A.G., Siggelkow, A.: A multiplier-less digital neural network. In: Digest 2nd Int. Conference on Microelectronics of Neural Networks, MicroNeuro 1991, Muenchen, Germany, pp. 281–290 (October 1991)

38. Zhang, L., Malki, S., Spaanenburg, L.: Intelligent camera cloud computing. In: IEEE International Symposium on Circuits and Systems, ISCAS 2009, Taipei, Taiwan, pp. 1209–1212, May 2009

39. Chalermsuk, K., Spaanenburg, R.H., Spaanenburg, L., Seuter, M., Stoorvogel, H.: Flexible-length fast Fourier transform for COFDM. In: Proceedings 15th IEEE International Conference on Electronics, Circuits and Systems, ICECS 2008, Malta, pp. 534–537, Aug 2008

40. Kari, J.: Theory of cellular automata: a survey. Elsevier Theoretical Computer Science **334**, 3–33 (2005)

41. Chua, L.O., Roska T.: The CNN universal machine – Part I: The architecture. In: Proceedings 2nd IEEE Int. workshop on Cellular Neural Networks and their Applications, CNNA-92, Muenich, Germany, pp. 110 (1992)

42. Roska, T et al.: A hardware accelerator board for Cellular Neural Networks: CNN-HAC. In: Proceedings IEEE Int. workshop on Cellular Neural Networks and their Applications, CNNA-90, Budapest, Hungary, pp. 160–168 (1990)

Chapter 17
Retinal Implants for Blind Patients

Albrecht Rothermel

Abstract Recently, very promising results have been obtained in clinical trials with eye-prostheses for the blind. There is a chance that advances in surgical techniques, microelectronics design, and material science may lead to the first really useful applications of retinal implants in the near future. This chapter will focus on the actual status of subretinal surgery and implant technologies.

Opportunities and limitations of the different technologies will be discussed in terms of patients benefit and technological challenges.

Finally, a vision on how the devices may work and look like in the future will be given.

17.1 Introduction: Causes of Blindness and Overview of Projects

Blindness is one of the most severe handicaps we can imagine. However, here we have first to distinguish between people who are blind from birth, on the one hand, and those who became blind by disease or accident.

People who are blind from birth completely lack any training of their visual system, and that means that probably there will never be any artificial vision for this group. However, people who have never seen in their life typically do not miss this ability too much.

Today's approaches for prostheses for the blind work exclusively with patients whose blindness is a result of diseases. Among them are age-related macular degeneration (AMD) and retinitis pigmentosa (RP), which are both related to photoreceptor degeneration [1, 2]. An important characteristic of these degenerative diseases lies in the fact that the optical nerve and most cell layers in the retina

A. Rothermel (✉)
University of Ulm, Institute of Microelectronics, Albert-Einstein-Allee 43, Ulm 89081, Germany
e-mail: info@albrecht-rothermel.de

B. Hoefflinger (ed.), *CHIPS 2020*, The Frontiers Collection,
DOI 10.1007/978-3-642-23096-7_17, © Springer-Verlag Berlin Heidelberg 2012

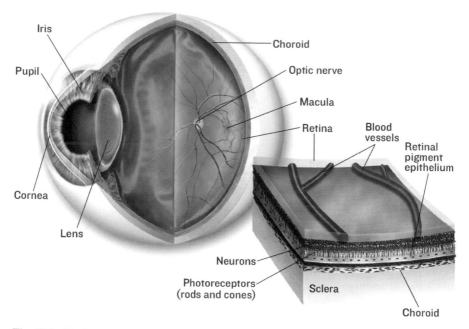

Fig. 17.1 The human visual system [3]

are still alive and functional. Another motivation for the development of electronic appliances to help RP patients derives from the complete lack of any medical treatment to stop the photoreceptor degeneration. A restoration of even rudimentary vision would be of significant benefit to these patients.

Figure 17.1 shows the main nerve layers in the retina of humans. Only the rods and cones are missing when RP patients have gone totally blind; the neurons, which include bipolar cells, horizontal cells, and ganglion cells, are still alive and are able to generate spike train action potentials.

Experiments have been carried out at the Natural and Medical Sciences Institute at the University of Tuebingen by Stett and Gerhardt. The retina was extracted from a Royal College of Surgeons (RCS) rat with inherited retinal dystrophy, which is a widely studied animal model of retinal degeneration. Samples of the extracted retina were placed on a microelectrode array. Pulse trains could be measured on ganglion cells as a response to stimulation pulses applied to the electrodes [4]. Figure 17.2 shows measured pulse trains for various stimulation pulse durations.

This kind of experiments verified that direct stimulation of retinal cells could generate so called phosphenes (visual perceptions) in patients [5]. With qualitative and quantitative results about the sensitivity of the retinal nerve cell layers, it was possible to design experiments on blind humans.

Four different locations of the stimulation electrodes are being investigated by different teams throughout the world.

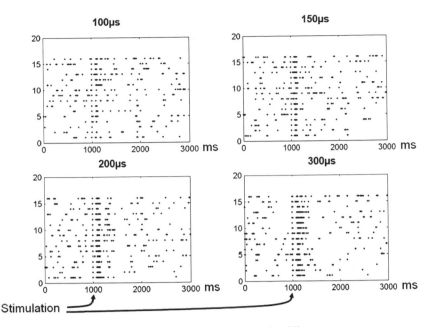

Fig. 17.2   Action potentials measured on a RP rat retina in vitro [4]

1. Epiretinal: Looking from the direction of the vitreous body in the eyeball, we approach the retinal layer from the ganglion cell side. The blood vessels and axons that form the optical nerve are located here. The epiretinal approach places the stimulating electrodes into the space of the vitreous body close to the ganglion cell layer. These devices typically separate the stimulation electronics from the electrodes themselves. The advantage is a high flexibility in electrode design and placement, the drawback the wiring required for each and every electrode. One very active team in this area is the company Second Sight [6]. In Europe, the same approach is followed by the company IMI [7] and the consortium Epiret. Figure 17.3 shows an example of such an approach. The picture information to control the electrodes is captured by an external camera, which allows for extensive image processing to optimize the stimulation signals. If the eye movements of the patient are to influence the stimulated pattern in a "natural" way, eye tracking is required.

2. Subretinal: Placing the electrodes at exactly the position where the photo-receptors were leads to the so-called subretinal approach. Most active in this area is a German group with the company Retina Implant AG driven by Zrenner [8, 9]. Also a group at the MIT follows this principle [10]. The advantage is that the stimulating electrodes are as close as possible to the location of the natural photoreceptors (Fig. 17.4). This arrangement promises the best stimulation sensitivity and best spatial resolution. The natural pressure of the retinal layer on the choroid guarantees the best proximity of electrodes and nerve cells. The MIT group [10] combines this approach with an independent design of

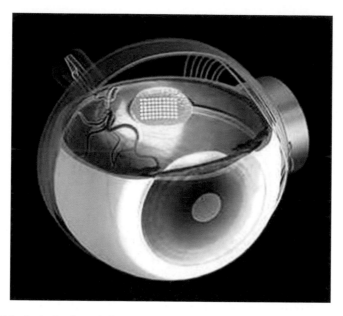

Fig. 17.3 Epiretinal stimulator design

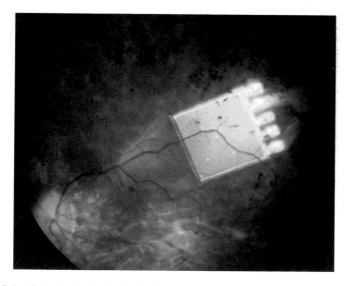

Fig. 17.4 Subretinal stimulation device in the human eye [11]

electrodes and electronics, allowing forcomplex electronics and high flexibility in the design, with the drawbacks of the wiring costs being proportional to the number of electrodes and the requirement of an external camera.

The German group based in Tuebingen [8] is the only one to the best of our knowledge that combines photosensors and electrodes in close proximity to use the natural optical path of the eye. The subretinal chip is a camera, signal processor, and stimulator all in one piece of silicon. Wiring this way is independent of the number of electrodes. In fact, the electrode number could be very high and is limited in practice by the limited spatial resolution and charge-transfer characteristics of the electrode material. The major drawbacks are the limited picture signal processing capabilities and limited flexibility of signal processing. Also the surgical procedure might be the most ambitious of the ones discussed here as the retinal layer, especially of RP patients, is a very delicate structure. However, retina manipulation is a quite common eye treatment surgery these days.

3. Subchoroidal: A little bit further towards the sclera is the position that an Australian consortium [12] is planning to use. The bionic eye project with a volume of Au$42 million uses an external camera again and places the electrodes between the choroid and the sclera. As far as we understand, the advantage lies in the fact that the eyeball does not have to be opened for this kind of surgery.

4. Suprachoroidal transretinal (STS): Going one step further, a Japanese group follows the STS approach, which means that the electrodes are attached to the sclera and again the eyeball does not have to be opened [13].

17.2 Organization of the Subretina Chip

The following sections mainly deal with a German project of the Tuebingen group. Driver of the project is Prof. Zrenner from the eye hospital in Tuebingen [14]. Commercial aspects are covered by the company Retina Implant AG. Research in this area started more than 10 years ago [15] together with the Natural and Medical Sciences Institute at the University of Tuebingen and the Institute for Microelectronics in Stuttgart.

The subretinal implant device as of today is wirelessly supplied by inductive energy and control signal transfer through the skin. A sealed ceramic box is used as the receiving device, similar to today's technology available for cochlea implants. Highly flexible wiring connects the subretinal chip with the receiver box (Fig. 17.5). This approach has been proved by the company Second Sight to allow a very satisfactory lifetime of several years [16].

With the German design, just the transmitting coil is attached to the outside of the patient's head, typically behind his ear, and magnetically secured. No other devices, no camera, and no glasses are essential for the operation of this type of device, which opens up chances for a very natural appearance of patients with this kind of prosthesis.

Two different chip designs are utilized for the subretinal stimulator. The earlier generation chip [17] was arranged in an array of about 1,500 pixels, each equipped with their own photosensor, amplifier, and stimulator. A more recent generation

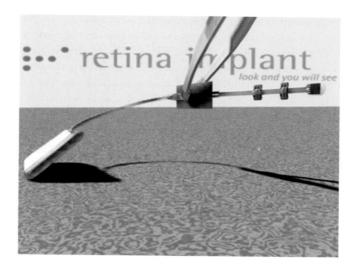

Fig. 17.5 Subretinal implant device

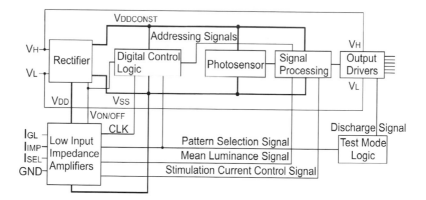

Fig. 17.6 Block diagram of subretinal chip [18]

[18] has been designed together with the University of Ulm, Germany. Figure 17.6 shows the block diagram. To improve the lifetime of the delicate structure, all terminals of the chip have been designed to work in a DC-free way. Control signals are supplied by control currents with an input voltage level clamped to GND. The supply voltages, however, are also DC-free. V_H and V_L are symmetrically oriented piecewise linear square wave voltages with peak-to-peak values of just ± 2 V. For internal operation, of course a rectifier is incorporated, however, to drive the stimulating electrodes, the polarity change of V_H is used directly at the output to generate a biphasic output stimulating current and voltage.

Not all output electrodes have to be addressed at the same time; they can be fired sequentially by a pattern selection signal to reduce peak currents in the supply and in the tissue.

Fig. 17.7 Chip micrograph

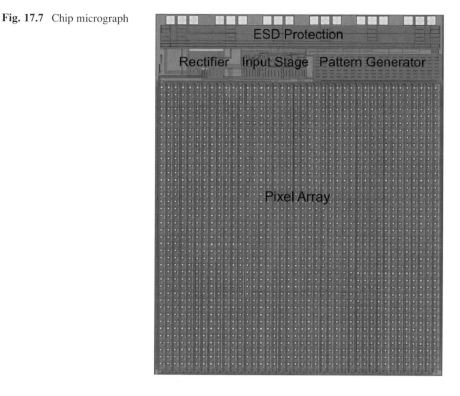

Figure 17.7 shows a micrograph of the subretinal chip with 40 × 40 pixels arranged on a 3 × 3 mm^2 surface. Three bond pads of usual size are available for every terminal in parallel. Although no DC supply voltages are available (except the GND potential) electrostatic discharge (ESD) protection of all input terminals is supported.

17.3   Function and Characteristics of the Pixel (Active Receptor)

Figure 17.8 shows the pixel cell schematic. A logarithmic photosensor is used [17] to obtain a maximized dynamic range of approximately seven decades. Although this kind of image sensor is not the first choice for lowest fixed pattern noise, the quality of the sensor is not the bottleneck in the overall stimulation homogeneity. The electrode properties and the interface to the tissue also cause a variation in the amount of charge transferred. In addition, it is known that the visual system itself has a temporal high-pass characteristic, which is assumed to cancel out local sensitivity variations to a certain extent. Power and space limitations make the mentioned photosensor technology the best choice for this kind of application.

Sensor output is amplified by a differential amplifier, which was adjusted by a reference voltage V_{GL} supplied externally in the earlier generation of implants.

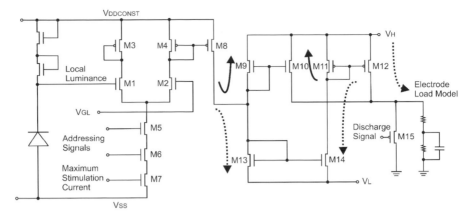

Fig. 17.8 Schematic of pixel cell

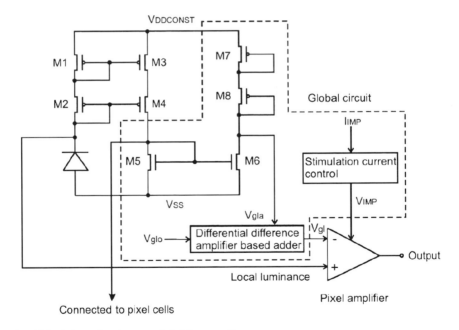

Fig. 17.9 Schematic of automatic brightness adjustment

The maximum stimulation current also is controlled externally and adjusted to the individual patient's sensitivity.

Addressing signals allow selective addressing of pixels as mentioned earlier. Transistors M_8 to M_{14} form the output-inverting driver. As V_H and V_L change polarity, the output driver polarity is automatically inverted. If V_H is positive, the dotted lines show how the output current of the differential amplifier is mirrored to a positive output current of the pixel cell. If V_H is negative and V_L positive, the solid arrows indicate the current mirror delivering a negative output stimulation polarity.

The most recent chip generation includes an automatic ambient-brightness adaptation, which is shown in Fig. 17.9 [19]. The signal of every fourth pixel in both spatial directions is read out, which gives 100 measuring points on the chip. An analog differential difference amplifier is incorporated to still allow fine adjustment of the sensitivity of the logarithmic sensor.

17.4 The Pixel–Retina Contact: Electronics and Physiology of the Interface Between Silicon and Nerve Cells

An advantageous property of the subretinal approach lies in the fact that the natural adhesion of the retinal tissue to the sclera is used to form the contact of the nerve cells to the stimulating electrodes (see Fig. 17.4). No additional mechanical measures are required. The surgical procedure to place the device beneath the retina requires specialized know-how, however, if the placement is successfully carried out, proximity of the tissue to the stimulator is very close.

The electrical contact between the electronics and the nerve cells is established by conducting ceramics such as titanium nitride or iridium oxide. Compared to approaches using a silicon dioxide interface [20], which is chemically inactive, the porous surface of the ceramics allows a much larger capacitance per unit area, which is essential for achieving a reasonable spatial resolution for the stimulation (Fig. 17.10) [9].

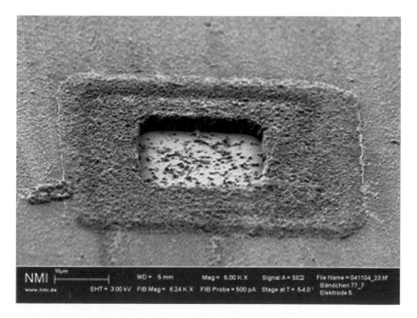

Fig. 17.10 Porous electrode for subretinal stimulation

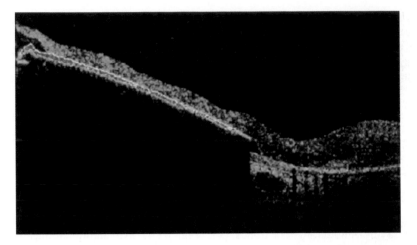

Fig. 17.11 Optical coherence tomography cross section showing the retinal layer in close proximity to the subretinal chip [11]

Experiments with in vitro retina from rats [4] and acute human experiments [8] have proven that with this kind of electrodes and stimulation voltages below 2 V visual sensations, called phosphenes, can be reliably generated.

To guarantee a long lifetime of the electrode material, charge balancing is an important electronic challenge. Many proposals have been made to control the charge transferred to the nerve tissue [21]. These include active charge balancing by measuring the residual voltage on the electrodes, controlling the injected charge, and active discharge of the electrodes after stimulation. As long as the electrode works in the safe voltage limits, it behaves like a capacitance and no ions are exchanged with the tissue layers.

In subretinal implants, it is essential that the device placed beneath the retina is as thin as possible. As can be seen in Fig. 17.11, the retina attaches itself closely to the implanted chip. No additional measures for holding the implant close to the nerve cells are required. The electrodes have to be flat therefore. However, maximizing the charge transfer calls for a rough structure, as depicted in Fig. 17.10.

17.5 The Surgical Procedure

Among the implantations of retinal prostheses, the subretinal surgery is possibly the most delicate one. Nevertheless, a series of 14 implantations have been carried out by the German team in Tuebingen without the occurrence of adverse effects. Therefore, the surgery can be considered to be safe and reproducible.

Figure 17.12 (from [22]) shows the major steps. First, the vitreous body has to be removed from the eyeball. The extraction of the vitreous body is a standard procedure in eye surgery manipulating the retina for various reasons. In a second

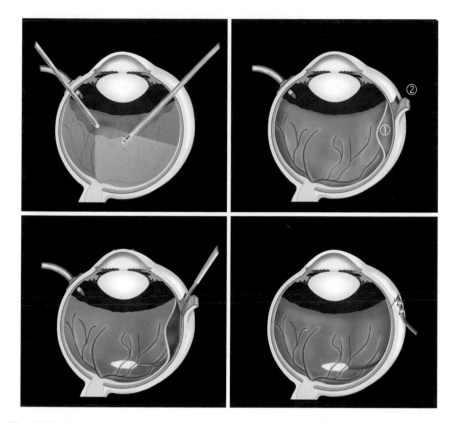

Fig. 17.12 Surgical procedure for subretinal implantation [22]

step, the retinal layer is detached from the background of the eyeball. Next, the subretinal implant can be safely inserted into the gap between the retina and the sclera. Finally, the retina is attached to the background of the eyeball again. Please note that most of the steps carried out here are surgical practices developed for different retinal diseases.

The ribbon that supplies the subretinal device is firmly attached to the outer part of the eyeball in such a way that no mechanical stress is introduced. The retina attaches the complete subretinal part of the implant to the background of the eyeball in a very natural way.

17.6   Results Achieved To-Date

A large variety of results have been achieved today with patients with subretinal implants. This includes the responses of blind patients to different amplitudes of stimulation, different pulse patterns, and different pulse frequencies. Generally, the perception is described as white to yellow in color. The intensity can reliably be varied, and maximum intensity is sometimes described to be very bright.

Different intensities, experimentally represented by sheets with a different gray color, can be distinguished very well. The orientation of a stripe pattern can be reliably recognized, as well as motion of stripes and other patterns.

One patient, who was blind for more than 10 years, demonstrated an extraordinarily high spatial resolution. Figure 17.13 shows examples of objects he was able to distinguish. He was even able to read the letters shown in the figure [8, 9]. Besides mug and plate, he could distinguish spoon and knife.

The only other group that consistently reports about visual perceptions in implanted patients, to the best of the author's knowledge, is Second Sight [16]. They show patients walking along a wide white line drawn on the floor, or they also show the perception of letters (Fig. 17.14).

Fig. 17.13 Blind patient with subretinal implant

Fig. 17.14 Letter recognition reported by the epiretinal approach (Second Sight Inc.)

17.7   Future Developments: What Can We Expect by the Year 2020?

Today, there are mainly two successful approaches, namely the epiretinal and subretinal approaches. Which of the two approaches will be a product in the future is difficult to predict, especially for somebody who is working for one of the two. There is also a certain chance that we may have two products that have different benefits for the different medical conditions of the patients.

It is very likely that we will have a fully self-contained implant that is completely located on the eyeball. Wireless inductive power and data transmission directly to the eyeball is available already today, e.g. in the Argus II prototype by Second Sight Inc. This type of power transmission requires a transmitting coil, which may be located on the frame of a pair of glasses the patient is wearing.

Epiretinal implants in all cases work with an external camera. To adjust the picture that is stimulated on the retinal ganglion cells to the gaze of the patient, eye tracking is required. Also for eye tracking, it will be necessary for the patient to wear a pair of glasses. A small camera tracking the eye position will be mounted on the frame. This means, in conclusion, that the epiretinal approach will always rely on the patient wearing a pair of glasses, and therefore it is very likely that a wireless transmission to the eyeball will be the solution for the future even in 2020.

For the subretinal approach, a similar setup is highly probable. The problem of power and data transmission is even relaxed, because only control data have to be transmitted, as the video data are generated by the active pixel array in the eye itself. In addition, the power consumption of subretinal approaches is relatively small up to now.

But there are additional, cosmetic considerations to be taken into account. For the subretinal approach, a pair of glasses is not essential and it might be discarded for some patients. The natural path of the light into the eyeball is still in use with subretinal prosthesis. Optical correction is not necessarily an issue, because typically the lens of retinitis pigmentosa patients is replaced anyway, as most of them suffer from cataract.

In conclusion, the subretinal approach has the potential of giving visual sensations to patients without them wearing any artifacts in front of their eyes. In that case, it might be of interest to supply the subretinal implant in the same way as is done today, by a wire from a receiving box located behind the ear. The transcutaneous transmission to that receiving box uses the well proven technology applied for the hundreds of thousands of cochlear implants in use today.

The experience of the subretinal group is that the wire does not cause any uncomfortable sensations to the patients. It is interesting, however, that also the epiretinal group with Second Sight Inc. has good long-term experience with the same approach with Argus I. At least two patients with such an implant have had no problems with the wired connection for 7 years now. Also the approach has proven to give a reliable long-term lifetime.

Please note that the optic nerve and muscles are connected to the eyeball anyway, therefore the additional wire does not disturb as much as one might suppose at first glance. Maybe in 2020 patients will have the choice between a device in the form of a pair of glasses, on the one hand, and an alternative device that requires a transmitter coil magnetically attached behind the ear, and nothing else.

In general, the subretinal approach has the limitation that the complete subretinal artifact has to have a certain form factor in order not to disturb the delicate retinal tissue with its supporting blood vessels too much. Epiretinal implants generally have a greater freedom of choice with respect to the shapes of wiring and electrodes. It is possible that the ability of the subretinal approach to cope with these form factor restrictions will finally define the winning approach.

In terms of improvements of electronics and electrode materials, we can expect the following developments. The most critical parameter in all retinal implants is the spatial resolution. Therefore, the teams following the epiretinal approach have continuously increased the number of electrodes (see [7] for example). As the stimulating chip and the electrodes have to be connected by wires, the wiring inside the eyeball turns out to be a critical issue. Argus II, which is the state-of-th- art prototype used by Second Sight Inc., uses 60 electrodes. Prototypes with higher numbers of electrodes have been developed also by Liu's group [23, 24]. Another idea to increase the number of electrodes is the use of multiplexers close to the electrodes. However, the technology of connections remains complicated.

The subretinal approach, however, has a different problem. The transferable charge per unit area has to be maximized to get the highest possible spatial resolution. On the other hand, it should be noted that the subretinal device does not have to be designed for a certain pixel density. We shall try to explain this surprising statement in the following.

The electrodes of a subretinal approach form an electrically active stimulating layer, which is organized like being cut into pieces. Assume the pieces were to be cut again, making the electrodes smaller. If the "fill factor" of electrode material remains constant, basically no restrictions due to the smaller individual electrodes arise. What happens is comparable to increasing the density of the two-dimensional spatial sampling pattern of the electrodes. The result of this consideration is that the electrodes themselves can be arbitrarily small without compromising the stimulation strength. Of course, not one electrode alone is able to elicit visual sensations in that case; a large enough light spot is required to generate enough charge to result in a visible object. Still, such a device would have the advantage that every patient would get the optimal resolution his retina is able to deliver to him.

Already today, subretinal implants use 1,500 electrodes, which allows a spatial resolution above the resolution of the electrode–tissue charge-transfer system we see today. It can be expected that the subretinal designs in 2020 will easily have electrode numbers in the range of 10,000, although not necessarily delivering the equivalent spatial resolution to all patients.

What should be solved by 2020 in addition is the increase of the "window size" of retinal implants open for the blind patients. Here, maybe the subretinal approach has the biggest surgical issues, as it requires a device that must be flexible in two dimensions to be placed beneath the retina and safely away from the optic nerve. Epiretinal stimulators may have easier access to the ganglion cells in the periphery, however, the wiring problem worsens.

Today, it is an open question whether stimulation in the periphery will be beneficial for retinitis pigmentosa patients. We know that the retinal signal processing in the area of sharpest vision and in the periphery differs significantly, therefore the spike trains of peripheral ganglion cells may be interpreted quite differently by the brain. Only more human experiments will answer those questions and help in designing the future implant devices.

Acknowledgments We acknowledge inspiration and continuous support by Prof. E. Zrenner from the University Eye Hospital Tuebingen and his team. We also acknowledge support by Dr. Wrobel, Dr. Harscher, and Mr. Kibbel from Retina Implant AG in Reutlingen, and Dr. Stett and Dr. Gerhardt from the Natural and Medical Sciences Institute at the University of Tuebingen.

References

1. Sharma, R.K., Ehinger, B.: Management of hereditary retinal degenerations: present status and future directions. Surv. Ophthalmol. **43**, 377 (1999)
2. Berson, E.L.: Retinitis pigmentosa: the Friedenwald lecture. Invest. Ophthalmol. Vis. Sci. **34**, 1659 (1993)
3. Pollack, J.: Displays of a different stripe. IEEE Spectrum **43**(8), 40 (2006)
4. Gerhardt, M., Stett, A.: Subretinal stimulation with hyperpolarising and depolarising voltage steps. In: Proceedings of the 6th International Meeting on Substrate-Integrated Micro Electrode Arrays, Reutlingen, Germany, pp. 144–147, July 2008
5. Stett, A., Mai, A., Herrmann, T.: Retinal charge sensitivity and spatial discrimination obtainable by subretinal implants: key lessons learned from isolated chicken retina. J. Neural Eng. **4**, 7 (2007)
6. Sivaprakasam, M., Liu, W., Humayun, M., Weiland, J.: A variable range bi-phasic current stimulus driver circuitry for an implantable retinal prosthetic device. IEEE J. Solid-St. Circ. **40**, 763 (2005)
7. Ortmanns, M., Rocke, A., Gehrke, M., Tiedtke, H.J.: A 232-channel epiretinal stimulator ASIC. IEEE J. Solid-St. Circ. **42**, 2946 (2007)
8. Zrenner, E., et al.: Blind retinitis pigmentosa patients can read letters and recognize the direction of fine stripe patterns with subretinal electronic implants. 2009 Association for Research in Vision and Ophthalmology (ARVO), Annu. Mtg., Fort Lauderdale, FL, D729, p. 4581, May 2009
9. Zrenner, E., et al.: Subretinal microelectrode arrays allow blind retinitis pigmentosa patients to recognize letters and combine them to words. In: 2nd International Conference on Biomedical Engineering and Informatics, 2009, BMEI '09. doi: 10.1109/BMEI.2009.5305315
10. Theogarajan, L., et al.: Minimally invasive retinal prothesis. In: IEEE International Solid-State Circuits Conference 2006, ISSCC 2006 Digest Technical Papers, San Francisco pp. 99–108, Feb. 2006
11. Rothermel, A., Liu, L., Pour Aryan, N., Fischer, M., Wuenschmann, J., Kibbel, S., Harscher, A.: A CMOS chip with active pixel array and specific test features for subretinal implantation. IEEE J. Solid-St. Circ. **44**, 290 (2009)
12. http://www.bionicvision.org.au/ Accessed April 2011

13. Fujikado, T., et al.: Evaluation of phosphenes elicited by extraocular stimulation in normals and by suprachoroidal-transretinal stimulation in patients with retinitis pigmentosa. Graefe's Arch. Clin. Exp. Ophtalmol. **245**, 1411 (2007)
14. Zrenner, E.: Will retinal implants restore vision? Science **295**, 1022 (2002)
15. Schubert, M., Stelzle, M., Graf, M., Stert, A., Nisch, W., Graf, H., Hammerle, H., Gabel, V., Hofflinger, B., Zrenner, E.: Subretinal implants for the recovery of vision. In: 1999 IEEE International Conference on Systems, Man, and Cybernetics, IEEE SMC '99 Conference Proceedings, Vol. 4, pp. 376–381 (1999)
16. Horsager, A., Greenberg, R.J., Fine, I.: Spatiotemporal interactions in retinal prosthesis subjects. Invest. Ophthalmol. Vis. Sci. **51**, 1223 (2010)
17. Dollberg, A., Graf, H.G., Höfflinger, B., Nisch, W., Schulze Spuentrup, J.D., Schumacher, K.: A fully testable retina implant. In: Hamza, M.H. (ed.) Proceedings of the IASTED International Conference on Biomedical Engineering, BioMED 2003, pp. 255–260. ACTA Press, Calgary, (2003)
18. Rothermel, A., Wieczorek, V., Liu, L., Stett, A., Gerhardt, M., Harscher, A., Kibbel, S.: A 1600-pixel subretinal chip with DC-free terminals and ±2 V supply optimized for long lifetime and high stimulation efficiency. In: IEEE International Solid-State Circuits Conference 2008, ISSCC 2008. Digest of Technical Papers, San Francisco, pp. 144–602 (2008)
19. Liu, L., Wunschmann, J., Aryan, N., Zohny, A., Fischer, M., Kibbel, S., Rothermel, A.: An ambient light adaptive subretinal stimulator. In: Proceedings of 35th European Solid-State Circuits Conference, ESSCIRC '09. pp. 420–423. doi: 10.1109/ESSCIRC.2009.5325980
20. Fromherz, P.: Joining microelectronics and microionics: Nerve cells and brain tissue on semiconductor chips. In: 33 rd European Solid-State Circuits Conference, ESSCIRC 2007. pp. 36–45. doi: 10.1109/ESSCIRC.2007.4430245
21. Sooksood, K., Stieglitz, T., Ortmanns, M.: Recent advances in charge balancing for functional electrical stimulation. In: 2009 Annual International Conference of the IEEE Engineering in Medicine and Biology Society, EMBC 2009, pp. 5518–5521. doi: 10.1109/IEMBS.2009.5333181
22. Wilke, R., Zrenner.E.: Clinical results, thresholds and visual sensations elicited by subretinal implants in 8 blind patients. In: The Eye and The Chip World Congress 2008, Detroit, June 2008. http://www.eyeson.org/documents/uploads/EyeandTheChipProgramBook_2008.pdf
23. Chen, K.F., Liu, W.T.: Highly programmable digital controller for high-density epi-retinal prosthesis. In: 2009 Annual International Conference of the IEEE Engineering in Medicine and Biology Society, EMBC 2009, pp. 1592–1595. doi: 10.1109/IEMBS.2009.5334120
24. Weiland, J.D., Fink, W., Humayun, M.S., Liu, W.T., Li, W., Sivaprakasam, M., Tai, Y.-C., Tarbell, M.A.: Systems design of a high resolution retinal prosthesis. In: IEEE International Electron Devices Meeting, 2008, IEDM 2008. doi: 10.1109/IEDM.2008.4796682

Chapter 18
Silicon Brains

Bernd Hoefflinger

Abstract Beyond the digital *neural* networks of Chap. 16, the more radical mapping of brain-like structures and processes into VLSI substrates has been pioneered by Carver Mead more than 30 years ago [1]. The basic idea was to exploit the massive parallelism of such circuits and to create low-power and fault-tolerant information-processing systems.

Neuromorphic engineering has recently seen a revival with the availability of deep-submicron CMOS technology, which allows for the construction of very-large-scale mixed-signal systems combining local analog processing in neuronal cells with binary signalling via action potentials. Modern implementations are able to reach the complexity-scale of large functional units of the human brain, and they feature the ability to learn by plasticity mechanisms found in neuroscience. Combined with high-performance programmable logic and elaborate software tools, such systems are currently evolving into user-configurable non-von-Neumann computing systems, which can be used to implement and test novel computational paradigms. The chapter introduces basic properties of biological brains with up to 200 Billion neurons and their 10^{14} synapses, where action on a synapse takes ~ 10 ms and involves an energy of ~ 10 fJ. We outline 10x programs on neuromorphic electronic systems in Europe and the USA, which are intended to integrate 10^8 neurons and 10^{12} synapses, the level of a cat's brain, in a volume of 1 L and with a power dissipation < 1 kW. For a balanced view on *intelligence*, we references Hawkins' view to first perceive the task and then design an intelligent technical response.

B. Hoefflinger (✉)
Leonberger Strasse 5, 71063 Sindelfingen, Germany
e-mail: bhoefflinger@t-online.de

B. Hoefflinger (ed.), *CHIPS 2020*, The Frontiers Collection,
DOI 10.1007/978-3-642-23096-7_18, © Springer-Verlag Berlin Heidelberg 2012

18.1   Some Features of the Human Brain

The human brain consists of about 200 billion neurons, each connected via thousands of synapses to other neurons, forming an incredibly powerful sensing, information, and action system that has

– Unparalleled intelligence, i.e., learning, comprehension, knowledge, reasoning, planning, and
– The unique capability of cognition, i.e., the power to reflect upon itself, to have compassion for others, and to develop a universal ethic code.

We settle for the intelligence aspect in this chapter, and we identify the following tasks as a collective motivation to understand and to use brain-like structures and operations:

– All kinds of feature- and pattern-recognition,
– Understanding many-body scenes, objects, people, traffic,
– Language understanding and translation,
– Building a relationship with care-bots.

We can imagine that the first three tasks can be handled by neural networks as described in [1] or in Chap. 16. Building a relationship with a care-bot requires a complexity of learning and adaptation plus the incorporation of feeling and compassion such that we have to look for the uniqueness of the brain to solve this task.

Figure 18.1 shows a micrograph of a part of a mouse brain with the central neuron green-enhanced. Each neuron consists of a nucleus (soma), dendrites, axons, and synapse barriers (Fig. 18.2; for an introduction see [4]).

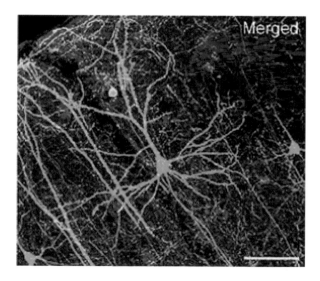

Fig. 18.1 A section of a mouse brain under the microscope. The scale bar is 100 μm. The *green* neuron in the center was made visible by a green-fluorescent protein [2]

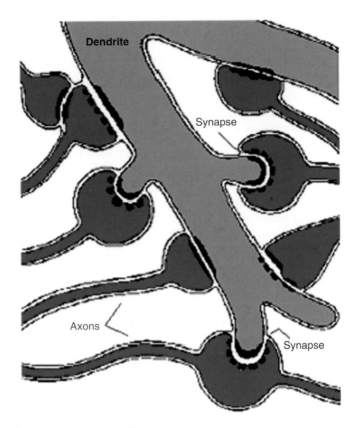

Fig. 18.2 Components of a neuron [3]

Nucleus: Receives and adds-up inputs from dendrites and sends outputs through axons. Has a nonlinear response curve with thresholds, above which a spike is generated.

Axons: Receive signal from nucleus and conduct ionic signal to synapse barrier.

Dendrites: Receive ionic signals from synapses.

Synapses: Consist of a presynaptic ending and a postsynaptic ending, separated by a gap of a few nanometers. Neurotransmitters are released across the gap depending on previous history and input signal, for which a simple model is the multiplication of the signal by a characteristic *weight* (see Fig. 16.1 for a comparison).

Contrary to the one-way feed-forward perceptrons in Chap. 16, two-way signal transport can occur in natural neurons, and new synapses with axons and dendrites can grow in a biological brain. Neurons are also specialized for certain tasks [5]:

– *Sensor neurons* sense inputs. The most prominent example is the retina (Chap. 17).
– *Inter-neurons* are the large majority operating inside the brain.

- *Memory neurons* are a more recently defined specialty of inter-neurons, which has strong reinforcement and long-term-potentiation (LTP) features.
- *Motor neurons* perform control and actuator functions.

The recent explosion of knowledge about the brain's functions is the result of the advancement of microelectronics and micro-sensing augmented by image processing and micro-physical chemistry. Out of the empire of acquired knowledge, we select a few topics, which are essential in developing *biomorphic Si brains:*

- *The key information* lies in the action potential, both in its height and in its frequency of occurrence. The action potential is a voltage spike 0–70 mV high and ~2 ms wide. The repetition rate can be 0.1–500 Hz.
- The resting potential across the gap of a synapse is 70 mV. When an ionic charge arrives, a specific *action potential* is generated, reducing the voltage difference.
- *The signal-transmission strength* is given by the spike density, which resembles a pulse-density-modulation code. The rate is <1 Hz to 500 Hz with an average of 10 Hz.
- *The synapse weight* determines to what degree the incoming signal at a synapse is transferred to the postsynaptic output onto a dendrite.

Table 18.1 is a performance guide to the human brain. The basis of this charge model is a charge density of 2×10^{-7} A s cm^{-2} per neuron firing, fairly common to biological brains [6] and a geometric mean for axon cross sections, which vary from 0.1 μm^2 to 100 μm^2.

We close this section with a brief summary on the circuit functions of the neural elements.

- **Synapses:** They receive as input signal a pulse-density-coded signal with a dynamic range of 0.1–500 Hz, equivalent to a 10-bit linear code, which could be log-encoded for efficiency and physical meaning. They transmit an output to the dendrite, which is weighted by the history of this connection, which means training, memory, recapitulation, and reinforcement, certainly the most significant and complex operation. Implementations require a multi-level, non-volatile, up-datable memory for the weight-function, which, in simple terms,

Table 18.1 Charge and energy model of the human brain

Number of neurons	2×10^{11}
Synapses/neuron	10^3
Ionic charges per neuron firing	6×10^{-11} A s
Mean cross section synaptic gap	30 μm^2
Charge/synaptic gap	6×10^{-14} A s $= 4 \times 10^5$ ions
Action potential	70 mV
Energy per synapse operation	4 fJ $= 2.5 \times 10^4$ eV
Energy per neuron firing	4 pJ
Average frequency of neuron firing	10 Hz
Average brain power	$(2 \times 10^{11}) \times (4 \times 10^{-12}) \times 10 = 8$ W

then has to be multiplied with the input signals. Thus, the electronic synapse is a complex, most-likely mixed-signal circuit with an *intelligent* memory.

– **Dendrites:** They perform the summation of their associated synapse-outputs, a fairly convenient electronic analog-circuit operation, but also common with digital accumulators.
– **Nucleus:** Its function is the firing decision, a nonlinear characteristic with hysteresis and historic adaptation (see the *digital* solutions in Chap. 16).

Beyond the dedicated technical implementations in Chap. 16, we introduce two generalist programs to build Si brains in the following.

18.2 3D Analog Wafer-Scale Integration of a Si Brain

In 2005, a consortium of 14 university and research centers in Europe started the FACETS project (Fast Analog Computing with Emergent Transient States) [7]. Within its broad scope of advancing neuromorphic computing, the hardware part is a very-large-scale, mixed-signal implementation of a highly connected, adaptive network of analog neurons. The basic element is a wafer-scale network of 50 million synapses, real-time configurable with

– 150 k neurons and 256 synapses each, to
– 2.3 k neurons and 16 k synapses each.

One key target is a 10^5-fold speed-up of the natural neuron-firing rate of 10 Hz. Since the communication of firing events is digital with a minimum of 16 bits/event, communication bandwidth is a major challenge for this system. The wafer is organized into 288 chips with 128 k synapses and 16 k inputs each, resulting in a total of 4.6 M inputs $\times$ 10^5 $\times$ 10 Hz = 4.6 Tera events/s = 74 Tb/s per wafer. This traffic is organized by a wafer-level, crossbar metal-interconnect, fabricated as a custom back-end-of-line structure. The manufacturing of this architecture would be a perfect example of future high-speed, maskless electron-beam lithography (Chap. 8). It is also a perfect example of at least three other challenges for the advancement of chip technology.

One is the power efficiency of chip/wafer-level interconnects. With the optimum efficiency of 1 mW/(Gb/s) in 2010 (Chap. 5), the resulting 74 W could be handled. The more likely level in the project at a less advanced technology node would be 1 kW. Scaling this up to mega-neurons clearly shows that power efficiency is the number 1 concern for these complex systems.

A second challenge is the implementation of a writable and non-volatile synapse weight. The FACETS hardware designers have opted for a multi-level (MLC) floating-gate memory, a nice example of utilizing the tremendous development activity of the industry in this direction (MLC Flash, Chap. 11). Providing this specific chip technology in congruence with the underlying synapse would make a natural application for 3D integration.

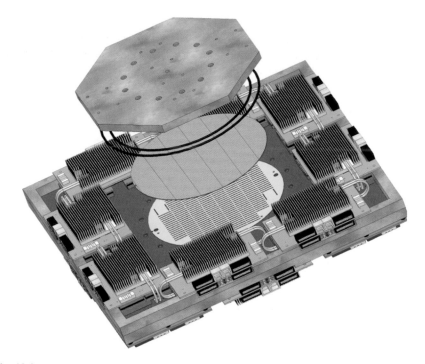

Fig. 18.3 View of the FACETS wafer-level mixed-signal brain-element with 150 k neurons and 50 M synapses [8]

This 3D integration is the third challenge for any Si brain. The memory Si layer on top of the mixed-signal Si layer can be achieved with a through-silicon-via (TSV) technology (Chaps. 3.7, 7, 12), and it is only one level in the task of building the whole system, because the global programmable digital control could be added on top. In the FACETS architecture, the digital control is implemented with FPGA (field-programmable logic array) chips on a printed-circuit board with the wafer in the center. An exploded view of the wafer-scale brain-element is shown in Fig. 18.3.

The chip area/synapse in the FACETS design is ~400 μm^2, comprising mostly analog circuits with about 100 devices (transistors, capacitors, resistors) operating at 1.8 V. The design and technology rules are given by the European research wafer service at the 130 nm level. As of 2011, Facets has been followed by BrainScaleS [8]. There is clearly room for scaling, and it will be interesting to follow the digital-versus-analog strategy, considering the alternative of digital 8 × 8 multipliers with 1 fJ and 100 μm^2 per multiplication (see Sect. 3.6).

The FACETS project has advanced high-speed vision tasks as a benchmark. Another application and a matter of cross-fertilization with the leaders in super-computer-based simulations of the human brain is the use of the FACETS system as a hardware accelerator for specific tasks. If a Blue Brain [9] simulation takes 100 times longer than a natural task, and the hardware could be 10^5 times faster than nature, this cooperation would accelerate the acquisition of know-how about some workings of the human brain quite significantly. This is also one of the aims of another large-scale project.

18.3   SyNAPSE: Systems of Neuromorphic Adaptive Plastic Scalable Electronics

This US program was launched by DARPA in 2009 [10], and it says in its description: "As compared to biological systems..., today's programmable machines are less efficient by a factor of 1 million to 1 billion in complex, real-world environments". And it continues: "The vision ... is the enabling of electronic neuromorphic machine technology that is scalable to biological levels. ... Biological neural systems (e.g. brains) autonomously process information in complex environments by automatically learning relevant features and associations."

Table 18.2 gives an overview of the capacity of biological brains and a comparison with two major electronics programs.

The SyNAPSE program is structured into four phases of ~2 years each:

1. The entry-level specification for synapse performance is:
   Density scalable to $>10^{10}$ cm^{-2}, <100 nm^2.
   Energy per synaptic operation <1 pJ, $<100\times$ nature.
   Operating speed >10 Hz (equivalent to nature).
   Dynamic range of synaptic conductance >10.
2. (~2010/2012) Specify a chip fabrication process for $>10^6$ neurons/cm^2, 10^{10} synapses/cm^2.
   Specify an electronics implementation of the neuromorphic design methodology supporting $>10^{10}$ neurons and $>10^{14}$ synapses, mammalian connectivity, <1 kW, <2 Liter (the final program goal).
3. (~2012/2014) Demonstrate chip fabrication $>10^6$ neurons/cm^2, 10^{10} synapses/cm^2.
   Demonstrate a simulated neural system of ~10^6 neurons performing at "mouse" level in the virtual environment.

Table 18.2 Brain characteristics[a]

Brain	Neurons	Synapses	Frequency [Hz] operations	Power	Volume weight	Refs.
Human	2×10^{11}	2×10^{14}	0.1–500	8 W		
Cat	10^8	10^{12}	10			
Mouse	10^6	10^{10}	10			
SyNAPSE Goal	10^8 Fab.	10^{12} Fab.				[10]
	10^{10} Des.	10^{14} Des.		1 kW	1 Liter	
Sim.2009	10^8	10^{12}	500 s / 5 s real	3 MW	227 to.	[9]
FACETS Goal	10^9	10^{12}	100 kHz			[8]

[a]Fab.: Fabricate a multi-chip neural system. Des.: Describe a high-level, conceptual electronics implementation

Table 18.3 Comparison of FACETS and SyNAPSE

	FACETS	SyNAPSE
Area/synapse	400 μm^2	<100 nm^2
Frequency of events	1 MHz	>10 Hz
Area/frequency	400 nm^2/Hz	<10 nm^2/Hz

4. (~2014/2016) Fabricate a single-chip neural system of ~10^6 neurons and package into a fully functional system.
   Design and simulate a neural system of ~10^8 neurons and ~10^{12} synapses performing at "cat"-level environment.
5. (~2016/2018) Fabricate a multi-chip neural system of ~10^8 neurons and instantiate into a robotic platform performing at "cat" level (hunting a "mouse").

SyNAPSE is a program with explicit specifications. It requires an existing system-simulation background (like Blue Brain in Lausanne [9] or C2 in Livermore) to assess the likelihood of realizing the milestones. One striking difference between FACETS and SyNAPSE is the density-speed figure-of-merit (FOM). See Table 18.3. SyNAPSE is an industry-driven, much larger program and will generate new hardware and software technologies. FACETS is based on existing chip technology, and it is an experiment on the large-scale, international cooperation of research institutes.

18.4 Outlook

The fascination associated with an electronic replication of the human brain has grown with the persistent exponential progress of chip technology. The present decade 2010–2020 has also made the electronic implementation feasible, because electronic circuits now perform synaptic operations such as multiplication and signal communication at energy levels of 10 fJ, comparable to or better than biological synapses. Nevertheless, an all-out assembly of 10^{14} synapses will remain a matter of a few exploratory systems for the next two decades because of several constraints:

Now that we are close to building an equivalent electronic core, we have moved the fundamental challenge to the periphery, namely the innumerable sensors and actuators (motors). In this outside-of-electronics domain, that of MEMS, remarkable progress has been made and continues with high growth (Chaps. 14 and 17). Furthermore, the critical conversion of analog sensor signals to digital signals is advancing rapidly (Chap. 4). And yet it is very early in the evolution of generic neuromorphic peripheries, so that all practical systems will be focused, application-specific solutions, certainly with growing *intelligence,* but with confined features of neural networks.

The other constraint is adaptation and control of brain-like architectures. Fuzzy and neuro-control have been practically available for 25 years, and yet their application is still limited to applications that are just feature-oriented and not critical for the technical performance of a product or service. It will take a major, globally consolidated effort involving many disciplines from reliability and ethics to social science to achieve a broad acceptance. The advancement of care-bots in various world regions will be the test-ground for this evolution of neuromorphic systems.

The third fundamental consideration is looking at the field of artificial intelligence (AI), where the Si brain is a major constituent, from the opposite end, i.e., not from the brain-side dealing with an environment, but from the environment, how to perceive it and how to deal with it. An outstanding leader with this viewpoint, to make predictions about the world by sensing patterns and then to respond intelligently, is Jeff Hawkins (the inventor of the first personal assistant, the PalmPilot) [11]. We refer to his work here in order to strike a balance with the hardware perspective on the Si brain offered in this chapter.

References

1. Mead, C., Ismail M.: Analog VLSI Implementation of Neural Systems, ISBN 978-0-7923-9040-4, Springer (1989)
2. www.wikipedia.org/Brain
3. Chudler, E.H.: Neuroscience for kids, http://faculty.washington.edu/chudler/synapse.html (2009)
4. Stufflebeam, R.: Neurons, synapses, action potentials, and neurotransmission. The Mind Project, www.mind.ilstu.edu/curriculum/neurons_intro (2008)
5. Martini, F.H., Nath, J.L.: Fundamentals of Anatomy and Physiology, Chapter 12: Neural Tissue, Prentice-Hall (2008)
6. Sengupta, B., Stemmler, M., Laughlin, S.B., Niven, J.E.: Action potential energy efficiency varies among neuron types in vertebrates and invertebrates. PLoS Computat. Biol (2010). DOI: 10.1371/journal.pcbi.1000840
7. www.facets-project.org
8. http://facets.kip.uni-heidelberg.de
   http://brainscales.kip.uni-heidelberg.de
9. www.bluebrain.epfl.ch
10. http://www.darpa.mil/Our_Work/DSO/Programs/Systems_of Neuromorphic_Adaptive_Plastic_Scalable_Electronics_%28SYNAPSE%29.aspx
11. Hawkins, J., Blakeslee, S.: On Intelligence. Times Books, Henry Holt, New York (2005)

Chapter 19
Energy Harvesting and Chip Autonomy

Yiannos Manoli, Thorsten Hehn, Daniel Hoffmann, Matthias Kuhl, Niklas Lotze, Dominic Maurath, Christian Moranz, Daniel Rossbach, and Dirk Spreemann

Abstract Energy harvesting micro-generators provide alternative sources of energy for many technical and personal applications. Since the power delivered by such miniaturized devices is limited they need to be optimized and adapted to the application. The associated electronics not only has to operate at very low voltages and use little power it also needs to be adaptive to the fluctuating harvesting conditions. A joint development and optimization of transducer and electronics is essential for improved efficiency.

19.1 Introduction

The progress in microelectronic components and battery development has provided us with portable equipment that is powerful and useful for our daily routine. Wireless communication makes information available at any time and at any place and has thus transformed the use of these devices. It is creating a whole new business where sensor systems can be employed in unforeseen ways. They can be used in buildings or other structures for improving maintenance, in remote areas for environmental observation, or in cars for better safety and comfort. In many cases this improved functionality also leads to savings in energy, e.g., fuel efficiency in cars or intelligent monitoring of buildings.

However, we are observing what can be seen as an energy gap [1]. Although microelectronic devices are becoming more power efficient, the demand for more

Y. Manoli (✉) • T. Hehn • M. Kuhl • D. Maurath • C. Moranz
IMTEK Microelectronics, University of Freiburg, Georges-Koehler-Allee 102, 79110 Freiburg, Germany
e-mail: manoli@imtek.de; maurath@imtek.de

D. Hoffmann • N. Lotze • D. Rossbach, • D. Spreemann
HSG-IMIT – Institute of Micromachining and Information Technology, Wilhelm-Schickard-Strasse 10, 78052 Villingen-Schwenningen, Germany

B. Hoefflinger (ed.), *CHIPS 2020*, The Frontiers Collection,
DOI 10.1007/978-3-642-23096-7_19, © Springer-Verlag Berlin Heidelberg 2012

performance is outpacing the progress in power density in batteries. Batteries have also the disadvantages of limited lifetime and the cost of replacement, maintenance, and waste generation. The use of harvesting concepts also in combination with rechargeable batteries can be a way to alleviate the power dilemma.

A great research effort has evolved in the last decade with the aim of improving power efficiency and reducing the size of harvesters as well as improving the power management electronics. We are extending the ideas of renewable resources to the miniaturized world, employing micro-engineering techniques to convert different forms of energy into electrical power. These energy sources can be thermal or mechanical or come in the form of light or as chemical or biological matter.[1] A number of industrially viable solutions are already present on the market [2, 3].

The next sections give an overview of available energy sources and show how dedicated energy management circuits can improve the system power efficiency.

19.2   Conversion Mechanisms

A number of different ambient energy sources, such as light, temperature gradients, motion and vibration, are available. This chapter does not attempt a comparison of these methods but provides a function and performance overview. What makes energy harvesting interesting for research but frustrating for the market development is the fact that it is always very application specific. It is very rarely the case that a task can be solved by employing a variety of these energy supplies. Environmental conditions will determine the choice of light, heat, or motion. Further conditions such as the allowed weight and size of the device will also constrain the options. Finally, the required system functionality will determine the power requirements and the selection of a solution. Should a choice of two mechanisms be actually possible, such as heat and vibration in machines and cars or vibration and light in industrial areas, then hybrid solutions can be considered. Since some of the sources might be sporadic and intermittent such a solution might be the only way out.

Although it might be questionable to include radio frequency (RF) scavenging in this discussion, we do deal with some of the electronic aspects since RF identification (RFID) concepts can be adapted to other areas. The same is true in the case of fuel cells, since most need energy storage. Emerging devices do not require this.

One can perhaps distinguish between generators such as thermal and solar or fuel cells that provide a DC voltage and those that deliver an AC voltage, as is the case for RF and vibration harvesters. The latter are often based on resonant systems and limited to narrowband applications. Thus, great interest exists in developing active tunable devices [4].

---

[1] This research has been funded in part by the German Research Foundation (DFG), the Federal Ministry of Education and Research (BMBF), the State of Baden-Wuerttemberg and industrial partners.

19.2.1   Motion and Vibration Transducers

When electrical energy is to be harvested from kinetic motion, in particular mechanical vibrations, the choice of a specific transducer mechanism (electromagnetic, piezoelectric, or electrostatic) is strongly dependent on the operating conditions (amplitude and frequency spectrum of the excitation source) and the available space given by the application environment.

The selection process may be further rationalized by considering the scaling behavior of electromagnetic and electrostatic forces [6]. The electromagnetic coupling coefficient scales at a different rate to the electrostatic coupling coefficient [7]. Therefore, when the transducer system decreases by a factor of 100 in size, the electrostatic coupling coefficient decreases by factor of 100, whereas the electromagnetic coupling coefficient decreases by a factor of 1000 [7]. Thus it follows that the electrostatic conversion mechanism is more efficient for transducer devices with a size of a typical MEMS (micro-electro-mechanical) device (<100 mm^3). On the other hand, an electrostatic transducer device may be outperformed by an electromagnetic transducer for larger device sizes (>1 cm^3).

Moreover, based on the fact that the technology for capacitive sensors (e.g., accelerometers, gyroscopes) is well established, it is beneficial to utilize the same standard MEMS process for energy harvesting devices. Thus, if miniaturization of a kinetic energy transducer is an important issue owing to little space available and if a large production scale is required, electrostatic MEMS transducers are the preferred choice. However, it must be considered that the output power of any inertial energy transducer is proportional to the oscillating mass. Consequently, miniaturized energy transducers with a very small proof mass will provide only low power levels in the range of microwatts. Therefore, electrostatic MEMS transducers require applications that are based on ultra-low-power devices.

In order to convert kinetic vibration energy into usable electrical energy by means of energy harvesting, two functional mechanisms are required:

- A transduction mechanism to convert mechanical energy into electrical energy
- A mechanical means of coupling ambient vibrations to the transduction mechanism.

The transduction mechanism enables the transformation of kinetic energy into the electrical domain. In this respect an electrical current or voltage is generated, which can be used to power electronic components. For the transduction mechanism to work effectively, a relative motion between two components must be available. A widespread principle for realizing inertial vibration generators is based on a mechanical resonator: A proof mass is supported by a suspension, which is attached to a frame. The oscillations of the proof mass can be damped by a suitable transducer mechanism and thus kinetic energy is converted into electrical power.

The mechanical oscillator should be adapted to the most energetic vibration frequency. However, there is great interest in increasing the bandwidth by using a piezoelectric harvester based on wideband vibrations [8] or to realize active tunable devices by tuning the harvester stiffness through piezoelectric actuators [9, 10]. The task in tunable devices is to be able to invest the power for the control system without limiting the converted energy too much.

19.2.2 Application-Oriented Design of Vibration Transducers

Electromagnetic resonant vibration transducers are already commercially available [3, 11, 12]. Commercial vibration transducers are typically add-on solutions, greater than 50 cm^3 and they fulfill standard industrial specifications or operation conditions (such as temperature range or shock limit).

Nevertheless, requests from the industry show that there is a great demand for application-specific solutions. This is because for each application the required output power, the available vibration level, and the overall mass and volume are significantly different. Moreover, it is often necessary to integrate the system into an existing subassembly. Especially for applications where the volume of the transducer is a critical parameter, these facts show that the required power can only be optimally extracted by application-specific customized developments. Thus, it can be guaranteed that the transducer is not over-dimensioned with respect to size, mass, and output performance.

In application-specific customized developments the underlying vibration source and the required output performance must be taken into account. Because real vibration sources are often stochastic in nature, the first task arising from this is to find the most energetic vibration frequency.

This is shown in the example of a real vibration source (car engine compartment) that has already been considered for energy harvesting applications (Fig. 19.1) [5]. An underlying typical acceleration profile and the corresponding frequency

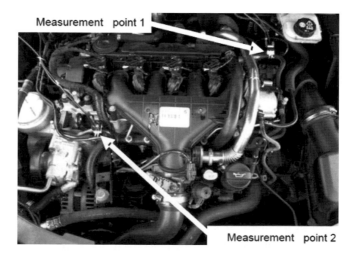

Fig. 19.1 Vibration measurements in the engine compartment of a four cylinder in-line diesel engine [5]

Fig. 19.2 (**a**) Acceleration
profile during country driving
route. (**b**) Frequency content
of the acceleration profile and
(**c**) expected output power for
an oscillation mass of 10 g
dependent on the resonance
frequency

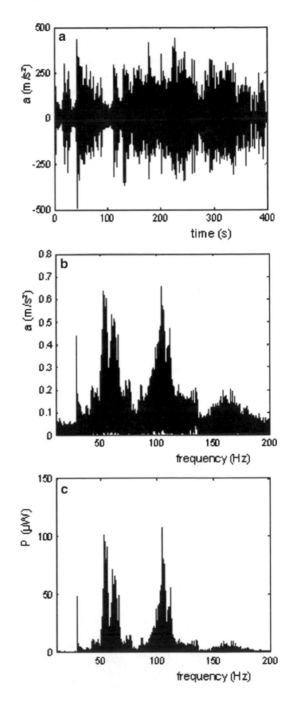

spectrum are shown in Fig. 19.2a, b. First of all, some predominant frequencies are
visible that depend on the number of cylinders and the revolutions per minute.
Based on the well-known analytic treatment of vibration transducers it can be

shown that the output power depends on the square of the acceleration amplitude but not on the vibration frequency. Hence the 2nd and 4th harmonic conversion will result in the same output power for this example vibration profile (Fig. 19.2c). Beside the amplitude, the bandwidth of the vibration needs to be taken into account. In other words, a narrow peak with high amplitude may be disadvantageous compared to a "plateau" with slightly smaller amplitudes. The mechanical oscillator should be adapted to the most energetic vibration.

19.2.2.1 Electromagnetic Generators

Since the earliest electromagnetic vibration transducers were proposed in the mid 1990s [13], a multiplicity of prototype transducers differing in size, electromagnetic coupling, excitation conditions, and output performance have been developed by numerous research facilities.

The data of published electromagnetic vibration transducer prototypes are widespread from 0.1 to 100 cm^3 and 1 μW to 100 mW. Voltage levels well below 1 V are in general barely applicable because of the dramatic drop in efficiency after rectification [10]. Microfabricated devices often have an output voltage well below 100 mV.

To date, no testing standard for vibration energy harvesting devices exists. Hence it is rather difficult to compare the transducers presented in the literature. The devices have been tested under different conditions and important parameters for comparison are often omitted. Nevertheless, an extensive review of electromagnetic vibration transducers has been presented [4] (Fig. 19.3), where the theoretical scaling law is observable even though there are orders of magnitudes between the normalized power densities (power density related to the excitation amplitude in

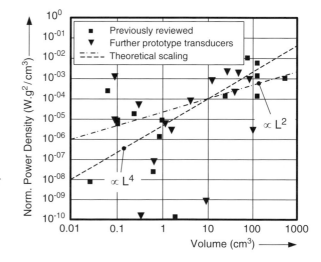

Fig. 19.3 Normalized power density of existing vibration transducer prototypes. The *dashed lines* indicate the theoretical scaling in the earlier review [4]

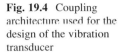

Fig. 19.4 Coupling
architecture used for the
design of the vibration
transducer

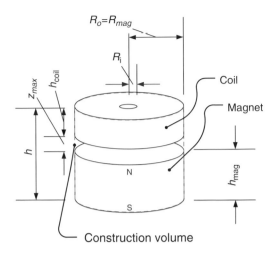

Construction volume

$g = 9.81$ m/s^2) of the published prototypes. At very small sizes L (linear dimension) the electromagnetic vibration transducers scale with L^4, and for large L, with L^2.

A possible reason why the published prototype transducers have significantly different output performances is that many different electromagnetic coupling architectures have been applied. A comparison of the maximum output performance of the coupling architectures (one shown in Fig. 19.4) is presented in [14], where the dimensions of the components have been optimized based on application-oriented boundary conditions. Such optimization calculations are necessary in order to achieve the highest conversion efficiencies. A further enhancement of the transducers would result with new magnetic materials. The development potential of $(BH)_{max}$ for permanent magnets is given in [1].

19.2.2.2   Piezoelectric Vibration Conversion

Piezoelectric materials are widely used in energy harvesters because of their ability to convert strain energy directly into electric charge and because of the ease with which they can be integrated into a harvesting system. The energy conversion process occurs because piezoelectric materials show a local charge separation, known as an electric dipole. Strain energy that is applied through external vibration results in a deformation of the dipole and the generation of charge, which can be transferred into a buffer capacitor, powering electronic devices.

Several piezoelectric materials appropriate for energy harvesting devices have been developed. Most common is a ceramic made of lead zirconate titanate (PZT). Although it is widely used in power harvesting applications, its brittle nature causes limitations in the strain that it can absorb without being damaged. Cyclic loading might cause the material to develop fatigue cracks, resulting in the device being irreversibly destroyed [13]. Thus, more flexible materials such as poly(vinylidene

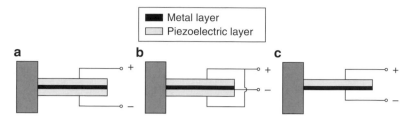

Fig. 19.5 Three types of piezoelectric sensors: (**a**) a series triple layer (bimorph), (**b**) a parallel triple layer (bimorph), and (**c**) a unimorph

fluoride) (PVDF), a piezoelectric polymer, have gained importance over the last few years. Thus, larger strains can be applied, increasing the amount of energy harvested from the piezoelectric device [15, 16].

Combining several piezoelectric layers is another strategy for increasing the amount of harvested energy. Instead of using a single piezoelectric layer in bending mode (unimorph), bimorphs consist of two piezoelectric layers connected electrically in series or in parallel. The different configurations are depicted in Fig. 19.5. Experiments have shown that for medium and high excitation frequencies, the output power of bimorph harvesters outperforms that of unimorphs [17].

19.2.2.3 Electrostatic Energy Conversion

The idea of electrostatic energy conversion goes back to 1976 with a rotary non-resonant electrostatic conversion system [18]. About 20 years later, the electrostatic and other conversion mechanisms were combined with a mechanical resonator in order to harvest electrical energy from kinetic motions [13]. Since then a number of research groups have been working on different electrostatic schemes for kinetic energy harvesting applications [7, 19–22].

The energy conversion mechanism of an electrostatic transducer is based on the physical coupling of the electrical and mechanical domains by an electrostatic force. The electrostatic force is induced between opposite charges stored on two opposing electrodes. In order to convert mechanical energy into electrical energy by means of the electrostatic transduction mechanism a variation of capacitance over time must occur [20, 23]. Despite the progress made in electrostatic energy harvesting devices, an energy-autonomous sensor system using this technology has not been demonstrated so far.

Figure 19.6 shows a microscope close-up view of a fabricated electrostatic transducer device, including one of the four mechanical suspension units and one of the ten electrode units. Each electrode unit comprises two comb electrodes, which consist of a number of interdigitated electrode elements.

Fig. 19.6 Microscopy image showing a detail of the electrostatic transducer structure, including one of the mechanical suspensions and the electrode units. The inset (*upper right*) shows a detailed view of the interdigitated electrodes, which implement an area-overlap characteristic

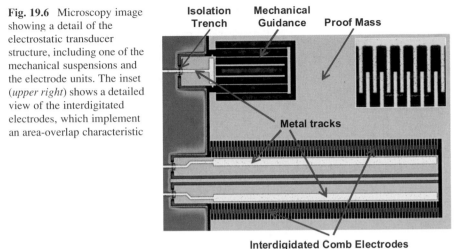

Isolation Trench   Mechanical Guidance   Proof Mass

Metal tracks

Interdigidated Comb Electrodes

19.2.3   Thermal Generators

Miniaturized devices with typically thin-film thermoelectric layers such as bismuth telluride-related compounds on silicon substrates in common vertical architecture are being successfully commercialized [24]. The performance data of these devices such as ΔT and heat flux densities are related to the thermoelectric performance of the materials, which are characterized by the dimensionless figure of merit ZT. Here, the figure of merit Z for thermoelectric devices is $Z = \sigma S^2/\kappa$, where σ is the electrical conductivity, S the Seebeck coefficient, and κ the thermal conductivity. Although there is no theoretical limit for ZT, thermoelectric bulk materials have not significantly exceeded $ZT \approx 1$ for almost 50 years. Typical characteristics of miniaturized devices at ΔT of 40–70 K are a few milliwatts [25].

19.2.4   Fuel Cells

At first glance fuel cells (FCs) do not seem to be energy harvesting devices in the classical sense, as they do not interact with the ambient medium to convert its energy into an electrical current. This might be true for high-power FCs utilizing fuel reservoirs, pumps, pressure reducers, and cooling devices with a power output of up to 300 W [26], but, concentrating on the micropower scale, several systems have been developed in recent years that are worth mentioning as energy harvesting devices – or at least in combination with them. Two designs of micro-FCs will be presented in detail in this section and some future prospects will be discussed.

19.2.4.1 Polymer Electrolyte Membrane Fuel Cells

Conventional PEM FCs consist of a polymer electrolyte membrane (PEM), two gas diffusion electrodes, two diffusion layers, and two flow fields. The reactants, e.g., hydrogen and oxygen, are supplied to the gas diffusion electrodes over feed pipes out of external tanks. The amount of supplied fuel is often controlled by active system periphery, such as pressure reducers and valves. As most of the energy harvesting devices are aiming towards autonomous microsystems, a new setup principle of PEM FCs was developed [27], enabling the miniaturization of its components. This work concentrates on the possibility of building FCs integrated into a well-established standard CMOS process. These chip-integrated fuel cells (CFCs) are made up of palladium-based hydrogen storage and an air-breathing cathode, separated by a PEM. The cross section of a CFC is depicted in Fig. 19.7. Advantages of the new approach are the omission of active devices for fuel supply and the reduction of system components such as flow fields and diffusion layers. Due to the simple assembly process, the fuel cells can be produced by thin-film or thick-film technologies, which fulfill all requirements for a post-CMOS fabrication. While the CFC itself has no conductive path to the CMOS chip, both the anode and the cathode can be connected to the CMOS circuit using wires and vias. Furthermore, the single fuel cells can be cascaded to generate multiples of the nominal 750 mV open circuit voltage. So-called integrated fuel cell cascades have been presented with an open circuit voltage of 6 V and a maximum power density of 450 μW/cm^2 [28] (Fig. 19.7).

19.2.4.2 Biofuel Cells

Other approaches to micro-FCs can be found in medical research. Biofuel cells oxidize biofuels, either with enzymatic, microbial, or non-biological catalysts, such as metal alloys and chelates, or with activated carbon. In the early literature, the term "biofuel cell" is used for fuel cells operated with biofuels such as glucose, whereas in more recent publications the term is used for biocatalyst-based fuel cells. Direct biofuel cells directly electro-oxidize the primary fuel at the anode, whereas in indirect fuel cells the fuel is first converted into a more reactive intermediate, which is then electro-oxidized at the anode.

   While the reactants inside of the previously presented CFC are clearly separated from each other by the PEM, there are scenarios without such a division for biofuel cells. Implantable fuel cells supplied by the surrounding body fluids (e.g., blood) for example have to pick the right components for the anode and cathode out of the same fluid by reactant selective electrodes. In [29] an implantable direct glucose fuel cell (GFC) is presented, in which the PEM is replaced by a combination of a selectively catalyzing oxygen reduction cathode and a porous membrane (Fig. 19.8).

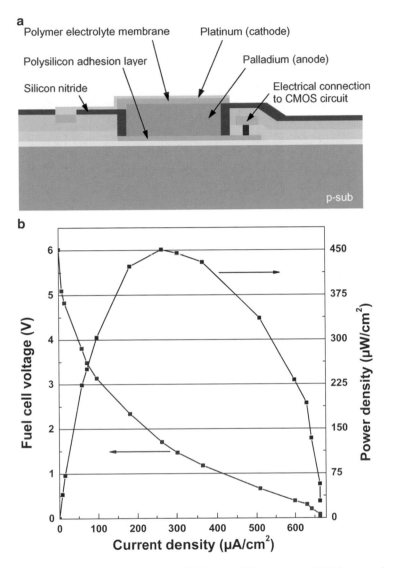

Fig. 19.7 Chip-integrated micro fuel cell (CFC). (**a**) CFC after post-CMOS processing. (**b**) Measured voltage and power density of a fuel cell cascade

This allows a setup with only one interface to the body fluids and enables thin-film GFCs to be attached to the front of an implant (e.g., a pacemaker). Such biofuel cells based on non-biological catalysts combine the advantages of biological safety, long-term stability, and biocompatibility. Under physiological conditions, the power output of these cells is around 2.5 $\mu W/cm^2$, sufficient to operate current pacemakers and many low-power MEMS implants.

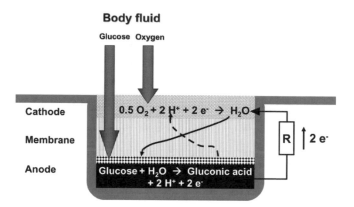

Fig. 19.8 Setup and operational concept of the direct glucose fuel cell (GFC)

19.2.4.3 Rechargeable Micro Fuel Cells

In combination with energy harvesting devices, the recharging of fuel cells will build a main focus in many laboratories. While the glucose cells are energy harvesting devices by themselves, the CFCs have to be supplied with hydrogen once the palladium reservoir is empty. As recharging can theoretically occur by electrolysis, the combination with vibration or solar harvesting devices will enable the CFCs to be used as alternatives to gold or super-capacitors. The discontinuously delivered power produced by the energy harvesters due to the changeable availability of environmental power can be stored within the CFCs. The CFC's capability as a long-term energy buffer will therefore promote its application as an uninterrupted power supply unit for autonomous microsystems. Examples of such systems may be found within cold-chain control or autonomous medical implants.

19.3 Power Management

In contrast to supplying power by battery or wire, using an energy transducer needs further consideration in order to achieve high efficiency. Besides rectification efficiency, voltage converter efficiency, and storage efficiency, the harvesting efficiency has also to be considered. There is a tremendous dependence on appropriate interfacing of the transducer, which is mainly achieved by input-impedance matching or adaptive voltage converters. Here, impedance matching not only includes the electrical load parameters but also considers the amplitude damping by the electromechanical feedback. That means the transducer's amplitude decreases as the applied load increases [30–32].

As depicted in Fig. 19.9, a transducer is typically connected to a particular low voltage rectifier (Sect. 19.3.2.2). Simply connecting a smoothing capacitor to a rectifier is not sufficient. A load adaptation interface to the transducer

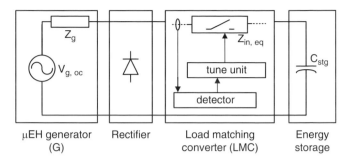

Fig. 19.9 Adaptive interface for optimal equivalent input impedance

becomes necessary, which can adapt its input impedance while converting the actual output voltage of the transducer into a (mostly higher) voltage level for the buffer capacitor or rechargeable battery (Sect. 19.3.3). Hence, the harvesting efficiency remains high over a wide range of excitation conditions and storage voltage levels [31, 33–36].

The design challenges are to create an interface that works at very low voltages, needs little power, and sets the load point (its equivalent input impedance) so as to obtain maximum output power. It should also convert the actual transducer output voltage to the voltage level of an energy storage device, e.g., a capacitor or battery.

The main limitation and difficulties of designing such an interface arise from the requirement of reducing the power consumption to a few microwatts while being able to accept fluctuant supply voltages – often in the sub-1-V regime. This restricts the applicable hardware components, e.g., processors, data converters, parameter monitoring, and communication devices. These restrictions include also the possible size, limiting the number of off-chip components [30, 34].

19.3.1 Power Multiplexing

A further step towards a more reliable power supply based on energy harvesting devices is power multiplexing, as illustrated in Fig. 19.10. That means several generators of complementary physical principles deliver power in a redundant manner to a central energy buffer. Thus, if one transducer is weakly excited other harvesters may be able to deliver power in order to keep the average power input above a minimum level [37, 38].

Similarly, power multiplexing is also reasonable at the output of the interface, e.g., for appropriate voltage scaling issues. Instead of one energy buffer, several buffers are present. Each buffer gets charged up to a certain voltage. Hence, different supply requirements of application components are accomplished more efficiently due to lower voltage conversion losses [36, 39, 40]. For example, sensors are often driven at a 3.3 V level while a radio transceiver needs only 1.8 V. Sensors need mostly small or moderate average currents compared to transceivers.

Fig. 19.10 Energy
harvesting and storage chain
with transducer interface

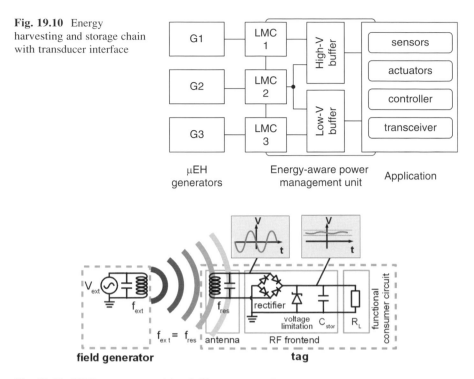

Fig. 19.11 RFID system comprising field generator and tag

Otherwise, if only one buffer is used, internal line regulators (low-dropout, LDO) typically provide the required voltages and efficiency decreases [39–41].

19.3.2 RFID and Signal Rectification

Strictly speaking, powering devices using radio frequencies such as in identification tags (RFID) cannot be regarded as energy harvesting, since the energy necessary for the operation of the system is not extracted from freely available environmental energy but from energy that was especially provided for this purpose. Anyhow, this discipline should be mentioned here, as there are several analogies concerning the circuit principles.

As described in [42], in passive RFID technology, a field generator creates an alternating high-frequency magnetic or electromagnetic field (Fig. 19.11), out of which battery-less independent sub-systems (tags) extract their energy. Such a subsystem essentially consists of an antenna, i.e., a receiver resonant circuit, and an RF frontend (energy conditioner). To convert the alternating voltage induced in the antenna into a constant supply voltage, a rectifying circuit is needed. Conventional CMOS integrated full-bridge rectifiers consist of four PMOS transistors connected as diodes. Since a forward-biased PMOS diode does not conduct any current until the threshold voltage is reached, the full-wave rectifier reduces the available voltage by two threshold voltages. In harvesting applications this leads to lower efficiency.

19.3.2.1 Floating Gate Diode

To minimize this voltage and the accompanying power loss, principles such as the rectifier using floating gate diodes presented in [43] can be used.

By applying a voltage (slightly larger than the usual operating voltage) between a MOSFET's control terminal and the channel, electrons can be transferred through the insulating oxide layer (Fowler–Nordheim tunneling, Fig. 19.12). If the circuit node receiving the charge carriers is isolated from surrounding circuit elements, i.e., is floating, a defined amount of charge can be deposited on it. This corresponds to a static bias voltage, which effectively reduces the voltage required to turn on the transistor. This technique needs an additional programming step after actual production.

19.3.2.2 Negative Voltage Converter

Another way to reduce the voltage loss is the use of the negative voltage converter followed by an active diode [10]. As illustrated in 19.13, the negative voltage converter also comprises only four transistors, but in contrast to the conventional full wave rectifier they are not connected as MOS diodes but as switches.

The switch creates a low-resistive connection between the alternating input voltage and the storage capacitor. Since a switch, unlike a diode, does not prevent

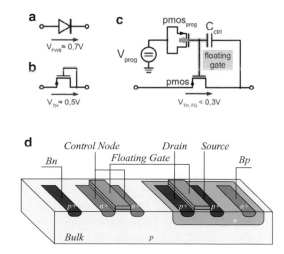

Fig. 19.12 Different types of diodes used in rectifiers: (**a**) PN diode; (**b**) PMOS diode; (**c**) floating gate diode. (**d**) Cross section of a single poly floating gate transistor

Fig. 19.13 Operation principle and transmission characteristic of the negative voltage converter with subsequent active diode

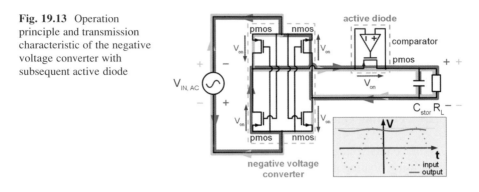

a possible backflow of charge that was already accumulated on the storage capacitor, a diode has to be attached to the negative voltage converter's output. In order to avoid reintroducing the typical voltage drop over the PMOS diode, one makes use of an active diode. As shown in Fig. 19.13, a voltage comparator evaluates the potentials at the input and output of the active diode and thus determines the direction of a potential current flow. It controls the PMOS switch so that only charge flowing onto the capacitor is permitted.

In order to achieve the optimum efficiency of the rectifier, the voltage comparator needs to be designed for very low power consumption. As a consequence its switching speed is limited and the rectifier is only suitable for systems with a frequency of a few-hundred kilohertz. To avoid the need of an active diode circuit and thus to make the rectifier suitable for higher frequency operation, [44] proposes that only two of the four diodes used in the conventional full-wave rectifier be replaced by switches. This topology requires a drop of only one threshold voltage between its input and output, while inherently preventing a current back-flow.

19.3.3 Input-Impedance Adaptive Interfaces

Instead of impedance matching, the term "load adaptation" between the generator and its electronics (Fig. 19.9) is sometimes preferred since typical conjugate impedance matching as normally understood is actually not sufficient. Therefore, the intention is to adjust the equivalent input impedance of the interface so as to maintain an optimal transducer output voltage to approximate the maximum power condition [33, 34, 36].

The following examples demonstrate this approach. One method continuously adapts to the actual open circuit voltage of the electromagnetic transducer, while two flying capacitor arrays are switched out of phase in order to best match the input voltage (Fig. 19.14). By switching the arrays, voltage conversion is achieved simultaneously [34]. Thus, non-harmonic and fluctuant transducer output voltages are also covered. As introduced in Sect. 19.2.1, rather low-frequency excitations are

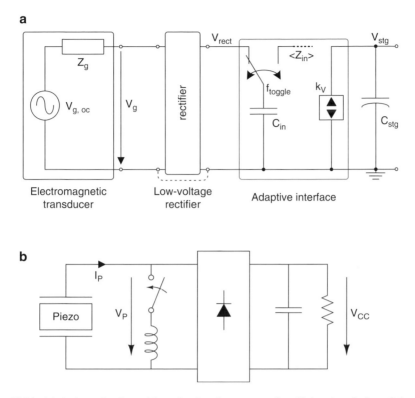

Fig. 19.14 (**a**) An input-load matching adaptive charge pump for efficient interfacing of electromagnetic transducers. (**b**) A SSHI energy harvesting circuit for efficient harvesting with piezoelectric transducers

converted into electrical power with highly inductive transducers. Thereby, the high resistive transducer coils appear as a quasi-resistive source impedance [10, 30, 45].

In a first step, interface circuits for piezoelectric harvesters, as for all AC generators, need to efficiently rectify the input voltage. Second, the voltage level, which is often very high, has to be down-converted to usable levels. One method keeps the piezoelectric harvester unloaded until the piezo voltage reaches a maximum, followed by a transfer phase extracting the total energy stored in the piezoelectric harvester temporarily into a coil and finally transferring it into a buffer capacitor. This principle is described as synchronous electric charge extraction (SECE) [46]. When tested experimentally against a linear impedance-based converter design, the SECE principle increases power transfer by over 400%. SECE circuits implemented in CMOS technology have been presented in [47] and [48].

In similar studies [35, 49–51] another nonlinear technique, called synchronous switch harvesting on inductor (SSHI), has been developed (Fig. 19.14). Unlike SECE, the coil–switch combination is placed between the piezoelectric harvester and the rectifier. A control circuit monitors the deflection of the piezoelectric harvester and triggers the switch at a deflection maximum, flipping the piezo voltage instantaneously. Thus, the rectifier is conducting most of the time, avoiding phases where the piezo voltage is below the voltage level of the buffer capacitor. Tests with discrete electronics show that the SSHI circuit is capable of extracting 400% more power than the standard rectifier circuit.

Since electrostatic transducers are capacitive, just like the piezoelectric ones, similar methods can be used. The initial approach for implementing electrostatic energy conversion was the implementation of a variable capacitor in a switched system [20]. Through the operation of switches a reconfiguration of the variable capacitor at different parts of the conversion cycle was induced, enabling the process of energy harvesting. A new approach based on charge transportation between two variable capacitors continuously connects the capacitors to the circuitry and therefore no switches are required [22]. To electrically bias the variable capacitors, precharged capacitors or polarized dielectric materials (electrets) can be used. Power outputs are predicted in the range of 1–40 µW based on theoretical analysis.

All these examples use very minimalistic, power-saving control and drive circuits. Compared to early harvesting and interfacing solutions, the tendency is clearly towards processorless and rather analog solutions [30, 52]. Light-load efficiency and high integration density in sub-milliwatt power ranges mark the direction of development [37, 41].

19.3.4   Circuitry for Fuel Cells

Both solar and fuel cells deliver very low output voltages, which need to be boosted for application using DC–DC converters. Furthermore, both the presented micro-FCs – the chip integrated fuel cells (CFCs) and the direct glucose fuel cell (GFC) – need to be monitored and controlled for efficient and safe operation. In [28] a stabilized power supply based on the CFCs with monitoring-and-control circuitry was realized within an extended CMOS process (Fig. 19.15). The electronic control circuitry consists of a low-dropout voltage regulator (LDO) [53], an on-chip oscillator, a programmable timing network, and a monitoring-and-bypass circuit for each CFC. All of these components are driven by the integrated CFCs. For the benefit of the system's lifetime, the core system is restricted to fundamental control elements such as an oscillator and an asynchronous timing network to avoid current peaks. It is powered by a redundant two-CFC-stack generating 1–1.5 V, depending on the charging level of the fuel cells.

For high reliability of the proposed power supply system all fuel cells within each cascade have to be checked for functionality before they are used to drive the LDO. Each cell's anode is compared to its slightly loaded cathode. In the case of defective or empty fuel cells, a small load current results in a significant voltage drop across the cell, enabling a dynamic comparator to classify this cell as "unusable". These defective or empty fuel cells are shorted by CMOS transmission gates

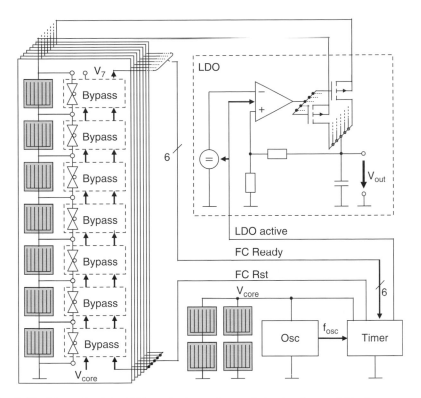

Fig. 19.15 Schematic of the CFC monitoring-and-control circuitry for a stabilized power supply with chip-integrated micro-FCs

to prevent loss of a complete cascade or to keep its input resistance as low as possible.

By stacking the fuel cells to cascades with an output voltage higher than 3.3 V for load currents up to 7 μA, a classical LDO was used to stabilize the output voltage to 3.3 V. As presented, six fuel cell cascades are integrated on chip. To increase the system's driving capability, these cascades are connected in parallel to the output by separate PMOS transistors of the LDO, used as pass elements, to avoid equalizing currents between the different cascades. Thus, the required load current is distributed to the cascades with respect to their driving capability. The complete circuitry is implemented without external components.

19.4   Ultra-Low-Voltage Circuit Techniques

The harvesting applications being focused on are sensor and data-acquisition devices and micro-actuators with wireless communication. Considering the previously discussed systems and architectures, the necessary building blocks are power interfaces for the transducers, voltage converters, and power management control units. All these modules include amplifiers, comparators, references, clock generators, and digital blocks. Owing to the limited energy resources, these circuits have to be designed compatible to the low voltages and power available rather than for performance. Feasibility, functionality, and reliability are seen as more critical and essential than absolute accuracy, bandwidth, and speed.

To handle the high voltages that might occur at the output of harvesting transducers and owing to the voltage requirements for driving certain sensors commonly used, bulk CMOS processes for energy harvesting processing modules remain rather constant at 0.18–0.35 μm technologies [54–58]. These processes, which usually include high voltage options, have relatively high threshold voltages, making the design of ultra-low-voltage circuits a challenging task.

19.4.1   Analog Design and Scaling Considerations

To enable analog integrated circuits to operate at low supply voltages, different strategies are available: (1) weak/moderate inversion operation, (2) bulk-driven stages [34, 59–61], (3) floating gate programming [62, 63], (4) input-level shifting, (5) field-programmable analog arrays (FPAAs), such as reconfigurable unity-gain cells [59, 61], and (6) digitally assisted architectures and sub-threshold designs [35, 64, 65].

Besides low supply voltages, the scaling of bulk CMOS technologies also has to be considered. As down-scaling reduces threshold voltages and minimum supply voltages [66], several degenerative effects arise. Most challenging are mismatch and reduced output resistance (gain), as well as lower dynamic range and signal-to-noise ratio [59, 67]. Often, speed has to be traded for noise margins. As an additional scaling result, the body effect will decrease, which reduces the potential of bulk-driven stages.

In conjunction with increased mismatch, unity-gain amplifiers with the option of reconfiguration become an interesting way to deal with these obstacles [68, 69]. Further, the technique of using digitally assisted architectures [65] helps to save power, while being useful for mismatch correcting and improving linearity.

19.4.1.1   Amplifiers and Comparators

For high amplification and current efficient operation the strong inversion regime might be omitted. The use of particular input-stages, e.g., bulk-driven MOS transistors or specially biased gate-driven stages, allows wide input common mode ranges while linearity remains high [59–61]. Fig. 19.16 gives further circuit examples.

Controlling the channel of MOS transistors by both the gate and the bulk terminals allows (exchangeable) use of either terminal as an input or for biasing.

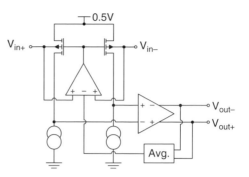

Fig. 19.16 Input-stage with two stacked transistors only [61]

Table 19.1 Low-voltage amplifiers with different design techniques: gate-driven (*GD*), bulk-driven (*BD*), multi-path (*MP*), multi-stage (*MS*), switched-capacitor (*SC*)

	GD[59]	BD[59]	BD[61]	MP[70]	MS[71]	SC[72]
V_{DD} [V]	0.5	0.5	0.5	2.0	1.0	1.0
P_V[μW]	75	100	1.7	400	10,000	1.8
GBW [MHz]	10	2.4	0.1	4.5	200	0.07
DC gain [dB]	62	52	93	100	84	90
Noise [nV/$\sqrt{\text{Hz}}$]	220	280	500	–	27	60
FOM [(MHz pF)/μW]	2.7	0.48	1.18	1.35	0.02	1.16
Load C_L [pF]	20	20	20	120	1 pF, 25kΩ	30
Technology [μm]	0.18	0.18	0.35	0.8	0.065	0.065

These bulk-driven stages make wide input ranges possible while biasing at the gate terminals guarantees high linearity [59]. However, low body transconductance adversely affects gain and noise margins, and an increased input capacitance limits the bandwidth [59, 61]. Prevention of latch-up is necessary and nMOS devices require triple-well processes [61].

The concept of switched biasing allows strong inversion operation [60]. This, however, requires a clock and continuous switching of the level-shifting input capacitors.

Floating gate programmed amplifiers are an interesting opportunity, as described in the rectifier section (Sect. 19.3.2). The threshold voltage and transconductance can be adjusted granularly, and device mismatch can be compensated [62]. But this technique is a research topic rather than an industrially applicable one, since each single device has to be carefully programmed.

Another emerging technique is using unity-gain cells and implementing them with reconfigurable options. Thus, mismatches could also be compensated and certain bandwidth and gain characteristics could be configured. This technique may especially help to compete with the challenging characteristic of short-channel transistors at process technologies with digital emphasis [62, 63]. For increasing accuracy or linearity, digital correction mechanisms are recommended [65, 73]. (Table 19.1)

For comparators, in addition to very efficient designs with regenerative or clocked latches, the use of asynchronous comparators is a reasonable choice [34, 65]. That is justified by the shallow clock edges and low slew rates if operation in strong inversion is not possible [35, 64], or if clocking is totally avoided, as in asynchronous digital circuits, as explained later [74]. In conjunction with active diodes, common-gate input stages allow a wide range of input and supply voltage levels. Even at low biasing currents adequate response and transition delay is maintained.

19.4.2 Ultra-Low-Voltage Digital CMOS Circuits

The requirements for digital blocks used in interface circuits for energy harvesting devices are typically very strict: Energy budgets are tight and the minimum supply voltage should be reduced as far as possible to allow for an early start-up of the device. Even though circuits operating at ultra-low supply voltage are typically also highly energy efficient, optimization for minimum energy consumption and minimum supply voltage are not equivalent targets.

In this section, we focus our discussion on supply voltage reduction for minimum power in digital circuits rather than energy-per-operation minimization as discussed, e.g., in [75], as the ultimate limit for energy harvesting devices typically is the startup supply voltage required by the operated electronics.

An early discussion about a theoretical lower supply voltage limit [76] states a minimum of kT/q based on noise considerations, more specifically for CMOS circuits a limit of $2(\ln 2)kT/q$ for inverters [77] and $2(\ln 5)kT/q$ for practical circuits [78], derived from the requirement that a logic gate needs more than unity gain.

Regarding practical circuit implementations, a very low supply voltage CMOS design was presented as early as the 1990s operating down to 125 mV [79] at room temperature. The operation occurs in saturation with transistor threshold voltages tuned to near-zero values though. As the convenience of such optimized process technology is not available for most applications, most ultra-low-voltage circuits operate the transistors in weak inversion. Examples are a hearing aid filter operating at 300 mV [80], a fast Fourier transform (FFT) processor operational at 180 mV [81], and a microcontroller optimized for the sub-threshold regime working at supplies as low as 168 mV [82]. Also memories with supply voltages down to 160 mV have been demonstrated [83]. These designs are optimized mainly for minimum energy consumption by supply voltage reduction. It is possible to achieve considerably lower supply voltages when optimizing specifically for this target though, as reported for a memory operational down to 135 mV [84], a FIR filter operational to 85 mV [85], and multiplier circuits remaining fully functional down to 62 mV [86].

Figure 19.17a gives an overview of a selection of publications regarding ultra-low-voltage digital circuits, relating the technology node used to the minimum supply voltage. It becomes clear that there is an obvious improvement in minimum

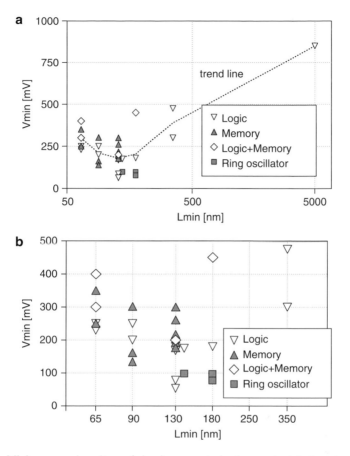

Fig. 19.17 Minimum supply voltage of circuits versus technology node: (**a**) a long-term interpolation, and (**b**) a zoomed view of recent technologies

supply coming from old technologies, but the supply voltage values level off or even increase for current ones. To understand this trend, a closer analysis of the reasons that limit the scaling of the supply voltage is necessary.

The ultimate limit for supply voltage reduction is imposed by a degradation of static noise margins (SNMs) to zero. More illustratively, the output levels of CMOS gates degrade due to the decrease in the "on" to "off" current ratio with decreasing supply voltage until the output level of some digital gate is falsely evaluated by the succeeding logic. Three main factors impact output level degradation: The circuit structure, process technology factors and process variabilities.

Regarding circuit structure, gate sizing has an important impact on minimum supply, as discussed in [87] and [81], requiring symmetric devices for minimum supply [88]. Furthermore, numerous circuit structures should be avoided, i.e., long transistor stacks (leading to an inherent worst-case asymmetry in CMOS), ratioed circuits [89] and dynamic circuits. All of these requirements though do not necessarily optimize energy consumption, as discussed for sizing in [87] and are therefore applied within limits in most circuits.

The major process technology factors affecting SNM are the sub-threshold slope factor and drain-induced barrier lowering (DIBL). DIBL corrupts SNM as it causes additional mismatch between "on" and "off" devices [90]. The sub-threshold slope defines the "on" to "off" current ratio of a transistor and therefore directly impacts SNM. It improves with shrinking oxide thickness, but then also degrades with decreasing gate length owing to short-channel effects. The fact that oxide thickness is scaled less than minimum gate length to limit gate leakage results in an overall sub-threshold slope factor deterioration in modern technologies [91].

Process variability impairs SNM as the imbalance of devices is increased, causing critical output level deviations at higher supply voltages. Threshold voltage variations are the major source of variability in sub-threshold circuits owing to their exponential impact. Many designs compensate global variabilities by the use of adaptive body biasing [82, 85, 92, 93], whereas local variability is dominated by random dopant fluctuations [94] and can therefore only be compensated by increasing device sizes following Pelgrom et al.'s model [95]. This approach is primarily used in designs optimized for ultra-low-voltage operation though (e.g., [86]) as it also increases the energy consumed per operation.

The increase in minimum supply voltage with shrinking technology in Fig. 19.17b is therefore easily explained by the fact that most of these circuits have a stronger optimization for minimum energy than for minimum supply and therefore suffer from increased variability, DIBL and sub-threshold slope factor. To profit from technology scaling, it is however possible to use larger than minimum-length devices [90, 91] to alleviate the DIBL and variability impacts and effectively profit from increased channel control due to the reduced oxide thickness.

If possible improvements in minimum supply voltage due to technology improvements are to be estimated, it needs to be noted that the only major process parameter relevant to minimum supply that can improve with scaling is sub-threshold slope. This factor is limited to approximately 60 mV/decade for physical reasons and has a value of 70–80 mV/decade in current processes. Using a first-order linear approximation, an ideal process might therefore improve supply voltage in a circuit highly optimized for minimum supply voltage operation from 62 [86] to 53 mV and in a design optimized for minimum energy per operation from 168 [82] to 144 mV. It is therefore obvious that for supply voltage reduction, circuit design innovations are more relevant than technology scaling. To achieve major technology-driven improvements for minimum supply voltage reduction, new device technologies are necessary, e.g., multi-gate FETs, improving channel control and removing random dopant fluctuations; band-to-band tunneling transistors [96] or carbon nanotubes [97], allowing considerably reduced sub-threshold slopes; or even devices such as the BiSFET [98], for which nominal supply voltages in the range of 50 mV are predicted.

19.5 Conclusion

The latest research on high-performance energy transducers and adaptive circuits has been presented. The challenge remains to close the gap between the energy generation capabilities of the harvesters and the energy consumption of the adaptive electronics used to optimize the output power of the harvesting devices. Whereas previous research focused mainly on characterizing the harvesting devices, in the future complete energy harvesting systems, including harvester, interface circuits, storage, and application electronics, need to be developed and optimized as a system.

References

1. Delagi, G.: International Solid State Circuits Conference Digest of Technical Papers, pp. 18–24. IEEE (2010). Reprinted as chap. 13 in the present book
2. EnOcean. http://www.enocean.com/
3. Perpetuum. http://www.perpetuum.com/
4. Arnold, D.: Trans. Magn. **43**(11), 3940–3951 (2007)
5. Spreemann, D., Willmann, A., Folkmer, B., Manoli, Y.: In: Proceedings of PowerMEMS, pp. 261–264 (2008)
6. Trimmer, W.: IEEE J. Solid-State Circuits **19**, 267 (1989)
7. Hoffmann, D., Folkmer, B., Manoli, Y.: J. Micromech. Microeng. **19**(094001), 2939, 11 pp (2009)
8. Ferrari, M., Ferrari, C., Guizzetti, M., And, B., Baglio, S., Trigona, C.: Proc. Chem **1**, 1203 (2009)
9. Eichhorn, C., Goldschmidtboeing, F., Woias, P.: J. Micromech. Microeng. **19**(094006) (2009)
10. Peters, C., Spreemann, D., Ortmanns, M., Manoli, Y.: J. Micromech. Microeng. **18**(10 104005), 1203 (2008)
11. Ferrosi. http://www.ferrosi.com/
12. KFC-Tech. http://www.kfctech.com/
13. Williams, C., Yates, R.: In: International Conference on Solid_State Sensors and Actuators, and Eurosensors IX, pp. 369–372 (1995)
14. Spreemann, D., Folkmer, B., Manoli, Y.: In: Proceedings of Power_MEMS (2008), pp. 257–260
15. Lee, C., Joo, J., Han, S., Lee, J.H., Koh, S.K.: Appl. Phys. Lett. **85**, 1841 (2004)
16. Lee, C., Joo, J., Han, S., Lee, J.H., Koh, S.K.: In: Proceedings of International Conference on Science and Technology of Synthetic Metal, pp. 49–52 (2005)
17. Ng, T.H., Liao, W.H.: J. Intell. Mater. Syst. Struct. **16**, 785797 (2005)
18. Breaux, O.P.: Electrostatic energy conversion system. 1976 U.S. Patent 04127804, 28 Nov 1978
19. Basset, P., Galayko, D., Paracha, A., Marty, F., Dudka, A., Bourouina, T.: J. Micromech. Microeng. **19**(115025), 12 pp (2009)
20. Meninger, S., Mur-Miranda, J., Amirtharajah, R., Chandrakasan, A., Lang, J.: IEEE Trans. VLSI Syst. **9**(1), 64 (2001)
21. Naruse, Y., Matsubara, N., Mabuchi, K., Izumi, M., Suzuki, S.: J. Micromech. Microeng. **19** (094002), 5 pp (2009)
22. Sterken, T., Baert, K., Puers, R., Borghs, G., Mertens, R.: In: Proceedings of Pan Pacific Microelectronics Symposium, pp. 27–34 (2001)
23. Williams, C.B., Yates, R.B.: In: Proceedings of Transducers/Eurosensors, pp. 369–372 (1995)
24. Bttner, H., et al.: IEEE J. Microelectromech. Syst. **13**(3), 414 (2004)

25. Micropelt. http://www.micropelt.com
26. Hopf, M., Schneider, K.: Fraunhofer ISE – Annual Report 2008. Fraunhofer ISE, Freiburg (2009)
27. Erdler, G., Frank, M., Lehmann, M., Reinecke, H., Mueller, C.: Sens. Actuators A **132**(1), 331336 (2006)
28. Frank, M., Kuhl, M., Erdler, G., Freund, I., Manoli, Y., Mueller, C., Reinecke, H.: IEEE J. Solid State Circuits **45**, 205 (2010)
29. Kerzenmacher, S., Ducre, J., Zengerle, R., von Stetten, F.: J. Power Sources **182**, 1 (2008)
30. Mitcheson, P., et al.: Proc. IEEE **96**, 1457 (2008)
31. Mitcheson, P.D., Green, T.C., Yeatman, E.M.: Proc. Microsyst. Technol. **13**(11–12), 1629 (2007)
32. Spreemann, D., Willmann, A., Manoli, B.F.Y.: In: Proceedings of PowerMEMS, pp. 372–375 (2009)
33. Lefeuvre, E., Badel, A., Richard, C., Petit, L., Guyomar, D.: J. Sens. Actuators, A Phys. **126**, 405416 (2006)
34. Maurath, D., Manoli, Y.: In: Proceedings of IEEE European Solid State Circuits Conference (ESSCIRC), pp. 284–287. IEEE (2009)
35. Ramadass, Y.K., Chandrakasan, A.: In: IEEE International Solid-State Circuits Conference Digest of Technical Papers, pp. 296–297. IEEE (2009)
36. Yi, J., et al.: In: Proceedings of IEEE International Symposium on Circuits and Systems (ISCAS), pp. 2570–2573 (2008)
37. Lee, S., Lee, H., Kiani, M., Jow, U., Ghovanloo, M.: In: IEEE International Solid-State Circuits Conference Digest of Technical Papers, pp. 12–14 (2010)
38. Li, Y., Yu, H., Su, B., Shang, Y.: IEEE Sens. J., 678–681 (2008)
39. Flatscher, M., Dielacher, M., Herndl, T., Lentsch, T., Matischek, R., Prainsack, J., Pribyl, W., Theuss, H., Weber, W.: In: IEEE International Solid-State Circuits Conference Digest of Technical Papers, pp. 286–28 (2009)
40. Guilar, N., Amirtharajah, R., Hurst, P., Lewis, S.: In: IEEE International Solid-State Circuits Conference Digest of Technical Papers, pp. 298–300 (2009)
41. Pop, V., et al.: In: Proceedings of PowerMEMS, pp. 141–144 (2008)
42. Finkenzeller, K.: RFID Handbook. Wiley, England (2003)
43. Peters, C., Henrici, F., Ortmanns, M., Manoli, Y.: In: IEEE International Symposium on Circuits and Systems (ISCAS), pp. 2598–2601 (2008)
44. Ghovanloo, M., Najafi, K.: IEEE J. Solid State Circuits **39**, 1976 (2004)
45. Kulah, H., Najafi, K.: In: Seventh IEEE International Conference on MEMS, pp. 237–240. IEEE (2004)
46. Lefeuvre, E., Badel, A., Richard, C., Guyomar, D.: J. Intell. Mater. Syst. Struct. **16**, 865876 (2005)
47. Hehn, T., Manoli, Y.: In: Proceedings of PowerMEMS, pp. 431–434 (2009)
48. Xu, S., Ngo, K.D.T., Nishida, T., Chung, G.B., Sharma, A.: IEEE Trans. Power Electr. **22**, 63 (2007)
49. Badel, D., Guyomar, E., Lefeuvre, C., Richard, J.: Intell. Mater. Syst. Struct. **16**, 889901 (2005)
50. Guyomar, D., Badel, A., Lefeuvre, E., Richard, C.: IEEE Trans. Ultrason. Ferroelectr. Freq. Control. **52**(4), 584594 (2005)
51. Lefeuvre, E., Badel, A., Richard, C., Guyomar, D.: In: Proceedings of Smart Structures and Materials Conference (2004), pp. 379–387
52. Ottman, G., Hofmann, H., Lesieutre, G.: IEEE Trans. Power Electron. **18**, 696 (2003)
53. Rincon-Mora, G., Allen, P.: IEEE J. Solid State Circuits **33**, 36 (1998)
54. Amirtharajah, R., Chandrakasan, A.: IEEE J. Solid-State Circuits **33**, 687 (1998)
55. Chao, L., Tsui, C.Y., Ki, W.H.: In: IEEE/ASM International Symposium on Low Power Electronics and Devices (ISLPED), pp. 316–319. IEEE (2007)
56. Man, T., Mok, P., Chan, M.: IEEE J. of Solid-State Circuits **44**(9), 2036 (2008)

57. Mandal, S., Turicchia, L., Sarpeshkar, R.: IEEE Pervasive Comput. **9**(1), 71 (2010)
58. Torres, E., Rincon-Mora, G.: IEEE Trans. Circ. Syst. I (TCAS I) **56**(9), 1938 (2009)
59. Chatterjee, S., Tsividis, Y., Kinget, P.: IEEE J. Solid State Circuits **40**(12), 2373 (2005)
60. Chiu, Y., Gray, P., Nikolic, B.: IEEE J. Solid-State Circuits **39**, 21392151 (2004)
61. Maurath, D., Michel, F., Ortmanns, M., Manoli, Y.: In: Proceedings of IEEE MWSCAS (2008)
62. Henrici, F., Becker, J., Trendelenburg, S., DeDorigo, D., Ortmanns, M., Manoli, Y.: In: Proceedings of IEEE International Symposium on Circuits and Systems, pp. 265–268. IEEE (2009)
63. Naess, O., Berg, Y.: In: IEEE International Symposium on Circuits and Systems (ISCAS), pp. 663–666. IEEE (2002)
64. Lotze, N., Ortmanns, M., Manoli, Y.: In: Proceedings of IEEE International Symposium on Low-Power Electronic Devices, pp. 221–224 (2008)
65. Murmann, B.: In: Proceedings of IEEE Custom Integrated Circuits Conference (CICC), pp. 105–112 (2008)
66. ITRS. The international technology roadmap for semiconductors. http://public.itrs.net (2009)
67. Vittoz, E., Fellrath, J.: IEEE J. Solid State Circuits **SC-12**, 224 (1977)
68. Becker, J., Henrici, F., Trendelenburg, S., Ortmanns, M., Manoli, Y.: IEEE Trans. Solid-State Circuits **43**(12), 2759 (2008)
69. Knig, A.: In: International Conference on Instrumentation, Communication and Information Technology (ICICI) (2005)
70. Lee, H., Mok, P.: IEEE J. Solid-State Circuits **38**(3), 511 (2003)
71. Sancarlo, I.D., Giotta, D., Baschirotto, A., Gaggl, R.: In: ESSCIRC 2008 – 34th European Solid State Circuits Conference, pp. 314–317. IEEE (2008)
72. Fan, Q., Sebastiano, F., Huijsing, H., Makinwa, K.: In: Proceedings of IEEE European Solid State Circuits Conference (ESSCIRC), pp. 170–173. IEEE (2009)
73. Witte, P., Ortmanns, M.: IEEE Trans. Cir. Syst. I: Reg. Papers **57**(7), 1500 (2010)
74. Lotze, N., Ortmanns, M., Manoli, Y.: In: Proceedings of International Conference on Computer Design (ICCD), pp. 533–540 (2007)
75. Calhoun, B., Wang, A., Chandrakasan, A.: IEEE J. Solid-State Circuits **40**(9), 1778 (2005)
76. Keyes, R., Watson, T.: In: Proceedings of IRE (Correspondence), pp. 24–85 (1962)
77. Swanson, R., Meindl, J.: IEEE J. Solid-State Circuits **7**(2), 146 (1972)
78. Schrom, G., Pichler, C., Simlinger, T., Selberherr, S.: Solid-State Electron. **39**, 425 (1996)
79. Burr, J.: In: Proceedings of IEEE Symposium on Low Power Electronics, pp. 82–83 (1995)
80. Kim, C.I., Soeleman, H., Roy, K.: IEEE Trans. Very Large Scale Integr. (VLSI) Syst. **11**(6), 1058 (2003)
81. Wang, A., Chandrakasan, A.: IEEE J. Solid-State Circuits **40**(1), 310 (2005)
82. Hanson, S., Zhai, B., Seok, M., Cline, B., Zhou, K., Singhal, M., Minuth, M., Olson, J., Nazhandali, L., Austin, T., Sylvester, D., Blaauw, D.: In: Proceedings of IEEE Symposium on VLSI Circuits, pp. 152–153 (2007)
83. Chang, I.J., Kim, J., Park, S., Roy, K.: IEEE J. Solid-State Circuits **44**(2), 650 (2009)
84. Hwang, M.E., Roy, K.: In: CICC 2008 – Custom Integrated Circuits Conference, 2008, pp. 419–422. IEEE (2008)
85. Hwang, M.E., Raychowdhury, A., Kim, K., Roy, K.: In: IEEE Symposium on VLSI Circuits, 154–155. doi:10.1109/VLSIC.2007.4342695 (2007)
86. Lotze, N., Manoli, Y.: In: ISSCC 2011 – IEEE International Solid-State Circuits Conference Digest of Technical Papers (to be published) (2011)
87. Kwong, J., Chandrakasan, A.: In: Proceedings of International Symposium on Low Power Electronics and Design, pp. 8–13 (2006)
88. Calhoun, B., Chandrakasan, A.: IEEE J. Solid-State Circuits **39**(9), 1504 (2004)
89. Calhoun, B.H., Chandrakasan, A.: IEEE J. Solid-State Circuits **41**(1), 238 (2006)
90. Bol, D., Ambroise, R., Flandre, D., Legat, J.D.: IEEE Trans. Very Large Scale Integr. Syst. **17**(10), 1508 (2009)
91. Hanson, S., Seok, M., Sylvester, D., Blaauw, D.: IEEE Trans. Electron Dev. **55**(1), 175 (2008)

92. Jayapal, S., Manoli, Y.: In: Proceedings of 13th IEEE International Conference on Electronics, Circuits and Systems ICECS'06, pp. 1300–1303 (2006)
93. Jayapal, S., Manoli, Y.: In: Proceedings of International Symposium on VLSI Design, Automation and Test VLSI-DAT 2007, pp. 1–4 (2007)
94. Roy, G., Brown, A., Adamu-Lema, F., Roy, S., Asenov, A.: IEEE Trans. Electron Dev. **53**(12), 3063 (2006)
95. Pelgrom, M., Duinmaijer, A., Welbers, A.: IEEE J. Solid-State Circuits **24**(5), 1433 (1989)
96. Koswatta, S., Lundstrom, M., Nikonov, D.: IEEE Trans. Electron Dev. **56**(3), 456 (2009)
97. Appenzeller, J., Lin, Y.M., Knoch, J., Avouris, P.: Phys. Rev. Lett. **93**(19), 196805 (2004)
98. Banerjee, S., Register, L., Tutuc, E., Reddy, D., MacDonald, A.: IEEE Electron. Device Lett. **30**(2), 158 (2009)

Chapter 20
The Energy Crisis

Bernd Hoefflinger

Abstract Chip-based electronics in 2010 consumed about 10% of the world's total electric power of $\sim$2 TW. We have seen throughout the book that all segments, processing, memory and communication, are expected to increase their performance or bandwidth by three orders of magnitude in the decade until 2020. If this progress would be realized, the world semiconductor revenue could grow by 50–100%, and the ICT industry by 43–66% in this decade (Fig. 6.1). Progress sustained at these levels certainly depends on investments and qualified manpower, but energy has become another roadblock almost overnight. In this chapter, we touch upon the life-cycle energy of chips by assessing the energy of Si wafer manufacturing, needed to bring the chips to life, and the power efficiencies in their respective operations. An outstanding segment of power-hungry chip operations is that of operating data centers, often called server farms. Their total operating power was $\sim$36 GW in 2010, and we look at their evolution under the prospect of a 1,000$\times$ growth in performance by 2020. One feasible scenario is that we succeed in improving the power efficiency of

- Processing 1,000$\times$,
- Memory 1,000$\times$,
- Communication 100$\times$,

within a decade.

In this case, the total required power for the world's data centers would still increase 4$\times$ to 144 GW by 2020, equivalent to 40% of the total electrical power available in all of Europe.

The power prospects for mobile/wireless as well as long-line cable/radio/satellite are equally serious. Any progression by less than the factors listed above will lead to economic growth smaller than the projections given above. This demands clearly that sustainable nanoelectronics must be minimum-energy (femtojoule) electronics.

B. Hoefflinger (✉)
Leonberger Strasse 5, 71063 Sindelfingen, Germany
e-mail: bhoefflinger@t-online.de

B. Hoefflinger (ed.), *CHIPS 2020*, The Frontiers Collection,
DOI 10.1007/978-3-642-23096-7_20, © Springer-Verlag Berlin Heidelberg 2012

20.1 Life-Cycle Energy

The tiny micro- and nano-chips are so ubiquitous today that they have become an issue on our overall environmental balance sheet. As a result, the science of a life-cycle analysis (LCA) of the chips from their conception to their disposal has been established. We focus here on two major aspects of their life cycle:

– The manufacturing of chips,
– The energy of data centers, the most energy-hungry operation of chips.

We simplify these highly complex topics by studying selected characteristics as they appear in 2010, and we project them to 2020 in two scenarios, a pessimistic one and an optimistic one.

The characteristics of manufacturing plants (fabs) are listed in Table 20.1. The projections in the table take into account that 3D chip-stacking will be introduced on a large scale, in order to achieve the expected advancements of performance per chip-footprint. We assume that the weighted average of stacked chip- or wafer-layers will be two layers. That is reflected both in the number of wafers and in the energy per wafer-stack and in the production energy per chip-stack. The required electrical power for the global chip production rises significantly to levels equivalent to 1/2–2/3 of the total electrical power used in Italy. The inflation in the total area of crystalline Si is a concern, too. However, thin wafer films have to be used anyway, which are the natural results of ELTRAN (Fig. 3.19) or CHIPFILM (Fig. 3.40), putting the usage of crystalline Si on a new, more favorable metric.

In any event, the energy is a major cost factor in the production of chips. At the price of 0.05 $/kWh, the cost breakdown in 2010 for a finished 30 cm wafer would be:

Energy/wafer:	$5,000
Plant/equipment	$1,700
Labor	$300
Total cost/wafer	$7,000

Table 20.1 Si wafer manufacturing 2010 and 2020

	2010	2020 conserv.	2020 optim.	Source
Revenue [$billion]	300	450	600	Fig. 5.1
Chip units [10^9]	150	225	300	
ASP [$]	2.00	2.00	2.00	
Wafers [30 cm diam. equiv., million]	60	180	240	
Total Si area [km^2]	4	12	16	
Plant and equipment [$billion]	80	120	160	
Elect. power [GW]	6	18	24	
Energy/wafer [MWh]	1	2	2	
Energy/chip/cm^2 [kWh]	1.7	3.4	3.4	
Cost/chip/cm^2 [$]	10	18	18	

Critical as this breakdown may be, the future trend of cost/wafer is unfavorable due to the rising cost of smaller feature sizes. We have seen in Chap. 8 that the cost of nanolithography makes up an increasing part not only of the total equipment cost but also of the total energy cost. The rule also holds that *the smaller the minimum dimensions on the product, the larger the overhead on plant volume and plant power for environmental control (temperature, air, media, vibration, etc.).* It is clear that this scenario is yet another inflection point for scaling down much beyond 16 nm (the 2015 node).

However, even with these limitations on the manufacturing roadmap, it is possible to foresee that, from 2010 to 2020, the much quoted and demanded 1,000-fold improvement for high-performance processors is possible.

20.2 Operating Performance and Energy

Here, we focus on the data centers (or server farms) because, not only do they operate with high-performance chips, *hot chips* [1], but also, as the digital brains of the global knowledge world and as the backbones of the internet, they are responsible for the largest part of the energy bill of the information and communication technology (ICT) industry. In 2010, worldwide *36 million servers operated and consumed ~36 GW* (roughly equivalent to the total electrical power of all of Italy). This power level is also six times the power consumption of all chip-manufacturing plants. And it is expected that the world's data centers improve their performance 1,000-fold in the present decade from 2010 to 2020.

We will try to analyze this 10-year perspective, starting from the present relation that *processors, memory and interconnects make up about 1/3 each of the total energy bill of the data centers.* We now look at these three entities and their potential based on our analyses in previous chapters. We start with processor performance in Table 20.2.

The 1,000-fold improvement in operations/s appears in the roadmap for supercomputers, Fig.6.4, and we have taken that as representative for the expectations on servers. In the two scenarios for 2020, the improvement of the power-efficiency by a factor of 100 is assumed in the mainstream of the industry, and it would increase the necessary electrical power by an order of magnitude in this decade. However, we have seen in Sect. 3.6 in the examples on digital multipliers that improvements in power-efficiency by more than an order-of-magnitude are possible with respect to the mainstream (Fig. 3.32). There are

Table 20.2 The advancement of processor performance and power consumption

Processors	2010	2020 pessim.	2020 optim.	Source
Rel. MOP/s	1	1,000	1,000	Fig. 6.4
Rel. power-efficiency	1	0.01	0.001	Fig. 3.32
Rel. power	1	10	1	

certainly many bottlenecks in processors (memory will be treated separately in the following) other than multipliers. However, it is also a strong point in Chap. 13 that the trend line of the significantly more power-efficient special processors for wireless applications (Fig. 13.8) is a sign for improvement by another factor of 10. Therefore, in the optimum scenario, a 1,000× improvement in operations/s is feasible at constant energy. We will find that this amount of progress is harder to predict for the other two constituencies of computers, namely memory and interconnects.

What is needed with processors is high-speed instruction and data memory, so that we now have to look at the progress of SRAM and DRAM with regard to their power-efficiency. We find in Fig. 11.6 that SRAM dynamic power-efficiency can advance to as low as 10 nW/Gb/s, while DRAM has reached its limits at 10 µW/Gb/s, but could gain another factor of 10 in a "differential DRAM", offering close to 1 µW/Gb/s. On the pessimistic side, we have to settle for a 100× advancement of memory power-efficiency, and an optimistic mix of SRAM and differential DRAM may achieve 1,000× (Table 20.3).

The third critical component is interconnects. In Chap. 5, we came to the conclusion that the power-efficiency will advance at best 100× in the decade 2010–2020 to 10 µW/(Gb/s) so that we obtain Table 20.4.

The development of the total electrical power of data centers is shown in Fig. 20.1. The pessimistic forecast would require 10× more power for the data centers, i.e., *360 GW needed in 2020 for data centers*, equivalent to the present total electrical power of all of Europe. Since this is not realistic, the performance improvement of data centers would slow down to 100×/decade, resulting in *a serious setback* for the expansion of computer and internet services. Even in the optimum scenario, the electrical power of data centers would increase 4× to 144 GW, which is hard to imagine from an investment and environment point-of-view. It was predicted in 2008 [2] that *energy costs could eat up to 40% of IT budgets.* Data centers are at the forefront of this escalation, and major developments are under

Table 20.3 The advancement of memory performance and power consumption

Memory	2010	2020 pessim.	2020 optim.	Source
Rel. Gb/s	1	1,000	1,000	
Rel. power-efficiency	1	0.01	0.001	Fig. 11.6
Rel. power	1	10	1	

Table 20.4 The advancement of interconnect performance and power consumption

Interconnects	2010	2020	Source
Rel. Gb/s	1	1,000	
Rel. power-efficiency	1	0.01	Fig. 5.2
Rel. power	1	10	

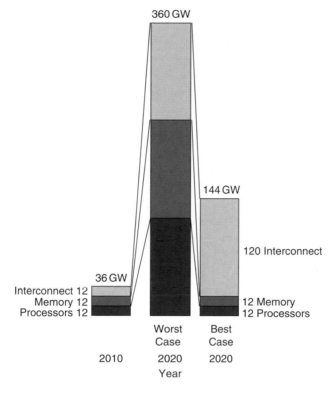

Fig. 20.1 The electrical power consumption of data centers in 2010 and 2020

way to alleviate this situation. Already, more than half of the energy bill of data centers is spent on cooling, so that recovering a major part of this by efficient water-cooling of the servers is an ongoing development. If the water is used for remote heating or industrial thermal processing, half of the original electrical energy could be recovered for other applications [3]. Sustainable growth will depend more and more on a serious commitment to ultra-low-energy architectures and circuits as well as integral optimization of the energy issue. Societies, regions, and future strategies will play an important role in view of the size of the energy issue. This is true as well for the two following domains.

20.3   Mobile and Wireless Energy Scenarios

The wireless and mobile interconnect/communication infrastructure has very specific energy constraints, which were covered in Chaps. 13 and 5. We recapitulate here that the quality of the mobile-communication world for billions of users is anything but certain, unless 1,000-fold improvements in processor power-efficiency can be realized. Among the four constituents, *processors*, *memory*, *displays*, and *transceivers*, we have good prospects for processors and memories to meet the challenge (Sect. 3.6, Chaps. 11 and 13). Displays have their own specific roadmap for power-efficiency, measured in mW/(Mpix/s), where a $10\times$ larger resolution and a $10\times$ reduction in drive-power/pixel would save the status quo in energy. However, the always-on, always-connected goal at constant or reduced battery energy will require highly intelligent snooze modes and on-board energy harvesting (Chap. 19). *The expected video and cloud-computing bandwidth causes an insatiable appetite for transceiver energy at the terminals as well as in the cellular grid.* The prospects for transceiver power-efficiency in Fig. 13.6 and according to Table 5.5, and for bandwidth in Fig. 5.4 put serious limits on a high-rate sustainable expansion. Breakthroughs in bit-rate compression and in the short-to-wide-range handshake will be needed. This makes us look at the long-distance evolution.

20.4 Long-Distance Fiber- and Satellite-Communication Energy

For decades, the installed long-line bandwidth, both cable and satellite, has been an invitation to invent ever more sophisticated web-based services. However, recently, the worldwide capacity has been filled up rapidly with the overwhelming expansion of high-resolution, high-speed video traffic. For its further potential expansion in the current decade 2010–2020, estimates could be:

- Low: (bit-rate $10\times$) $\times$ (users $100\times$) $= 1,000\times$
- High: (bit rate $100\times$) $\times$ (users $1,000\times$) $= 10^5\times$

Capital investment, energy, and environment will dictate how much of this potential can be sustained, and business creativity will be the surprise factor pushing this expansion.

The other major new-use component is cloud-computing and web-supported large-scale education (Chap. 22). More and more regions worldwide provide web-based educational materials and library and laboratory access to their schools – free-of-charge. Estimates of growth for these services are difficult because they depend on regional and global policies as well as new organizational structures for these services.

In any event, the challenges on providing the electrical energy needed for the hardware in the internet, as well as at the user sites, have dimensions similar to those of the data centers, that is *hundreds of gigawatts in 2020*. These communication nodes and user facilities are all candidates for renewable energy.

20.5   Conclusion

We have seen in the technology and product chapters that we have the innovation potential to increase significantly all chip functions, processors, memories, communication, which would lead to >1/3 of the total world electric power being used, a level that would not be economical or sustainable. In any event, to support reasonable growth, large-scale development, driven by serious requirements for systems-on-chip with minimum energy-per-function is needed, starting from proven prototypes and meeting significant milestones in short periods of time.

References

1. www.hotchips.org
2. McCue, A.: Energy could eat up to 40% of IT budgets. www.businessweek.com (2006). 7 Nov 2006
3. Taft, T.K.: IBM building zero-emission data centers. www.eweek.com (2009). 6 Nov 2009

Chapter 21
The Extreme-Technology Industry

Bernd Hoefflinger

Abstract The persistent annual R&D quota of >15% of revenue in the semiconductor industry has been and continues to be more than twice as high as the OECD
definition for *High-Technology Industry*. At the frontiers of miniaturization, the
Cost-of-Ownership (COO) continues to rise upwards to beyond 10 billion $ for
a Gigafactory. Only leaders in the world market for selected processors and
memories or for foundry services can afford this. Others can succeed with high-
value custom products equipped with high-performance application-specific
standard products acquired from the leaders in their specific fields or as fabless
original-device manufacturers buying wafers from top foundries and packaging/
testing from contract manufacturers, thus eliminating the fixed cost for a factory.
An overview is offered on the leaders in these different business models. In view of
the coming highly diversified and heterogeneous world of nanoelectronic-systems
competence, the point is made for global networks of manufacturing and services
with the highest standards for product quality and liability.

21.1 Extreme Technology

The term *high technology* arose in the 1950s to denote industries with a significant
commitment to research and development (R&D). Since the 1970s, the OECD
has classified industries as *high-tech*, if their R&D budget exceeds 7% of their
revenues. A ranking of such high-tech industries is shown in Table 21.1.

As another reference, electrical machinery and apparatus with 3.6% R&D is
classified as medium-high-technology. A comparison with the R&D percentage of
the semiconductor industry, shown in Fig. 21.1, demonstrates why the classification
extreme-technology industry is justified. The R&D percentage has been above

B. Hoefflinger (✉)
Leonberger Strasse 5, 71063 Sindelfingen, Germany
e-mail: bhoefflinger@t-online.de

B. Hoefflinger (ed.), *CHIPS 2020*, The Frontiers Collection,
DOI 10.1007/978-3-642-23096-7_21, © Springer-Verlag Berlin Heidelberg 2012

Table 21.1 R&D quota in high-tech industries (% of revenues)

Pharmaceuticals	10.5%
Air and space	10.3%
Medical, precision, optical	9.7%
Radio, TV, communication	7.5%
Office and computing machines	7.2%

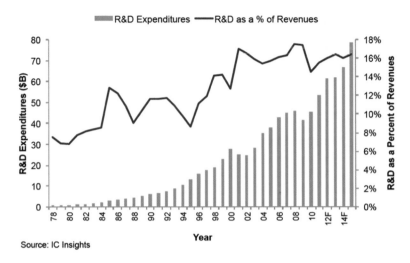

Fig. 21.1 Integrated-circuits industry R&D expenditure [1]. © IC Insights

15% for the IC industry since the mid-1990s, twice as high as the closest electronic-equipment industry, TV, communication and computing. The leaders in the industry have spent and are spending closer to 20% or even higher.

This is only the R&D effort. In a way more extreme is the investment in facilities and equipment, which has consistently exceeded the R&D. The semiconductor industry has had and continues to have unique dynamics in its evolution and continuing expansion. The global picture is well represented by the installed wafer capacity (Fig. 21.2).

In 2010, investment in facilities and equipment ran high at ~$60 billion, arriving late after a cautious 2009 and being effective only in 2011. Therefore, wafer capacity in 2010 did not meet demand, causing healthy prices and great revenues for 2010. However, the next unfavorable ratio of demand/capacity can be predicted for 2012. These extreme cycles and the sheer magnitude of required capital led to three distinct business models:

– The original-device-manufacturer model
– The foundry model
– The fabless-manufacturer model

Together, the top-10 representatives of these models serve well over 50% of the world market. However, we look at the remaining players as well, because they are

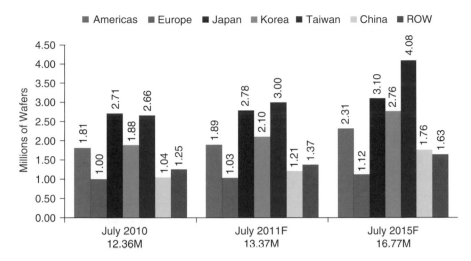

Fig. 21.2 The installed capacity in wafers/month (200 mm equivalents), July 2010, July 2011, and July 2015 [2]

often leaders in their field with application-specific integrated chips (ASICs) or chip-systems. This type of customization is also being served by the PCB (printed-circuit-board) industry, which merits attention for that reason.

21.2   The Original-Device-Manufacturer Model

In the early years of microelectronics, the 1960s, chips were produced by semiconductor divisions within vertically integrated electronics enterprises, with some notable exceptions such as Intel (Chap. 2). Intel has been the world leader since the early 1990s and continues to hold that position, as shown in Fig. 21.3.

Besides Intel, five others have a history throughout most of the 50 years of microelectronics. The Koreans and Taiwanese started in the 1980s. There are many dramatic stories behind these leaders. The striking observation about the 5-year data is that the memory heavyweights grew much faster than the group 5-year average of 35%, and that the world's largest foundry also did much better than this average.

21.3   The Foundry Model

By the early 1980s, the business cycles and the necessary critical-size investments led several leading chip manufacturers to reduce their risk by securing some wafer capacity through shares in a *contract manufacturer* for wafers, which later received the label *silicon foundry*. With his formidable track record at Texas Instruments,

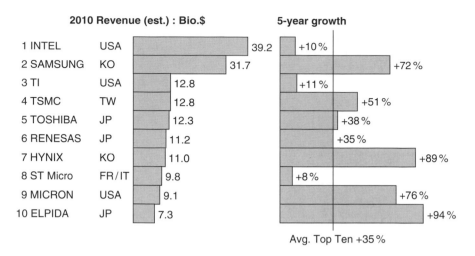

2010 Revenue (est.) : Bio.$ **5-year growth**

1 INTEL	USA	39.2	+10%
2 SAMSUNG	KO	31.7	+72%
3 TI	USA	12.8	+11%
4 TSMC	TW	12.8	+51%
5 TOSHIBA	JP	12.3	+38%
6 RENESAS	JP	11.2	+35%
7 HYNIX	KO	11.0	+89%
8 ST Micro	FR/IT	9.8	+8%
9 MICRON	USA	9.1	+76%
10 ELPIDA	JP	7.3	+94%

Avg. Top Ten +35%

Total Top 10 : 157.2 Bio.$ ≈ 50% World 2010

Fig. 21.3 The world top-10 chip manufacturers in 2010 and their recent 5-year growth

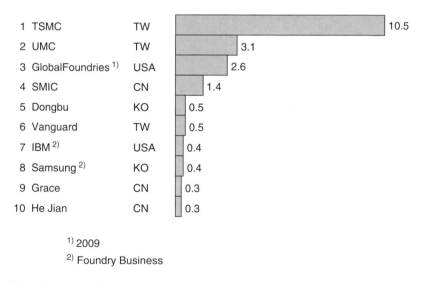

Si Foundries 2008 Revenue (Bio.$)

1	TSMC	TW	10.5
2	UMC	TW	3.1
3	GlobalFoundries [1]	USA	2.6
4	SMIC	CN	1.4
5	Dongbu	KO	0.5
6	Vanguard	TW	0.5
7	IBM [2]	USA	0.4
8	Samsung [2]	KO	0.4
9	Grace	CN	0.3
10	He Jian	CN	0.3

[1] 2009
[2] Foundry Business

Fig. 21.4 The leading Si foundries with their 2008 revenue

Morris Chang became the most famous leader, returning to his home country Taiwan and founding TSMC (Taiwan Semiconductor Manufacturing Corporation). Offering top chip technology totally to outside customers became in a way the *silicon open source*, like *open-source* code in software, launching a tsunami of chip innovations starting in the 1980s, continuing through today and becoming even more important in the future, now that the super-foundries are at the forefront in Si nanotechnology.

The top-10 Si foundries are listed in Fig. 21.4 with their 2008 revenue, with the exception of Global Foundries, which is listed with 2009 data after Chartered Semi-conductor (Singapore), the global number 3 until 2008, joined in the second year of this new enterprise, which also includes now the foundry business of IBM. IBM and Samsung's foundry business are examples of world leaders with the mixed portfolio of original, proprietary products, and outside services (even for potential competitors), a seemingly touchy constellation, which IBM has practiced since the mid-1990s.

The dynamics in the foundry scene show that the tremendous Si technology challenges are mastered with adaptive business models, and, in some prominent cases, with a radical decision from the start or early-on. This leads us to the group of fabless chip suppliers.

21.4   The Fabless Model

Companies with a fables business model contract Si-wafer manufacturing, packaging, and testing to outside sources, namely to foundries and contract manufacturers. Their 2009 ranking is shown in Fig. 21.5.

It is interesting to note that many of these are leaders in their fields, Qualcomm and Broadcomm in very-high-speed transceivers, Xilinx and Altera in field-programmable arrays, and that they require from their foundries advanced technologies with special features. They demonstrate that high-performance chips

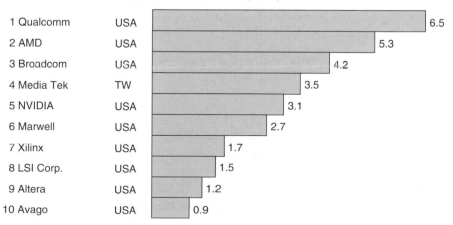

Fabless IC Suppliers Revenue 2009 (Bio.$)

1 Qualcomm	USA	6.5
2 AMD	USA	5.3
3 Broadcom	USA	4.2
4 Media Tek	TW	3.5
5 NVIDIA	USA	3.1
6 Marwell	USA	2.7
7 Xilinx	USA	1.7
8 LSI Corp.	USA	1.5
9 Altera	USA	1.2
10 Avago	USA	0.9

Fig. 21.5 Fabless IC suppliers in 2009

can be brought to the market without an in-house wafer-manufacturing line, and that the ecosystem for the production of high-performance chips has a lot to offer.

Competition for the next node on the roadmap and global alliances have enabled a large-scale sharing of progress for all three models. What may future strategies bring at the end of the roadmap?

21.5   Manufacturing Strategies Towards the End of the Roadmap

In the past, a definitive competitive strategy for the leaders in chip manufacturing has been to reduce the minimum features on a wafer faster than the competitor. The consensus is growing rapidly that this linear strategy will come to an end between 16 nm (2015 node) and 10 nm (2018 node)

– Economically because of diminishing returns and
– Physically because of atomic variance and vanishing amplification.

At the same time,

– Power-efficiency of processing, memory and communication can be improved 100–1,000× (femtoelectronics), and
– The use of power and, really, energy is becoming the no.1 global force for progress – and economic growth.

Therefore, the future performance questions for the chip industry will be:

– Which femtojoule (10^4 eV) and attojoule (10 eV) functions/bit do we offer in our portfolio?
– By how much do we reduce the energy/year necessary for the production of a chip function from sand to product?

This will amount to a life-cycle assessment of the chips and ultimately of the systems into which they are integrated, and of their operating life. With this perspective, it is worthwhile looking at the whole network of facilities involved in the creation and operation of micro- and nanochip systems.

21.6   Networks for the Creation of Micro- and Nanochip Systems

Our attention to the giga- and megafactories in the previous section and, regarding their energy and investment characteristics, in Chaps. 8 and 20 is well justified because of their challenging dimensions. We now consider these factories as backbones in an elaborate network of facilities, which are all essential in the design and production of optimum-performance and high-quality chip-based products for highly diversified customers, as illustrated in Fig. 21.6.

The world's leading and aspiring regions in nanoelectronics have to be concerned about the total quality of this network and its facilities, regionally and globally. We highlight the diversity of facilities in this network in Table 21.2 by looking at the global characteristics of three types of manufacturers, namely those for

– Standard chips
– Custom chips and microchip modules (MCMs)
– Printed-circuit boards (PCBs)

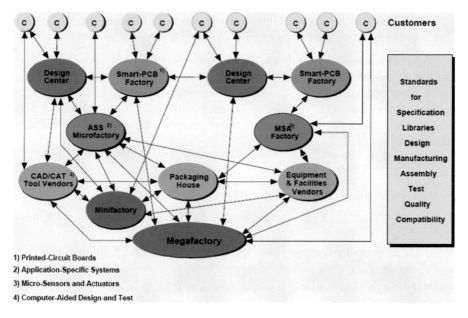

1) Printed-Circuit Boards
2) Application-Specific Systems
3) Micro-Sensors and Actuators
4) Computer-Aided Design and Test

Fig. 21.6 Networks and value chains for the design and production of chip systems

Table 21.2 Scenario in 2000 of chip and PCB manufacturers

	Standard chips	Custom chips and micromodules	PCBs
World market [$ billion]	160	40	60
No. of companies	200	400	4,000
No. of companies for 90% of the market	40	200	1,500
Top-10 revenue share	50%	5%	3%
Avg. revenue of top 10 [$ billion]	10	0.2	0.2
No. of design starts/year	30 k	200 k	2 mio.
Avg. volume [units/product]	5 mio.	10 k	5 k

We include the PCB industry because their role is changing fast with the level of integration on chips and systems-on-chip (SOCs).

Contrary to the importance and to the dominance of the top-10 suppliers of standard chips in their segment, the supply of custom chips and chip systems, including PCBs, is much more distributed in market shares and, at the same time, characterized by millions of design starts/year, without which standard chips would not reach their large volumes.

This constituency of manufacturers sees a continuing and often disruptive shift both in their product complexity as well as in their technology complexity:

- The progress of integrating more functionality on a chip continues to be driven by cost, performance, and energy.
- SOCs and 3D chip stacks mean more customization, more specific services, and fewer units/product.
- The PCB industry has to follow a roadmap, synchronizing itself with the advancing interconnect hierarchy (Fig. 3.39 and Chap. 12).

With the increasing importance of 3D integration of chip stacks, where the chips, processors, memory, transceivers, MEMS, etc. come from different vendors, the diversity and specialization in services will increase, and entirely new levels of total quality management (TQM) have to be established. The new SOCs and SIPs (systems-in-package) with chips from many sources and complex value chains require

- New standards,
- New test strategies,
- New compliance rules,
- New warranty and liability rules,
- Life-cycle assessment.

At the end of the scaling roadmap and in the era of the energy roadmap, *a diminishing part of the total investment will go into the gigafactories, and an increasing part will go into the most effective network to produce the best 3D multi-chip systems.*

21.7   Conclusion

Chips have become the basis of 50% of the world's gross product, and this share will rise. The ingenuity of the concept has triggered and feeds an unprecedented research, development, and factory investment of over 30% of annual revenue, and that with exponential growth for half a century. The R&D expenditure in the chip industry was more than $400 billion over the past 10 years, and the accumulated know-how in this industry is equivalent to over three million man-years, not counting the millions from the concurrent and essential software industry. It will take a lot of manpower and many years to make any measurable impact on the continuing silicon innovation wave with alternative technologies such as molecular, bio-, or quantum electronics.

The move of the mega and gigafactories to Asia continues. Soon, more than 80% of the world's chips will come from the Far East (Fig. 21.2).

In spite of Europe's investment in megafabs between the late 1980s and the early 2000s, Europe's share of the world production of chips never exceeded 10%. Therefore, Europe has been a net importer of finished chips and of custom-processed wafers for a long time. The only surviving wafer fabs are high-value-added fabs for proprietary smart chips with integrated sensing and actuating. Substantial employment can result only from creative, early entry into sizable markets such as the internet, mobility, and health care in the high-value-added early phase before the rapid worldwide commoditization. This requires a highly qualified work force, highly innovative companies with global alliances for marketing and standardization, and super-critical investment to reach the market in time.

The US share of global chip production has steadily declined from 50% in the 1980s to 15% in 2010. However, the US has been and still is the single largest provider of manufacturing know-how and the largest customer for the Far-East silicon foundries, preferably in Taiwan and Singapore. The US has by far the largest integrated network power to open and serve the world's dominant markets such as the internet and personal computing. Even so, the implementation of their strategies, such as interactive and internet TV, has been delayed due to a lack of manpower for product development rather than by a lack of chip technology. In order to be a part of the global competition for qualified manpower, US corporations and universities have established a vast cooperative network with the many rising universities in Taiwan, South Korea, India, and China, which have grown since the 1980s at a rate much faster than the established academia in the USA or Europe and Japan. Both the return of US-educated compatriots and the establishment of affiliated campuses in those countries will secure a super-critical force for innovation. However, the concern over qualified manpower and over the rate of innovation persists in the USA and everywhere else where nanoelectronics is considered to be essential, as we will assess in the following chapter.

References

1. The McClean report . www.icinsights.com (2010)
2. IC Insights, after D. McGraw: Taiwan to beat Japan wafer capacity. http://www.eetasia.com/ ART_8800626005_480200.HTM. Nov. 16, 2010

Chapter 22
Education and Research for the Age of Nanoelectronics

Bernd Hoefflinger

Abstract Nanoelectronics has great potential for further, sustainable growth, and this growth is needed worldwide, because new chips provide the technology foundation for all those products and services that shape our lives. However, the concern is justified that this truth is not the perception of the public in the first decades of the new millennium. How can we work towards a broad, sustained commitment to an innovation ecosystem involving education, research, business, and public policy?

Reminding ourselves of the 10x programs invoked in Chap. 2 to describe major milestones in advancing microelectronics towards today's nanoelectronics, we notice that all of them demanded requirements-driven, top-down research with ambitious, often disruptive targets for new products or services. Coming closer to the end of the *nanometer focus,* the new task of global proportion should be a *femto-Joule focus* on minimum-energy nanoelectronic systems research.

22.1 Chips in High School

The children in our schools are the critical long-term resource for a creative, healthy society. However, in spite of ICT (information and communication technology) being the largest contributor to global economic growth in the past 40 years, the interest of the public, and with it within schools, in engineering and microelectronic engineering, in particular, has declined since the early 1990s. For an excellent review of important studies on this subject, see John Cohn's lecture "Kids today – Engineers Tomorrow?" [1]. His statement "What we need is a new Sputnik to galvanize the next generation of students. . ." (and we might add the public), makes us reflect on the waves in electronics innovation, which we covered in Chap. 2.

B. Hoefflinger (✉)
Leonberger Strasse 5, 71063 Sindelfingen, Germany
e-mail: bhoefflinger@t-online.de

B. Hoefflinger (ed.), *CHIPS 2020*, The Frontiers Collection,
DOI 10.1007/978-3-642-23096-7_22, © Springer-Verlag Berlin Heidelberg 2012

It was indeed Sputnik and Apollo that launched the rise of integrated circuits in the 1960s, prominently in the US. This investment waivered in the 1970s in the US for well-known reasons, letting the Japanese take over the lead, and it needed the renaissance of 1981 to generate a worldwide engagement in microelectronics as an industrial – and, very significantly – as a public policy, a phenomenon unparalleled since then. It was in a 1984 meeting at MCNC (Microelectronics Center of North Carolina) that their communication expert showed me the Lego transistor model, which she used in her classes with elementary-school children. I have used Lego since that day, and you will find examples in Figs. 2.9 and 3.5, no longer as a toy, but now, due to their buttons, as a visualization of individual boron and arsenic atoms in a nanotransistor.

Certainly, the internet and educational-robotics programs have risen to a formidable offering for children in the meantime, and a listing of proven programs is contained in [1]. At a more advanced level and in the serious concern to carry microelectronics to high-school juniors in time for their professional inclinations, we launched a 2-week intensive lab course on the making of microchips in 1992, which has since been offered annually (Table 22.1 and Fig. 22.1).

The course program, although challenging, with its subject matter reaching from freshman- to junior-year material, has been very well received by the high-school students. Over the past 18 years, more than 600 students from some 60 high schools participated, among them 30% girls. Returning to their schools, the students reported in their classes about their experience, thereby multiplying the effect of this program. The experience with that material was used to establish a vocational training program in collaboration with regional SMEs (small and medium enterprises). This program and its certificate have been accepted by the regional association of electronics enterprises as an additional professional qualification for electronics technicians.

22.2   University Programs in Micro- and Nanoelectronics

The teaching of microelectronics has been promoted throughout its history for 60 years by monographs and textbooks written by authors based in the US (and Canada). It is important to note that the originators of new material in the early years were the world leaders in industrial research:

– William Shockley: *Electrons and Holes in Semiconductors*, 1950 (Bell Laboratories)
– Simon Sze: *Physics of Semiconductor Devices*, 1970 (Bell Laboratories),
– The Series on Bipolar ICs, ~1967 (Texas Instruments and Motorola) and MOS ICs, 1970 (Texas Instruments)

The leading universities, especially Berkeley and MIT, were quick to structure and present these new topics in textbooks, including important sections on problems (and solutions), and the competitive US university system established

Table 22.1 SMC (Schüler und Schülerinnen machen Chips) – High-School Juniors Make Chips, a 10-day laboratory course

	Day 1
C1	Microchips for health, mobility, and entertainment
C2	The electron, force, energy, potential
C3	Resistors, capacitors, voltage and current sources
L1	Tour of the institute
C4	Ideal and real amplifiers
C5	Semiconductors – the shockley garage model
	Day 2
C6	PN-junction and diode
C7	MOS transistor
C8	Analog and digital signals, codes
C9	Logic gates and standard cells
C10	CMOS logic – the inverter, its gain and noise margin
M1	Meeting with engineering students
	Day 3
C11	CMOS gates and counters
C12	Field-programmable arrays
L1	Design of a digital clock (group B)
	Day 4
C13	Logic synthesis with VHDL I
C14	Logic synthesis with VHDL II
C15	Logic synthesis with VHDL III
L1	Design of a digital clock (group A)
	Day 5
C16	Digital circuits testing
C17	CMOS process technology front-end
C18	CMOS process technology back-end, parameter test
L2	Design of a digital counter (group B)
M2	Reunion with previous classes
	Day 6
L2	Design of a digital counter (group A)
L3	Chip manufacturing line (group A)
L4	Counter in VHDL (group B)
	Day 7
L5	Programming of the FPGA (group B)
L3	Chip manufacturing line (group B)
L4	Counter in VHDL (group A)
	Day 8
C19	ASICs – from specification to test
C20	Test program generation
C21	Quality and reliability
L5	Programming of the FPGA (group A)
	Day 9
L6	Clock printed-circuit board assembly and test (group B)
L6	Clock PCB assembly and test
	Day 10
M3	Professional development – an industry representative
C22	High-dynamic-range CMOS video cameras
M4	Certificates and course review/feedback

Fig. 22.1 High-school juniors make chips, class of 2000 (www.ims-chips.de)

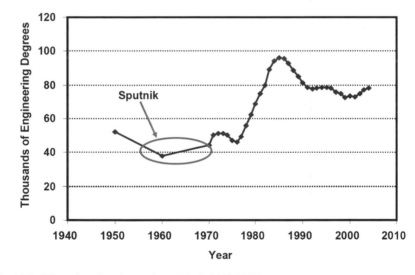

Fig. 22.2 US engineering degrees/year [1]. © 2009 IEEE

the expansion within the electrical-engineering curricula, supported by a national accreditation program, executed by the AAEE (American Association for Engineering Education), whose teams consist of peer professors as well as industry leaders. This system, fed by the world's best students from both home and abroad, enabled the annual number of US engineering degrees to rise from 40,000 in 1960 to a peak of almost 100,000 in 1985 (Fig. 22.2) and led to the sustained world leadership of US universities.

A worldwide ranking of engineering programs generated by Shanghai Technical University [2], which is quoted widely, shows among the top 50 engineering programs:

34 from the US (among them 15 at the top)
4 from the UK
4 from Japan
3 from Canada
1 from Switzerland
1 from Singapore
1 from Hong Kong
1 from Israel
1 from Taiwan

Table 22.2 shows more details of this global ranking:

In spite of this sustained leadership, Fig. 22.2 shows that the US system has stagnated below 80,000 engineering degrees/year since 1990. At the same time, China has strongly increased its engineering output so that by 2008 their number of young engineering professionals became three times that in the USA (Fig. 22.3).

The concern about US leadership is quoted here because, in the past, it has been the largest force to revive and to re-focus the innovation system, not only in the US but also globally. The concern has been voiced in the US since 2001 with increasing intensity, and it has been addressed as a *perfect storm:* "There is a quiet crisis in US science and technology...This could challenge our pre-eminence and capacity to innovate..." (Shirley Ann Jackson, President of Rensselaer Polytechnic Institute, 2004, quoted from [4]). The global financial and economic crisis of 2008/2009 has made it even clearer that the exit strategy has to be focused innovation for global markets and services, an Apollo program as the answer to Sputnik (where the present crisis is much larger than Sputnik). Apollo was the largest requirements-driven program ever, and there will certainly not be a simple analogue today. However, we will look at the present situation in nanoelectronics research and technology transfer with special attention to the requirements that have been set before the scientists and which they have promised to meet.

Table 22.2 Top 10 university engineering programs USA/ROW 2007 [2]

	USA		Rest of world (ROW)
1.	MIT	16	Cambridge U., UK
2.	Stanford U.	17	Tohoku U., Japan
3.	U. of Illinois, Urbana-Champaign	19	U. Toronto, Canada
4.	U. of Michigan, Ann Arbor	25	Kyoto U., Japan
5.	U. of California, Berkeley	27	Imperial College, UK
6.	Penn State U.	28	EPF Lausanne, Switzerland
7.	Georgia Tech	28	Tokyo Institute of Technology, Japan
8.	U. of Texas, Austin	32	Singapore U., Singapore
9.	U. of California, San Diego	37	Hong Kong U. of Science and Technology, Hong Kong
10.	Purdue U., IN	38	Technion, Haifa, Israel

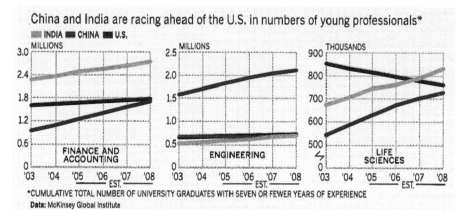

Fig. 22.3 The number of young professionals in the US, China, and India [3]

22.3 Large-Scale Nanoelectronics Research

The microelectronics age has had its two prominent international scientific meetings for well over half a century, and they continue to be the most representative yardsticks for the global development of micro- and now nanoelectronics:

- The International Electron-Devices Meeting (IEDM), held annually in December, as the name says with a focus on new devices, their physics, technology and characterization.
- The International Solid-State-Circuits Conference (ISSCC), held annually in February, the no.1 event for the presentation of new chips, and most often quoted in this book.

To get a view of the global participation in 2010, we show the origins of the scientific papers by country in the categories *universities and research institutes* and *industry* in Fig. 22.4. As a foretaste, we note that, among the more than 450 papers, 85 are the result of industry–university/institute cooperations, and of these 46 are international/global ones, a remarkable new development since ~2000. We have counted these cooperations in both categories to weigh in their significance.

The US lead in papers remains significant, and it has gained a visible cooperative note with well over 20 cooperations. Japan is holding a solid second place. Significantly, Taiwan and Korea have become key contributors to global innovation. Chinese universities appear on the map as well. The single most remarkable force is IMEC in Belgium (the Inter-University Microelectronics Center), which has become the world's largest nanoelectronics research center. One other message is the success of the university–industry network of Grenoble, France, and northern Italy, around the major locations of ST Microelectronics. This has given France/Italy a significant scientific output in nanoelectronics.

Fig. 22.4 Origins of the scientific papers presented at the 2010 ISSCC and IEDM

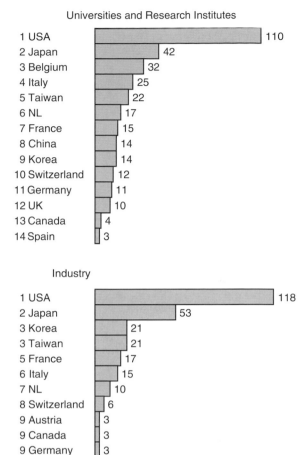

Universities and Research Institutes

1 USA	110
2 Japan	42
3 Belgium	32
4 Italy	25
5 Taiwan	22
6 NL	17
7 France	15
8 China	14
9 Korea	14
10 Switzerland	12
11 Germany	11
12 UK	10
13 Canada	4
14 Spain	3

Industry

1 USA	118
2 Japan	53
3 Korea	21
3 Taiwan	21
5 France	17
6 Italy	15
7 NL	10
8 Switzerland	6
9 Austria	3
9 Canada	3
9 Germany	3

The ranking in Table 22.2 reflects very closely the 2010 presence of US universities at ISSCC and IEDM. However, the ROW scenario of top universities regarding nanochip innovations is marked by great dynamics, which is supported by the world's leading chip manufacturers (Chap. 21). Therefore, on a global map of nano-chip research regions, we have to look at the following:

USA

- NANO NY, Albany NY with IBM, AMD, and Global Foundries.
- The universities in Table 22.2, related institutes with the US companies among the top 10.

Japan

- Tokyo-Kanagawa Pref. with the Japanese companies among the top 10.

Taiwan

 – Taipeh-Hsinshu with TSMC, UMC, and MediaTek.

Korea

 – Seoul-Taejon with Samsung and Hyundai.

Europe

 – IMEC, Leuven, Belgium, cooperating with most of the world top 20 companies.
 – Grenoble, France, and Lombardy, Italy, with ST Micro, Micron, and Numonix.

These top regions and their governments have highly diverse cultures in their ultimate task: Transfer of nanoelectronics innovations.

22.4   Requirements-Driven Research and Innovation Transfer in Nanoelectronics

The concern in this section is the timely question of how to energize the innovation constituency of academia, public research institutes, and industrial research divisions with a sustained drive out of the 2008/2009 world crisis and consecutive critical conditions, accepting that in many parts of the world, notably the *West,* that is North America and Europe, in the words of the US Commerce Secretary [5], "*The Innovation Ecosystem is broken.*" In his speech on January 15, 2010, before the PCAST, the President's Council of Advisors on Science and Technology, he said: "Just a decade or two ago, the US had the unquestioned lead in the design and production of things like semiconductors, batteries, robotics and consumer electronics. No more. Our balance of trade in advanced technology products turned negative in 2002, and has shown little signs of abating."

The European balance of trade in semiconductors has always been negative, with the notable exception of automotive semiconductors, and the situation of Germany and Great Britain, as illustrated by Fig. 22.4, shows that the US concern should be a concern in Europe as well. The research community, together with the research policy makers in governments, has to commit itself to initiate a sustained recovery with tangible, real-world solutions.

As far as nanoelectronics is concerned, approaching the end of the *roadmap* (an effective, simple requirement in the past) and after two decades of too much scientific optimism about quantum-nano solving all problems, it is time to ask which top-down requirements nanoelectronics should meet with clear milestones and with a tight time-frame.

We have seen throughout this book that *energy efficiency in processing, memory and communication* is the single most important task and the most effective one generating new performance and added value. And let us remember again: power efficiency tells us the atomistic value of how much energy we need for a single operation, but not how much energy we need to solve a task such as compressing one HDTV frame to a minimum of bits for communication.

It is an interesting test of relevance to search recent major research programs for power efficiency or, better, energy-efficiency goals. We covered the *Silicon Brains* program SyNAPSE [6] under the leadership of DARPA, and, as a case in point, it is demanded there that the electronic synapses perform an operation with an energy <10 pJ. Another DARPA program, Ubiquitous High-Performance Computing [7], also sets clear goals: *Build a PetaFLOP-class system in a 57 kW rack* [7]. In a simple model, like the one we used in Chap. 20, where we assumed that processing, memory, and interconnects each take about 1/3 of the total power, this would leave 19 kW for 10^{15} floating-point operations/s or 19 pJ/FLOP, which is mostly a 54×54-bit multiply. This is one tenth that of the best comparable graphics processor in 2010 (200 pJ) and a significant challenge if we consider the data on short-word-length multipliers in Fig. 3.32.

By contrast, if we read through the program statements of ENIAC, the European Nano Initiative [8], there is nothing other than nanometer nodes of the ITRS and "more-than-Moore" commonplaces. The 2010 update of the European 7th Programme for ICT for the tasks 2011–2012 is the most generic catalog of hit-words without any target other than at least asking for "radically improved energy efficiency"[9]. While energy efficiency is the right performance goal, why not demand from the research community a *reduction in the energy for recording, compression, communication, decompression, and delivery of a single HD frame to a display by at least a factor of 10,000*. This would make a serious $10\times$ program for the 2010s, considering that video-bandwidth and -energy is the no.1 roadblock for high-quality video, not only for entertainment but, more seriously, in professional applications such as health care and public safety.

While system energy is a requirement forcing a top-down approach (researching functions, components, and physics at the bottom), the higher levels of system applications and benefits for the public should be focused on. The ones with the biggest public interest and with broad government obligations are

- Health and care
- Education
- Safe mobility
- Communication

Here again, test questions have to be asked regularly:

- What is the dedication in funded research on these subjects,
- Their requirements on nanoelectronics,
- Their benefits from nanoelectronics and, most important and most often forgotten,
- Their sustained program of implementation?

22.4.1   Nanoelectronics in Health and Care

The potential of intelligent, portable, potentially energy-autonomous diagnostic and therapeutic nanochips is boundless so that we can only name a few representative ones: The development of a superficial, intelligent heart-monitoring coin for all patients with potential heart problems would benefit hundreds of millions of patients as well as their health-care and health-insurance systems. Supporting the introduction of this device through public health insurance would accelerate its development and its approval, and it would push low-energy chip technology in the semiconductor industry with spin-offs into many other new products.

22.4.2   Nanochips for Safe Mobility

Saving lives and preventing injuries and life-long handicaps in health care and traffic are tasks with a tremendous social significance, certainly as great as providing energy compatible with the environment. In the latter domain, we see billions of dollars being spent in subsidies for renewable power generation such as solar and wind. Avoiding fatal accidents on public roads caused by tired or distracted truck and bus drivers has been proven by hundreds of thousands of kilometers of test drives by the automotive industry, and as yet no legal requirement for these safe-distance and automatic-braking systems has been issued by the governments and supported with some public cost sharing for their rapid introduction.

The rapidly advancing urbanization of our world presents formidable challenges and opportunities for nanoelectronics-enabled ICT to support public and personal mobility. The intelligent-transportation-systems (ITS) programs in Europe, Japan, and the USA merit a review and a new, sustained effort in view of the new era of mobility driven by environmental and life-cycle energy-efficiency issues.

22.4.3   Education with and for Nanoelectronics

The Texas Instruments Speak & Spell system introduced in the late 1970s was a pioneering technical achievement with its linear-predictive-coding (LPC) voice encoder, and it was the most effective toy ever to advance language skills. Educational electronics today could be incredibly more powerful than Speak & Spell, and yet it is a domain with very patchy success given the base of billions of potential customers from kindergarten to old age. In Chap. 6, on requirements and markets, we covered broad issues such as "Open Libraries for Practically all Children" and carebots as personal tutors and assistants in rehabilitation. These examples raise the extremely challenging issue of producing and supporting content for the application of nanoelectronic systems – and for the training of knowledge in nanoelectronics.

Education and continuing-education (CE) programs in microelectronics were a vital part in corporate CE programs and nationally as well as internationally (like COMETT in Europe) funded R&D programs. In spite of the boundless opportunities with the internet, there are no critical-size public programs today. Several world-class universities have made internet course offerings into a business model, and a truly interactive mode should receive sustained support. This requires standards and certification, as were set up in the USA by the AAEE and the NTU (National Technological University) for satellite-broadcast programs.

22.5   Conclusion

Overall, it is worthwhile reconsidering the public research strategy. The advent of the transistor and of the integrated circuit has not been the result of financing large national research labs for the advancement of physics. It was rather driven by ambitious system requirements, in system language: not bottom-up, but top-down. First came the radar requirement and then the massive-computing requirement of the military establishment in the 1940s and 1950s. Soon after, in the 1960s, the requirement by the Japanese industry of a single-chip desk-top calculator changed the world, again a system requirement forcing a creative solution with the best, albeit "available" technology, not a search for new physical phenomena. ARPANET, the predecessor of the internet, is another example of the significance of an ambitious government requirement. Similarly, the European success of the digital cellular phone based on the GSM standard became possible with the long-term strategic planning of the still functional public telecom authorities requiring and pushing a joint standard, while the USA was occupied with taking its telecom system, majestically guided by ATT, apart.

The lesson for our public and governments in their efforts to support innovation for growing markets and employment should be to assign more resources to realize ambitious systems and services requirements with top-down R&D, noting that there is a lot in place and sitting at the bottom.

References

1. Cohn, J.: Kids today! Engineers tomorrow? In: IEEE ISSCC (International Solid-State Circuits Conference) 2009, Digest of Technical Papers, pp. 29–35, Feb 2009, San Francisco, USA
2. http://www.arwu.org/fieldENG
3. BusinessWeek, Special Report: China 6 India, 22 August 2005
4. Friedman, T.L.: The World Is Flat, Chaps. 7 and 8. Farrar, Straus and Giroux, New York (2005)
5. US Department of Commerce News, Feb 2010

6. http://www.darpa.mil/Our_Work/DSO/Programs/Systems_of_Neuromorphic_Adaptive_Plastic_
   Scalable_Electronics_%28SYNAPSE%29.aspx
7. http://www.darpa.mil/Our_Work/I2O/Programs/Ubiquitous_High_Performance_Computing_
   (UHPC).aspx
8. www.eniac.eu
9. www.cordis.europa.eu/fp7/ICT/Objective ICT-2011.9.8, p.106.

Chapter 23
2020 World with Chips

Bernd Hoefflinger

Abstract Although we are well advised to *look at the future 1 day at a time*, we have seen in the chapters of this book, and they necessarily could cover only a selection on the features and applications of those tiny chips, that their potential continues to grow at the exceptional rates of the past. However, the new commitment has to be towards Sustainable Nanoelectronics, guided by creating sensing, computing, memory, and communication functions, which move just a few electrons per operation, each operation consuming energy less than one or a few femtojoule, less than any of the 10^{14} synapses in our brains.

At these energy levels, chips can serve everywhere, making them ubiquitous, pervasive, certainly wireless, and often energy-autonomous.

The expected six Billion users of these chips in 2020, through their mobile, intelligent companions, will benefit from global and largely equal access to information, education, knowledge, skills, and care.

23.1 Users of 2020 Chips

The six billion users in 2020 will use, on average, about 40 new chips per year, on top of the hundreds they will have had in use statistically, giving everyone the power of a supercomputer. Most of these chips will go unnoticed, from the terabit memory in the MP3 ear plug to the heart monitor in your shirt. The always-on mobile companion will be ready to perform trillions of multiplies per second to provide its owner with high-quality video and animated information.

The access to information, education, knowledge, and skills will be global and largely equal. The number of users in Asia will be almost ten times from the number in North America and Europe.

B. Hoefflinger (✉)
Leonberger Strasse 5, 71063 Sindelfingen, Germany
e-mail: bhoefflinger@t-online.de

B. Hoefflinger (ed.), *CHIPS 2020*, The Frontiers Collection,
DOI 10.1007/978-3-642-23096-7_23, © Springer-Verlag Berlin Heidelberg 2012

The most visible products powered by minimum-energy chips will be paper-thin e-books and tablets for information and education, and, at least in Japan, Korea, Taiwan, and China, personal robots for children and for the elderly and handicapped. These shocking new companions will change human lives more than the phone, the car, the TV, and the internet. Health monitoring, diagnostics, therapy, and telemedicine will be most significantly advanced by 2020, again enabled by high-resolution and large-bandwidth imaging and video. Personal and public, safe and effective mobility can be greatly enhanced by intelligent nanoelectronic systems. Both health and mobility are a public concern – and obligation.

23.2   Regional and Global Policies for Sustainable Nanoelectronics

Given the potential and power of chips in 2020, a valid question is whether this progression happens just because of market forces such as competition and new-product appeal or as a result of long-term strategies such as those pursued in Asia or by DARPA in the USA.

R&D strategies should focus on system requirements in health, education, and mobility, and they should be accompanied by certification policies to assure the beneficial introduction of these new, *intelligent* nanosystems such as *silicon brains.*

23.3   The Makers of 2020 Chips

The re-focusing from nanometers to femtojoules is another one of the more fundamental turning points in the history of microelectronics, such as the transitions from bipolar to NMOS and NMOS to CMOS, however, this time, not from one technology to another, but from technology to functional energy.

Certainly, the broad introduction of a 10-nm Si process technology is a sufficiently large challenge for a decade. But to stop pushing the nanometer competition further and to invest instead in energy, and to get this done overnight and in parallel with the remaining nanometer work, is very tough. The time scale is short indeed, if we recall the energy crisis exemplified for data centers in Chap. 20. However, the focus on nanoelectronics as a sustainable, minimum-energy, even self-sustained technology also has the potential to interest the public and the young, in particular [1].

We identified the interconnect- and communication-energy problem in Chap. 20, and, if we turn our attention all the way back to Fig. 1.4, we see the magnitude of the problem. Correspondingly, since 2005, the 3D integration of thinned chips with TSVs has appeared as one measure to shorten interconnects. This heterogeneous process with heterogeneous, multiple chip suppliers is reshaping the industry and one sure sign of change.

With the large-scale wafer manufacturing – >80% of the world total – in Asia, the West and the rest of the world (ROW) have to realize their high-value-added share on finished chip-systems in highly innovative ways. By 2020, this will have evolved largely on the basis of available skilled, creative manpower.

The two energy worlds of processors, general-purpose as opposed to application-specific, minimum-energy, well described in Chap. 13, will have hit each other decisively by 2020. Even more fundamentally, the professionals have to come to grips with the more radical, adaptive, and self-learning biomorphic chip systems regarding their certification and fields of applicability. Here, the chip industry has to work closely with the public and with policy makers.

23.4 Beyond 2020

Will a 10-nm chip industry have enough potential to grow at > 4%/a beyond 2020? The answer is yes, because all of the developments described in this book and all those based on the same large-scale Si technology but not described here will just be achieved by a few demonstrators and by selected leaders in their field, and the 2020s will be marked, in the first place, by the widespread insertion of these achievements in a very large available global market.

Of course, there are longer-term research issues, and to keep these focused, we identify some fundamentals for future large-scale electronics in a natural environment with operating temperatures >70° C, advancing, by at least two orders of magnitude, the 2020 performance of Si nanoelectronics projected in this book:

For programming, writing, reading, and communication within the natural environment, electrons, currents and voltages are a must. In 10nm transistors, MOS or bipolar, currents are generated by single electrons, and the 2020 performance levels are as follows:

Digital logic and high-speed memory:

– Differential, full-swing regenerator
– Footprint < 60 × 60 nm^2
– Dynamic energy <7 eV
– DC energy <7 eV.

Tenfold improvements on footprint and energy are a tough challenge and/or limited by the electron thermal energy of 25 mV at room temperature.

In any more radical innovation of high-speed, low-energy nanometer-logic, the fundamental component still has to be a differential, amplifying regenerator such as a cross-coupled inverter, with a variance per chip compatible with the power supply. In a carbon-based technology, this means complementary n- and p-type carbon nanotube (CNT) or graphene transistors. A 3D vertical, doped CNT structure with a topography as in Fig. 3.42 may be better than a thin-film graphene implementation. In any event, a harmonious mix of, most likely short-range, vertical and (long-range) horizontal integration should be achieved. Self-assembly by selective, doped crystal growth should receive sustained development because of its density, quality, and efficiency.

A different scenario exists for non-volatile (NV), field-programmable memory, because here the selection of the token for a memory bit has a high priority. Charge- and resistance-based memories such as multi-level per cell (MLC) Flash, phase-change memory (PCM), and resistive RAM (RRAM) are ruling for the time being. They have intrinsic energies per cell per bit > 100 eV and require large programming energies.

A most fascinating token for a bit is the spin of the electron so that *spintronics for NV memories* continues to be an important research area. Programming (writing) and reading require electronics resulting in unfavorable energy and area requirements (see Table 11.4). The write and read energy remain the fundamental issue of NV memories. It is here that the single-electron nanodevice, the *SE device-chain with Coulomb confinement*, built possibly with direct, focused-ion-beam doping, has fundamental potential, if, as a thought experiment, we revive the charge-coupled-device (CCD) serial-memory concept [2] and imagine a single electron (or several electrons, for that matter) per bucket in a CCD, either holding the electrons in the retention state without power dissipation, or reading them out serially, coupled with a re-write. In such a multi-phase, clocked architecture, write and read energies have to be assessed carefully.

The fantastic progress in understanding ionic and organic information processing in nature will continue with exciting insights and discoveries, many of them achieved with ever more powerful electronic sensing, processing, and pattern-recognition functions. This is of significant merit for diagnostics, therapy, learning, and for the interaction of electronic systems with functions and components of the human body via surface contacts, implants, and prosthetics.

However, R&D with the aim of replacing the age of micro- and nanoelectronics by a new age of molecular or organic computing is ill conceived, as summarized in the final section.

23.5 Conclusions

- The electronic circuits being built in the present decade 2010–2020 perform the operation of a neuron's synapse with less energy and one million times faster than nature.
- Electronics is a fundamental technical invention.
- Electronic circuits are for information and communication what wheels and motors are for mobility.
- We should improve these with all our creativity and energy, but not try to fly like birds!

References

1. Cohn, J.: Kids today! Engineers tomorrow? IEEE ISSCC (International Solid-State Circuits Conference) Digest Technical Papers, pp. 29–35 (2009)
2. Boyle, W.S., Smith, G.E.: Charge Coupled Semiconductor Devices, Bell Sys. Tech. J. Vol. 49, pp. 587–593, Apr. 1970.

Index

Titles in this Series

Quantum Mechanics and Gravity
By Mendel Sachs

Quantum-Classical Correspondence
Dynamical Quantization and the Classical Limit
By Josef Bolitschek

Knowledge and the World: Challenges Beyond the Science Wars
Ed. by M. Carrier, J. Roggenhofer, G. Küppers and P. Blanchard

Quantum-Classical Analogies
By Daniela Dragoman and Mircea Dragoman

Life - As a Matter of Fat
The Emerging Science of Lipidomics
By Ole G. Mouritsen

Quo Vadis Quantum Mechanics?
Ed. by Avshalom C. Elitzur, Shahar Dolev and Nancy Kolenda

Information and Its Role in Nature
By Juan G. Roederer

Extreme Events in Nature and Society
Ed. by Sergio Albeverio, Volker Jentsch and Holger Kantz

The Thermodynamic Machinery of Life
By Michal Kurzynski

Weak Links
The Universal Key to the Stability of Networks and Complex Systems
By Csermely Peter

The Emerging Physics of Consciousness
Ed. by Jack A. Tuszynski

Extreme States of Matter
on Earth and in the Cosmos
By Vladimir E. Fortov

Searching for Extraterrestrial Intelligence
SETI Past, Present, and Future
Ed. by H. Paul Shuch

Essential Building Blocks of Human Nature
Ed. by Ulrich J. Frey, Charlotte Störmer and Kai P. Willführ

Mindful Universe
Quantum Mechanics and the Participating Observer
By Henry P. Stapp

Principles of Evolution
From the Planck Epoch to Complex Multicellular Life
Ed. by Hildegard Meyer-Ortmanns and Stefan Thurner

The Second Law of Economics
Energy, Entropy, and the Origins of Wealth
By Reiner Kümmel

States of Consciousness
Experimental Insights into Meditation, Waking, Sleep and Dreams
Ed. by Dean Cvetkovic and Irena Cosic

Elegance and Enigma
The Quantum Interviews
Ed. by Maximilian Schlosshauer

Humans on Earth
From Origins to Possible Futures
By Filipe Duarte Santos

Evolution 2.0
Implications of Darwinism in Philosophy and the Social and Natural Sciences
Ed. by Martin Brinkworth and Friedel Weinert

Probability in Physics
Ed. by Yemima Ben-Menahem and Meir Hemmo

Chips 2020
A Guide to the Future of Nanoelectronics
Ed. by Bernd Hoefflinger

From the Web to the Grid and Beyond
Computing Paradigms Driven by High-Energy Physics
Ed. by René Brun, Federico Carminati and Giuliana Galli Carminati

Printing: Ten Brink, Meppel, The Netherlands
Binding: Stürtz, Würzburg, Germany